I0041748

Numerical Methods and Analysis with Mathematical Modelling

What sets this book apart are the modelling aspects utilizing numerical analysis (methods) to obtain solutions. The authors cover first the basic numerical analysis methods with simple examples to illustrate the techniques and discuss possible errors. The modelling prospective reveals the practical relevance of the numerical methods in context to real-world problems.

At the core of this text are real-world modelling projects. Chapters are introduced and techniques are discussed with common examples. A modelling scenario is introduced that will be solved with these techniques later in the chapter. Often, the modelling problems require more than one previously covered technique presented in the book.

Fundamental exercises to practice the techniques are included. Multiple modelling scenarios per numerical methods illustrate the applications of the techniques introduced. Each chapter has several modelling examples that are solved by the methods described within the chapter.

The use of technology is instrumental in numerical analysis and numerical methods. In this text, Maple, Excel, R, and Python are illustrated. The goal is not to teach technology but to illustrate its power and limitations to perform algorithms and reach conclusions.

This book fulfills a need in the education of all students who plan to use technology to solve problems whether using physical models or true creative mathematical modelling, like discrete dynamical systems.

Textbooks in Mathematics

Series editors:
Al Boggess, Kenneth H. Rosen

An Introduction to Optimization with Applications in Data Analytics and Machine Learning
Jeffrey Paul Wheeler

Encounters with Chaos and Fractals, Third Edition
Denny Gulick and Jeff Ford

Differential Calculus in Several Variables
A Learning-by-Doing Approach
Marius Ghergu

Taking the "Oof!" out of Proofs
A Primer on Mathematical Proofs
Alexandr Draganov

Vector Calculus
Steven G. Krantz and Harold Parks

Intuitive Axiomatic Set Theory
José Luis García

Fundamentals of Abstract Algebra
Mark J. DeBonis

A Bridge to Higher Mathematics
James R. Kirkwood and Raina S. Robeva

Advanced Linear Algebra, Second Edition
Nicholas Loehr

Mathematical Biology: An Introduction to Differential Equations
Christina Alvey and Daniel Alvey

Numerical Methods and Analysis with Mathematical Modelling
William P. Fox and Richard D. West

www.routledge.com/Textbooks-in-Mathematics/book-series/CANDHTEX
BOOMTH

Numerical Methods and Analysis with Mathematical Modelling

William P. Fox and Richard D. West

CRC Press
Taylor & Francis Group
Boca Raton London New York

CRC Press is an imprint of the
Taylor & Francis Group, an **informa** business

A CHAPMAN & HALL BOOK

First edition published 2025
by CRC Press
2385 Executive Center Drive, Suite 320, Boca Raton, FL 33431

and by CRC Press
4 Park Square, Milton Park, Abingdon, Oxon, OX14 4RN

CRC Press is an imprint of Taylor & Francis Group, LLC

© 2025 selection and editorial matter, William P. Fox and Richard D. West individual chapters, the contributors

Reasonable efforts have been made to publish reliable data and information, but the author and publisher cannot assume responsibility for the validity of all materials or the consequences of their use. The authors and publishers have attempted to trace the copyright holders of all material reproduced in this publication and apologize to copyright holders if permission to publish in this form has not been obtained. If any copyright material has not been acknowledged please write and let us know so we may rectify in any future reprint.

Except as permitted under U.S. Copyright Law, no part of this book may be reprinted, reproduced, transmitted, or utilized in any form by any electronic, mechanical, or other means, now known or hereafter invented, including photocopying, microfilming, and recording, or in any information storage or retrieval system, without written permission from the publishers.

For permission to photocopy or use material electronically from this work, access www.copyright.com or contact the Copyright Clearance Center, Inc. (CCC), 222 Rosewood Drive, Danvers, MA 01923, 978-750-8400. For works that are not available on CCC please contact mpkbookspermissions@tandf.co.uk

Trademark notice: Product or corporate names may be trademarks or registered trademarks and are used only for identification and explanation without intent to infringe.

Library of Congress Cataloging-in-Publication Data
Names: Fox, William P., 1949– author. | West, Richard D., author.
Title: Numerical methods and analysis with mathematical modelling /
    William P. Fox and Richard D. West.
Description: First edition. | Boca Raton, FL : CRC Press, 2025. |
    Series: Textbooks in mathematics | Includes bibliographical references and index.
Identifiers: LCCN 2024006546 | ISBN 9781032697239 (hardback) |
    ISBN 9781032703688 (paperback) | ISBN 9781032703671 (ebook)
Subjects: LCSH: Numerical analysis—Textbooks. | Numerical analysis—
    Data processing—Textbooks. | Mathematical models—Textbooks.
Classification: LCC QA297 .F658 2025 | DDC 518—dc23/eng/20240412
LC record available at https://lccn.loc.gov/2024006546

ISBN: 978-1-032-69723-9 (hbk)
ISBN: 978-1-032-70368-8 (pbk)
ISBN: 978-1-032-70367-1 (ebk)

DOI: 10.1201/9781032703671

Typeset in Palatino
by Apex CoVantage, LLC

Dedicated to our wives: Hamilton Dix-Fox and Mary West

Contents

About the Authors

Dr. William P. Fox is an emeritus professor in the Department of Defense Analysis at the Naval Postgraduate School. Currently, he is a visiting professor in the Department of Mathematics at the College of William and Mary. He received his PhD in industrial engineering from Clemson University. He has taught at the United States Military Academy, Francis Marion University, and the Naval Postgraduate School. He has many publications and scholarly activities, including more than 20 books, 24 chapters of books and technical reports, 150 journal articles, and over 150 conference presentations and mathematical modelling workshops.

Richard D. West is a professor emeritus of Francis Marion University and a retired colonel of the U.S. Army. He received an MS in applied mathematics from the University of Colorado in Boulder, which launched his teaching interest in numerical analysis, and earned his PhD in college mathematics education from New York University. After a 30-year career in the army, he taught at Francis Marion University in Florence, South Carolina, where he served as a professor of mathematics.

Preface

Welcome to *Numerical Methods and Analysis with Mathematical Modelling*. This endeavor has been a 20-plus-year development. We fully believe that there are many real-world applications that require numerical analysis, and we want to illustrate those applications. We use mathematical modelling as our thread.

Our original intent is to introduce numerical analysis and numerical methods through mathematical modelling and mathematical modelling projects. We have both evolved in our mathematical thinking and teaching. We spend less time on mathematical theory and notation and more time on the use of numerical algorithms to answer modelling problems. Numerical analysis is a course requirement for computer science majors and a nice advanced focus for math majors. Through our modelling experience, we have become more interested in the problems. In the meantime, computers and technology have become commonplace. And our experience has adapted to this regime. Our hope is that our book fulfills very well a need in the education of all students who plan to use technology to solve problems whether using physical models or true creative mathematical modelling, like discrete dynamical systems.

Audience

This book is designed for a one- or two-semester course in numerical analysis or numerical methods in applied mathematics. Students in computer science, mathematics, applied mathematics, and operation research should take this course.

Numerical Methods and Analysis with Mathematical Modelling

Each chapter has several modelling examples that are solved by the methods described within the chapter. Often the modelling problems require more than one previously covered technique presented in our book.

Chapters are introduced and techniques for the chapter are discussed with common examples. A modelling scenario is introduced that will be solved with these techniques later in the chapter.

For example, in Chapter 7, we model bungee jumping and solve as a first-order initial value problems (IVP) in terms of velocity. We use numerical methods such as Euler's and Runge–Kutta 4 to solve. We are interested in finding the terminal velocity, which is straightforward. We are interested in the velocity at a time where we have free-fallen 200 feet. We use numerical integration methods to find these values as well as roots-finding methods. In Chapter 11, we revisit this same bungee problem but as a boundary value problem in position, a second-order ordinary differential equation (ODE). Here, we need results from Chapter 7 as well as shooting point and finite difference to solve the position equation and interpret our results. In the scenario, there are questions we need to answer, and some of these require previously covered numerical methods. Extensions are usually in the projects at the end of the chapter.

In this way we cover new material as well as revisit previously covered material.

In Chapter 7, we solve $dv/dt = 0.18095v - 32.17$, $v(0) = 0$ using numerical methods.

a. If the force due to the wind resistance is 0.9 times the velocity of the jumper, then use Newton's second law ($\Sigma F = MA$) to write a differential equation that models the fall of the jumper. Be sure to include the initial conditions for the jumper for your differential equation. (Hint: This problem can be formulated as a second-order differential equation [DE] in position or as a first-order DE in velocity—use the first-order DE model in this section initially.)

b. Solve this differential equation and find a function that describes the jumper's velocity (as a function of time) using numerical methods.

c. We use numerical differentiation from Chapter 6 to describe the jumper's position (as a function of time) numerically. We need to find the time it takes the jumper to fall 200 feet.

d. We determine the velocity of the jumper after the jumper has fallen 200 feet.

e. We determine the terminal velocity of the jumper, if any.

After a bit more research, you have found that the force due to wind resistance is not linear, as assumed earlier. Apparently, the force due to wind resistance is more closely modelled by $0.9v + 0.0009v^2$.

In the projects section, we modify the air resistance from $0.9v$ to $0.9v + v^2$ and have them resolve. This is a much more difficult problem.

Write a new differential equation governing the velocity of the jumper (prior to the bungee cord coming into effect).

You should notice that this DE is no longer easy to solve. Nonetheless, you are determined to find the velocity of the jumper after 4 seconds by using a numerical technique. Use Euler's method with a step size of 0.5 and estimate the velocity of the jumper after 4 seconds.

Find the terminal velocity, if any exists.

How do your results compare with those found in requirement one under the linear assumption?

What is the velocity of the jumper after 200 feet? (HINT: you can find this from your numerical table).

In Chapter 11, we use Newton's Second Law to develop a second-order ODE, now that considering the bungee cord has acted like a spring. We need to numerically find the new equilibrium for our jumper in terms of feet fallen as well as the jumper's velocity.

With boundary conditions that match what we did in Chapter 7, we then require other numerical methods to find (a) the maximum height on the bungee rebound, (b) whether the jumper hits the bottom of the gorge, and so on.

In this manner, again we cover new material and require a review of previously covered material.

What makes our approach so different are the modelling aspects that utilize numerical analysis (methods) to obtain a solution. We must cover the basic numerical analysis methods first with simple examples to illustrate the techniques and discuss errors. The modelling perspective is novel and shows the practical relevance of the numerical methods in context to real-world problems, not just made-up exercises.

Multiple modelling scenarios per numerical methods in each main chapter help illustrate the applications of the techniques introduced.

For example:

Chapter 1 is a review of calculus with no modelling examples.

Chapter 2 covers the modelling process, and some of the latter models are introduced here.

Chapter 3 focuses on discrete dynamical systems and systems of systems. There are many real-world applications in the chapter, including prescription drug dosage, a mortgage payment example, the growth of bacteria in a culture, the spread of a contagious disease, and the competitive hunter, predator–prey, and SIR models. Discrete dynamical systems model many numerical analysis techniques presented in a later chapter.

Chapter 4 (root-finding methods) has a design of a ship illustration model and a new car buying illustration (interest rate) model.

Chapter 5 (interpolation and polynomials) has a telemetry model with garbled data and a water tank problem with missing data to model.

Chapter 6 (numerical integration and differential) has three models present: a car traveling models and revisits the earlier telemetry and water tank models to answer other key modelling questions and distance, speed, acceleration, and volume.

Chapter 7 (numerical solutions to ODE IVP) has a parachute-jumping model, a bungee-jumping model, population modelling, and a spread-of-a-contagious-disease model.

Chapter 8 (direct and indirect methods in matrix algebra) has a bridge design model, an economic Leontief model, and a cubic spline interpolation model.

Chapter 9 (single variable [SV] numerical search methods) has exhaustive coverage of methods: unrestricted, dichotomous, bisection, golden section, Fibonacci, and Newton's method directed to finding extrema. Application models include an oil-rig location example to minimize costs and an inventory model.

Chapter 10 (multi-variable [MV] search methods) includes steeped ascent, deepest descent, and MV Newton–Raphson methods. Modelling applications include a computer central-placement model and a harbor design model.

Chapter 11 (boundary value ODE) has models being revisited for the parachute-jumping model, the bungee-jumping model, motorcycle suspension models, and a beam deflection model.

Chapter 12 (approximation theory) has Kepler's law model; a spring-mass system; a production, planning, and control model; and machine interpretation of a cosine oscillator.

Chapter 13 (numerical solutions to partial differential equations [PDEs] has two heat equation problems.

Exercises/Projects

At the core of this text are real-world modelling projects. Each chapter has fundamental exercises to practice the techniques covered in the chapter. Many projects require techniques from previous chapters to complete as we show with some of our illustrious examples.

Technology

The use of technology is instrumental in numerical analysis and numerical methods. In our text, we illustrate Maple, Excel, R, and Python. Our files are available upon request to wpfox1973@gmail.com

Maple is the only package that is not free, but many colleges and universities have Maple available.

One cannot adequately do mathematical modelling, let alone numerical analysis, without technology. We illustrate various technologies with the book. Our goal is not to teach the technology but to illustrate the power and limitations of technology to perform algorithms and reach conclusions. The technologies mentioned in this book include the Ti-84 calculator, Excel, R, Python, and Maple.

Maple

Maple has a new internal section called The Student[NumericalAnalysis] Package. We used this callout often in the text to illustrate algorithms in numerical analysis. According to Maple, the Student:-NumericalAnalysis subpackage is designed to help teachers present and help students understand the basic material of a standard course in numerical analysis. There are three principal components to the subpackage: computation commands, visualization commands, and interactive routines. These components are described in the following sections and help commands. To access Student:-NumericalAnalysis tutors, select Tools>Tutors>NumericalAnalysis.

Each command in the Student:-NumericalAnalysis subpackage can be accessed by using either the long form or the short form of the command name in the command calling sequence.

The long form, Student:-NumericalAnalysis:-command, is always available. The short form can be used after loading the package.

Many of the commands and tutors in the Student:-NumericalAnalysis package can be accessed through the context menu. These commands are consolidated under the Student:-NumericalAnalysis name.

The following is a partial list of commands available from Maple's help.

Computation
• The computation commands in the `Student:-NumericalAnalysis` subpackage implement standard numerical analysis operations.
For more information on this functionality, see Student[NumericalAnalysis][ComputationOverview].
The computation commands are:

AbsoluteError	AddPoint	ApproximateExactUpper Bound
ApproximateValue	BackSubstitution	BasisFunctions
CubicSpline	DataPoints	Distance
DividedDifferenceTable	ExactValue	ForwardSubstitution
Function	Interpolant	InterpolantRemainderTerm
IsConvergent	IsMatrixShape	IterativeFormula
LeadingPrincipalSubmatrix	LinearSolve	LinearSystem
MatrixConvergence	MatrixDecomposition	NevilleTable
NumberOfSignificantDigits	Polynomialinterpolation	RateOfConvergence
RelativeError	RemainderTerm	SpectralRadius
TaylorPolynomial	UpperBoundOfRemainderTerm	VectorLimit

We recommend, if using Maple, go to the Student[NumericalAnalysis] help page and execute the Example Worksheets. These are useful worksheets. For examples using the Student:-NumericalAnalysis subpackage, see Student:-NumericalAnalysis Example Worksheet.

Python

Executing commands in Python requires the ability to understand Python and write code. Almost every section of numerical methods in our book has Python code listed, illustrated, and the output shown.

R

R has many libraries for numerical methods. Many useful ones are illustrated within this book.

Excel

We illustrate some algorithms using Excel as Excel is useful for numerical mathematical calculations and iterations. We find Excel useful for dynamical systems and numerical solutions to differential equations, as well as numerical search methods.

William P. Fox
Richard D. West

Acknowledgements

We want to thank everyone who was involved in the development of Interdisciplinary Lively Applications Projects. This National Science Foundation grant, run by Dr. Richard D. West, was instrumental in developing applications across many disciplines in applied mathematics. Some are illustrated within this book.

We thank Frank R. Gordano for the opportunities he gave us that assisted us along the way in our careers.

1

Review of Differential Calculus

1.1 Introduction

Nearly all problems can be approximated using continuous function, so calculus is extremely important for deriving approximate numerical method and verifying the solutions. The calculus definitions provided in this text are fundamental, and we will see many again in our discussions of the numerical algorithm in the text.

There are two important concepts when applying numerical methods to solve problems. The first is to obtain a "good" approximation, and the second is to obtain approximations with some degree of precision or accuracy.

1.2 Limits

The limit is an important concept in calculus and in mathematical modelling. Although often misunderstood, the limit is basic to the study of calculus and provides necessary information. As we allow the independent variable to approach a value, c, we see where the dependent variable goes. In modelling as we let the independent variable approach ∞, we can also understand the long-term behavior in systems that are an important concept in mathematical modelling. The definition of a limit is

as $t \to t_0$ if the function approaches a finite limit, L, and then we say

$$\lim_{t \to t_0} f(t) = L.$$

This is depicted in Figure 1.1

The idea of a limit is one of the most basic ideas in calculus. The equation $\lim_{x \to a} f(x) = L$ means that as x gets closer to a (but not equal to a), the value of $f(x)$ gets arbitrarily close to L. It is also possible that the $\lim_{x \to a} f(x)$ will not exist. Limits can be viewed analytically, graphically, and by numerical tables.

DOI: 10.1201/9781032703671-1

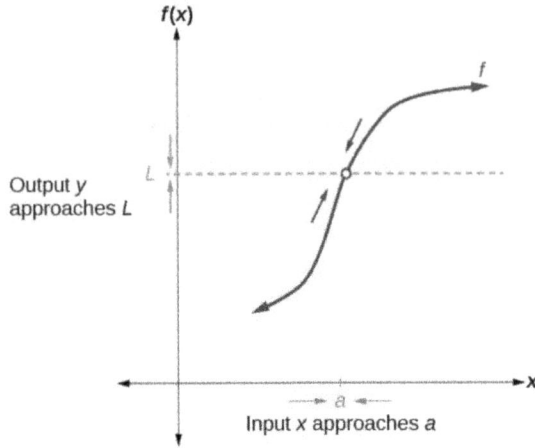

FIGURE 1.1
Graphical depiction of a limit.

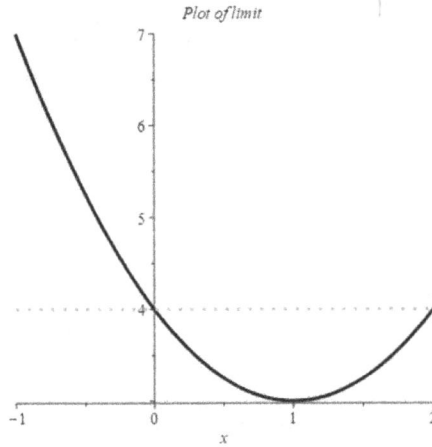

FIGURE 1.2
Plot of $x^2 - 2x + 4$ as x approaches 2.

Let's illustrate with some examples.

Example 1. Consider the $\lim_{x \to 2}\left(x^2 - 2x + 4\right)$

a. Analytical: we substitute $x = a$ into $f(x)$ to determine whether $f(a)$ exists and is a real value.

$\lim_{x \to 2}\left(x^2 - 2x + 4\right) = 2^2 - 2(2) + 4 = 4$.

Since $\lim_{x \to 2}\left(x^2 - 2x + 4\right) = 4$, the limit exists. As $x \to 2$, $f(x)$ approaches 4.

b. Graphically. We see that Figure 1.2 shows as x approaches 2 that $f(x)$ approaches 4.

c. Numerical table. We see in Table 1.1 that as x approaches 2, from the left, the values get closer to 4, and as x approaches 2 from the right, $f(x)$ approaches 4. We must allow x to approach 2 from both the left and from the right in the limiting process to determine whether the limit exists.

Example 2. Consider $\lim_{x \to 0} \dfrac{1}{x}$

a. Analytical. We substitute $x = 0$ for x and see that $1/0$ is not defined. Therefore, we might conclude that limit does not exist (LDNE). We might check to see if the function can be reduced or simplified before we reach this conclusion as shown in Example 3.

b. Graphically. In Figure 1.3, we see that as we approach 0 from the left and the right the function approaches different quantities, $\pm\infty$. We conclude the LDNE.

TABLE 1.1

Limit by Numerical Values

x	$f(x)$	x	$f(x)$
1.9	3.81	2.1	4.21
1.95	3.9025	2.05	4.1025
1.99	3.9801	2.1	4.21
1.995	3.990025	2.005	4.010025
1.999	3.998001	2.001	4.002001
1.9999	3.99980001	2.0001	4.00020001

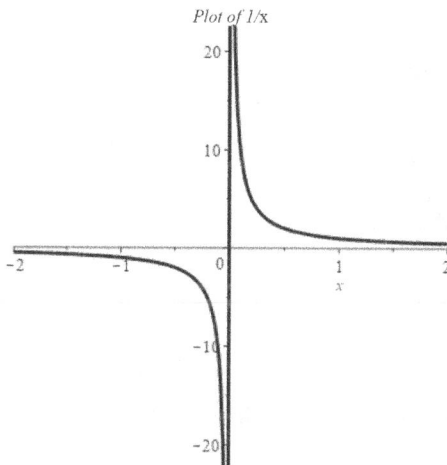

FIGURE 1.3
Plot of $1/x$.

TABLE 1.2

Numerical Table for Example 2

x	$f(x)$	x	$f(x)$
−1	−1	1	1
−0.95	−1.052631579	0.95	1.052631579
−0.9	−1.111111111	0.9	1.111111111
−0.5	−2	0.5	2
−0.1	−10	0.1	10
−0.05	−20	0.05	20
−0.0001	−10,000	0.0001	10,000

   c. Numerical table. In Table 1.2, we clearly see that the values from the
      left and right are not tending toward the same values.

Example 3. Consider the $\lim_{x \to 2} \dfrac{(x^2 - 4)}{(x - 2)}$

   a. Analytical. If we merely substitute, we get 0/0 which is an indetermi-
      nate form. We do not want to conclude the LDNE until we exhaust the
      following rules. If we have an indeterminate form such as 0/0 or ∞/∞,
      then we might try simplification of the functions or use L'Hôpital's
      rule. L'Hôpital's rule states that

$$\lim_{x \to a} \frac{f(x)}{g(x)} = \lim_{x \to a} \frac{f'(x)}{g'(x)}.$$

Therefore, if we employ L'Hôpital's rule, then we have $\lim_{x \to 2} \dfrac{2x}{1} = 4$. The
limit does exist and it is $f(2) = 4$.

   b. Graphical, shown in Figure 1.4.

   In Figure 1.4, we see that as we approach 2 from the left or the right, we
approach $f(x) = 4$.

   c. Numerical table

Table 1.3 shows that as x approaches 2, $f(x)$ approaches 4.

1.3 Continuity

We begin with a definition of continuity.

A function, $f(x)$, is continuous at a point a, if $\lim_{x \to a} f(x) = f(a)$.

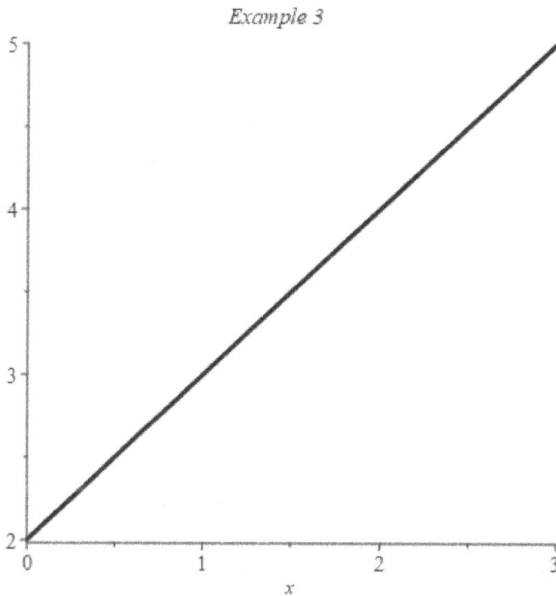

FIGURE 1.4
Plot of $x^2 - 4/(x - 2)$ as x approaches 2.

TABLE 1.3

Limit of $x^2 - 4/(x - 2)$ as x Approaches 2 Numerically

x	$f(x)$	x	$f(x)$
1.9	3.9	2.1	4.1
1.95	3.95	2.05	4.05
1.99	3.99	2.1	4.1
1.995	3.995	2.005	4.005
1.999	3.999	2.001	4.001
1.9999	3.9999	2.0001	4.0001

If $f(x)$ is not continuous at $x = a$, we say that $f(x)$ is discontinuous (or has a discontinuity) at a. Recall from your study of functions that often discontinuities are points *not* in the domain of the input variable x.

The above definition requires three conditions hold for $f(x)$ to be continuous at a.

1. The function, $f(a)$, is defined (a is in the domain of x).
2. The $\lim_{x \to a} f(x)$ exists.
3. The $\lim_{x \to a} f(x) = f(a)$.

Also recall the following two facts from calculus.

a. Polynomials are continuous everywhere. That is, they are continuous on the open interval $(-\infty, \infty)$.

b. Rational functions are continuous wherever they are defined. That is, they are continuous over the domain of x.

Sometimes, we might restrict the domain of a function so that we can make the functions continuous over the restricted domain.

Example 4. Show That the Function $f(x) = x^2 + \sqrt{13-x}$ Is Continuous at $x = 4$

$$f(4) = 19$$
$$\lim_{x \to 4} x^2 + \sqrt{13-4} = 19$$

Since, $f(4) = \lim_{x \to 4} x^2 + \sqrt{13-4} = 19$, then $f(x)$ is continuous at $x = 4$.

Example 5. Determine if $f(x) = ln(x-2)$ Is Continuous at $x = 2$

Since $f(2) = ln(0)$ that is not defined. Therefore, $f(x)$ is not continuous at $x = 2$. Recall from precalculus that the $ln(x)$ is only defined for $x > 0$.

Example 6. Given a Step Function

Consider the cost function $c(x) = \begin{cases} 25x + 5 & 0 \le x \le 100 \\ 15x + 3 & x > 100 \end{cases}$.

Is $c(x)$ continuous at $x = 100$?

We find $c(100) = 2505$.

$$\lim_{x \to 100^-} c(x) = 2505$$

$$\lim_{x \to 100^-} c(x) = 1503$$

Since the limits from the left and right are not equal, the LDNE as $x \to 100$. The function is not continuous at $x = 100$.

1.4 Differentiation

The derivative of a function, $f(x)$, is defined as the limit quotient:

$$\frac{df}{dx} = f'(x) = \lim_{\Delta x \to 0} \frac{f(a + \Delta x) - f(a)}{\Delta x}.$$

If the LDNE, then the function has no derivative at $x = a$.

The geometric interpretation of $f(a)$ is that it represents the slope of the tangent line to $f(x)$ at the point, $x = a$. The derivative is also the instantaneous rate of change. Recall from precalculus that the concept of average rate of change between two points a and b is

$$\text{Average rate of change} = \frac{f(b) - f(a)}{b - a}.$$

As we allow the difference $b - a$ to approach 0, the average rate of change becomes the instantaneous rate of change: let $b - a = \Delta x$, we have the definition of the derivative:

$$\lim_{\Delta x \to 0} \frac{f(a + \Delta x) - f(a)}{\Delta x}.$$

Example 7. Consider the Function $f(x) = (0.5 - x)^2 + 4$

a. Determine the average rate of change from $x = 2$ to $x = 8$.

$$f(8) = 60.25, f(2) = 6.25$$

The average rate of change is 9.

b. Determine the instantaneous rate of change at $x = 4$.

$$f'(4) = 1 - 2(4) = -7.$$

c. Determine what this instantaneous rate of change means.

In Figure 1.5, we see the plot of the function and its tangent line at $x = 4$. The slope of the tangent line is negative (recall is just found that the slope was –7).

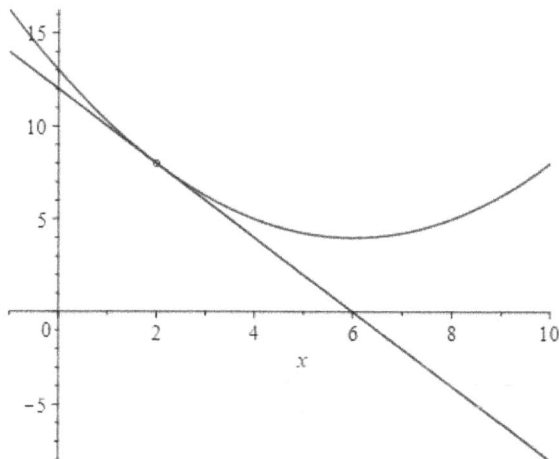

At $x = 2$, for the function $f(x) = (0.5x - 3)^2 + 4$, a graph of $f(x)$ and a tangent line.

FIGURE 1.5
Plot of the function and its tangent line at $x = 4$.

TABLE 1.4

Derivative Rules from Calculus

$\dfrac{d}{dx}(a)=0$	$\dfrac{d}{dx}[\ln u]=\dfrac{d}{du}[\log_e u]=\dfrac{1}{u}\dfrac{du}{dx}$
$\dfrac{d}{dx}(x)=1$	$\dfrac{d}{dx}[\log_a u]=\log_a e\,\dfrac{1}{u}\dfrac{du}{dx}$
$\dfrac{d}{dx}(au)=a\dfrac{du}{dx}$	$\dfrac{d}{dx}e^u=e^u\dfrac{du}{dx}$
$\dfrac{d}{dx}(u+v-w)=\dfrac{du}{dx}+\dfrac{dv}{dx}-\dfrac{dw}{dx}$	$\dfrac{d}{dx}a^u=a^u\ln a\,\dfrac{du}{dx}$
$\dfrac{d}{dx}(uv)=u\dfrac{dv}{dx}+v\dfrac{du}{dx}$	$\dfrac{d}{dx}(u^v)=vu^{v-1}\dfrac{du}{dx}+\ln u\;u^v\dfrac{dv}{dx}$
$\dfrac{d}{dx}\left(\dfrac{u}{v}\right)=\dfrac{1}{v}\dfrac{du}{dx}-\dfrac{u}{v^2}\dfrac{dv}{dx}$	$\dfrac{d}{dx}\sin u=\cos u\dfrac{du}{dx}$
$\dfrac{d}{dx}(u^n)=nu^{n-1}\dfrac{du}{dx}$	$\dfrac{d}{dx}\cos u=-\sin u\dfrac{du}{dx}$
$\dfrac{d}{dx}(\sqrt{u})=\dfrac{1}{2\sqrt{u}}\dfrac{du}{dx}$	$\dfrac{d}{dx}\tan u=\sec^2 u\dfrac{du}{dx}$
$\dfrac{d}{dx}\left(\dfrac{1}{u}\right)=-\dfrac{1}{u^2}\dfrac{du}{dx}$	$\dfrac{d}{dx}\cot u=-\csc^2 u\dfrac{du}{dx}$
$\dfrac{d}{dx}\left(\dfrac{1}{u^n}\right)=-\dfrac{1}{u^{n+1}}\dfrac{du}{dx}$	$\dfrac{d}{dx}\sec u=\sec u\tan u\dfrac{du}{dx}$
$\dfrac{d}{dx}[f(u)]=-\dfrac{d}{du}[f(u)]\dfrac{du}{dx}$	$\dfrac{d}{dx}\csc u=-\csc u\cot u\dfrac{du}{dx}$

We provide some basic rules for finding derivatives of a function in Table 1.4.

It is important to state that *differentiability implies continuity.*

Increasing and Decreasing Functions

If $f(x) > 0$, then $f(x)$ is increasing, and if $f(x) < 0$, then $f(x)$ is decreasing.

Example 8. Increasing–Decreasing Functions

If a company charges a price, p, for a product, then it can sell $5000e^{-p}$ thousands of items. Then $f(p) = 5000p * e^{-p}$ is the company's revenue if it charges a price, p.

1. For what values of p, will $f(p)$ be increasing, and for what values will $f(p)$ be decreasing?
2. Suppose the price is \$4 and the company is considering an increase in price of \$0.10. How much would revenue change?

$$f(p) = 5000p^* \, e^{-p}$$
$$f'(p) = 5000(1 - p) \, e^{-p}.$$

3. From the derivative, if $p < 1$, then $f(p) > 0$, and our function is increasing. If $p > 1$, then $f'(p) < 0$, and our function is decreasing.

4. $f'(4) = 5000(-3)e^{-4} = -274.73$, and we find $0.10 * f(4) - 27.47$.

In actuality, $f(4) - 366.71$ and $f(4.1) = 339.74$. So, $f(4.1) - f(4) = -26.57$.

Higher Derivatives

We define a higher derivate as $f^{(n)}(x) \dfrac{d^n y}{dx^n}$.

For example, consider $f(x) = -2x^2 + 3x - 4$. Then $f^{(2)}(x) = f''(x) = -4$.

The second derivative provides very useful information, If $f'(a) < 0$, then the function is concave at $x = a$. If $f'(a) > 0$, then the function is convex at $x = a$. If $f''(x) = 0$, then we investigate if $x = a$ is an inflection point.

1.5 Convex and Concave Functions

Convex and concave functions play an important role in the study of numerical analysis. We begin with a few definitions and theorems that we will need in subsequent chapters.

Let $f(x_1, x_2, \ldots, x_n)$ be a function defined for all points $(x_1, x_2, \ldots, x_n)$ in a convex set. A convex set is the set of all points for which all points on a line segment for x' to x'' are members of the set S.

A function $f(x_1, x_2, \ldots, x_n)$ is a convex function on a convex set S if for any $x' \in S$ and $x'' \in S$,

$$f[c \, x' + (1 - c) \, x''] \le c \, f(x') + (1 - c) f(x') \text{ holds for all } 0 \le c \le 1.$$

A function $f(x_1, x_2, \ldots, x_n)$ is a concave function on a convex set S if for any $x' \in S$ and $x'' \in S$,

$$f[c \, x' + (1 - c) \, x''] \ge c \, f(x') + (1 - c) f(x') \text{ hold for all } 0 \le c \le 1.$$

To gain some additional insights, let's view these geometrically. Let $f(x)$ be a function of a single variable. In Figure 1.6 and the previous definitions, we find $f(x)$ is convex if and only if for any line segment, the line segment is always above the curve. In Figure 1.7, and the previous definitions, we find $f(x)$ is concave if and only if for any line segment, the line segment is always below the curve.

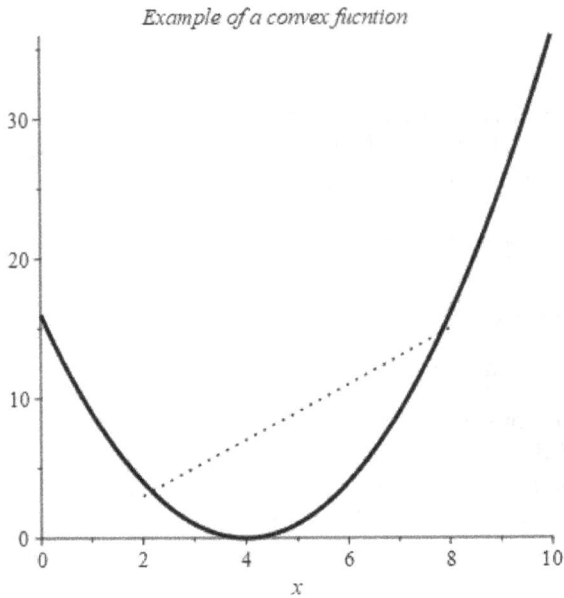

FIGURE 1.6
Example of a convex function.

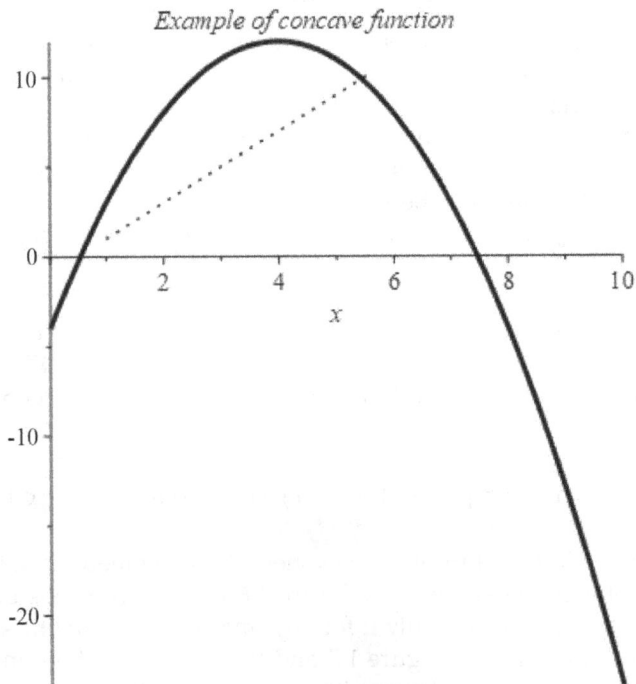

FIGURE 1.7
Example of a concave function.

Example 9. Determine the Convexity of $f(x) = x^2$

Plot of x^2

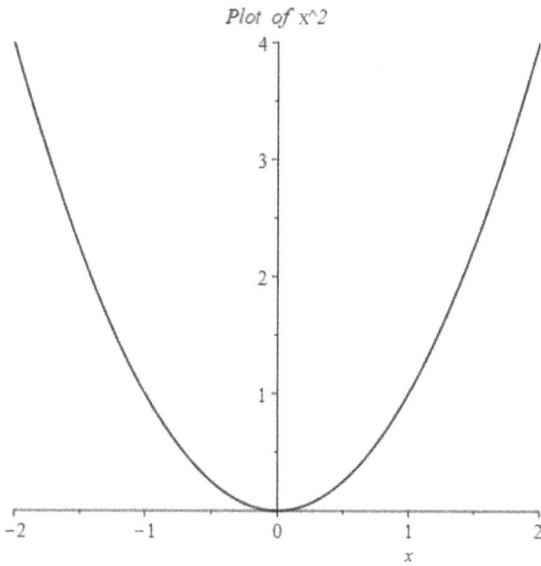

FIGURE 1.8
Plot of $f(x) = x^2$.

We see in Figure 1.8 the function is convex.

Example 10. Determine the Convexity of e^x

Plot of exp x

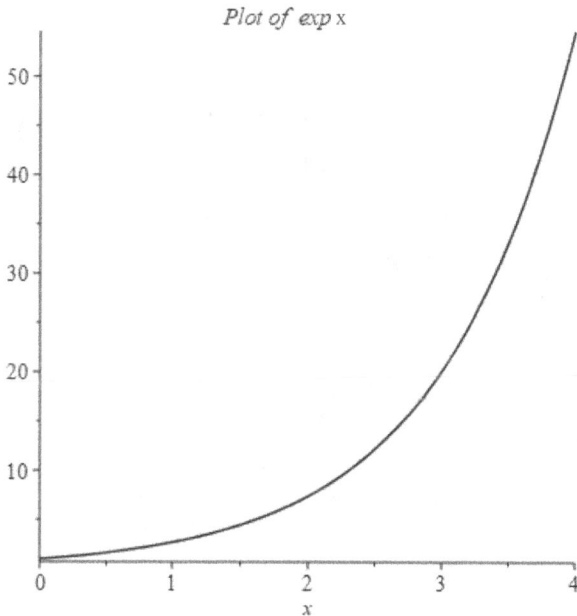

Example 11. Determine the Convexity of ln(x)

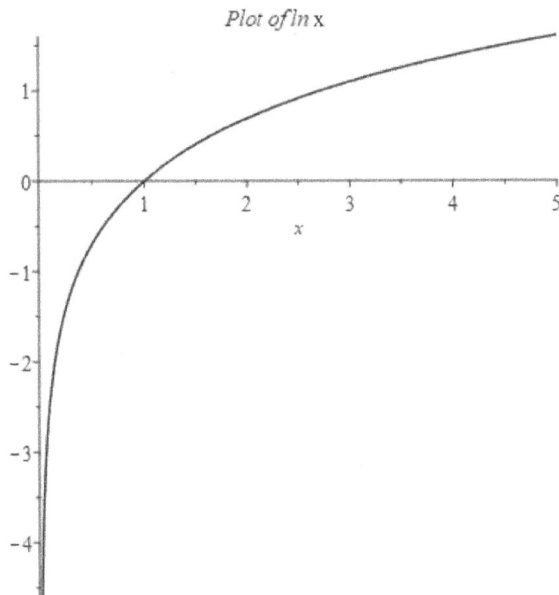

Plot of ln x

Example 12. Determine the Convexity of x^3

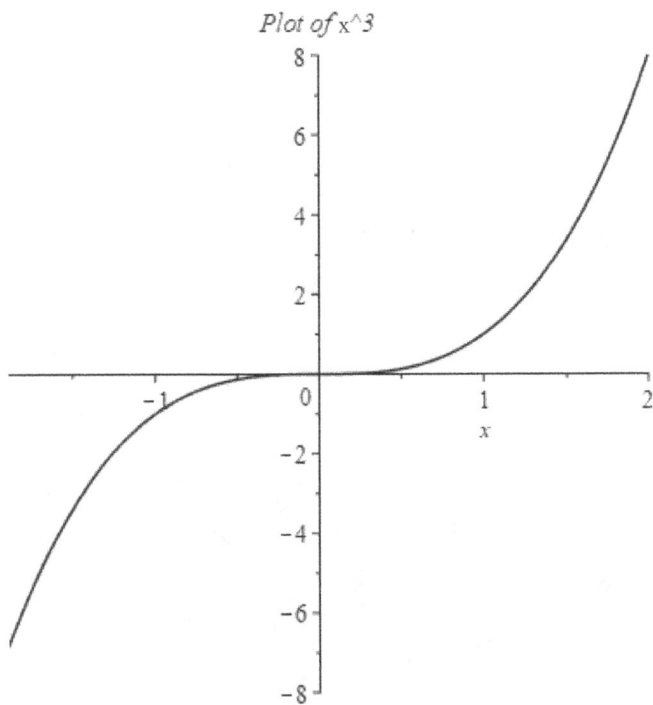

Plot of x^3

This function is both convex and concave. It is convex for $x > 0$ and concave for $x < 0$. There is an inflection point at $x = 0$.

We note that a line, $y = mx + b$, is both convex and concave.

The following theorem is presented from calculus.

Theorem 1.1 (1) Suppose $f''(x)$ exists for all x in a convex set S. Then $f(x)$ is convex on S if and only if $f''(x) \geq 0$ for all x in S.

(2) Suppose $f''(x)$ exists for all x in a convex set S. Then $f(x)$ is concave on S if and only if $f''(x) \leq 0$ for all x in S.

Example 13. The Second Derivative Theorem

Use the second derivative theorem to determine the convexity of the following functions:

a. $f(x) = -2x^2 + 2x$

b. $f(x) = 5 - 4x$

c. $f(x) = 3\,x3$

Solution

a. $f'' < -4 < 0$, so by Theorem 1, the function is concave.

b. $f'' = 0$, so the derivative is both ≥ 0 and ≤ 0. The function is both convex and concave.

c. $f'' = 18x$. Since the sign of f'' depends on the sign and value of x, then this function is neither convex nor concave over R^2. We could restrict the domain to have a convex $(x > 0)$ or concave region $(x < 0)$.

Exercises

Problems 1–5: Find each limit (if it exits).

1. $\lim\limits_{x \to 4} \left(x^3 + 2x - 21 \right)$

2. $\lim\limits_{x \to \infty} \dfrac{x^2 + x}{x}$

3. $\lim\limits_{t \to 1} \dfrac{t^2 - 1}{t - 1}$

4. $\lim\limits_{y \to 0} \dfrac{\tan(y)}{y}$

5. $\lim\limits_{x \to 0} \dfrac{|x|}{x}$

6. Given some function $f(x)$, state and give an example (i.e., a graph) of three fundamental types of discontinuities.

7. Differentiate $y = \sin(x^2)$

8. Differentiate $y = (\sin^2 x)$

9. Find all first- and second-order partial derivatives for $f(x,y) = \exp(xy^2)$.

10. Show that the following functions are convex using the definition of convexity. Verify your result by using the second derivative test.

   a. $3x + 4$

   b. $-x^{1/2}$ for $x \geq 0$

11. Characterize the function $x^3 - 9x^2 + 24x - 10$ in terms of any convexity and/or concavity properties over R, by any method.

12. Show $e^x - x - 1$ is convex using the definition of convexity. Verify your result using the second derivative test.

1.6 Accumulation and Integration

The integral from calculus is used a lot for numerical approximation. We need to understand and consider the Riemann sum so that we can employ either Simpson's rule or the trapezoidal rule:

$$\int_a^b f(x)\,dx = \lim_{\Delta x_i \to 0} \sum_{i=1}^n f(z_i)\Delta x_i,$$

where $x_0 = a < x_1 < x_2 \ldots < x_n = b$,
$\Delta x_i = x_i - x_{i-1}$ for $i = 1, 2, \ldots, m$ and z_i in $[a, b]$.

Figures 1.9 and 1.10 illustrate the graphical and numerical solutions to $f(x) = e^{x^2}$ over [0,1] using Maple.

Exercises

Using the Reimann sums, the trapezoidal method, and Simpson's method on each of the following to approximate the area under the curve:

1. $f(x) = \cos(x)$ $0 \leq x \leq 1$
2. $f(x) = -x^2$ $-1 \leq x \leq 1$
3. $f(x) = \ln(x)$ $1 \leq x \leq 2$

with(*Student*[*Calculus1*]) :
RiemannSum (*f*, *x* = 0. ..1.0, *method* = *midpoint*)
$$1.460393091$$

RiemannSum (*f*, *x* = 0 ..1, *method* = *midpoint*, *output* = *plot*)

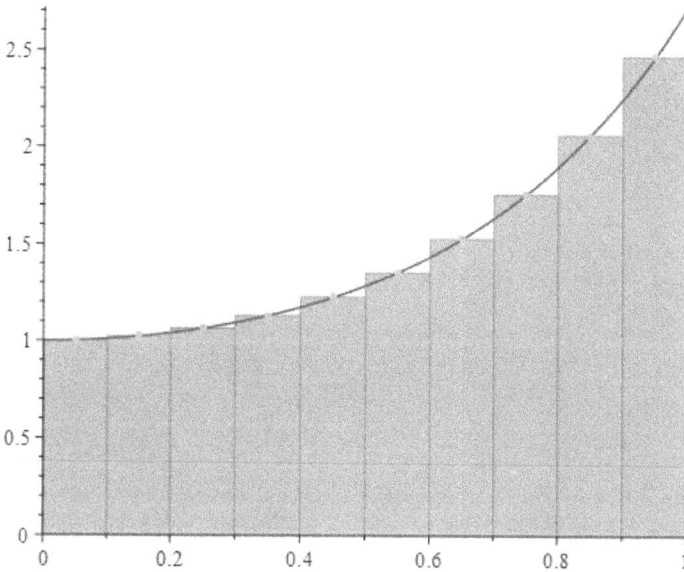

A midpoint Riemann sum approximation of $\int_0^1 f(x) \, dx$, where $f(x) = e^{x^2}$ and the partition is uniform. The approximate value of the integral is 1.460393091. Number of subintervals used: 10.

FIGURE 1.9
Midpoint Reimann sum.

1.7 Taylor Polynomials

Many numerical techniques are derived directly from the Taylor series. Therefore, the Taylor series forms a firm foundation for approximation by polynomials of functions both known and unknown. Any function that is continuous and has continuous derivatives may be expanded into a Taylor expansion.

Taylor's theorem: Suppose *f* is a continuous function on a closed interval [*a*, *b*] and has *n* + 1 continuous derivatives on the open interval (*a*, *b*).

> $evalf(ApproximateInt(f, x = 0 ..1, method = simpson))$;
> 1.462653625
> $ApproximateInt(f, x = 0 ..1, method = simpson, output = plot)$

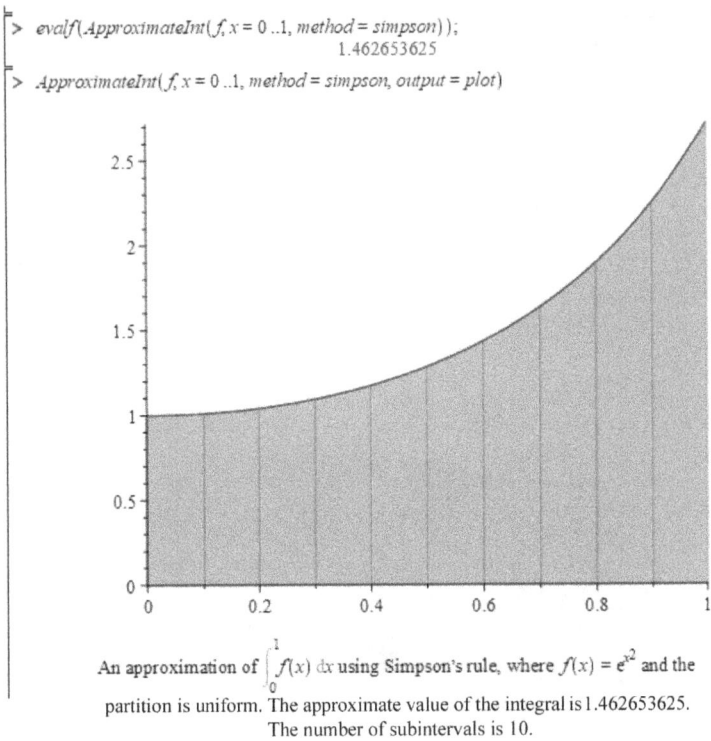

An approximation of $\int_0^1 f(x)\, dx$ using Simpson's rule, where $f(x) = e^{x^2}$ and the
partition is uniform. The approximate value of the integral is 1.462653625.
The number of subintervals is 10.

FIGURE 1.10
Simpson's rule approximation.

If x and c are points in the interval $[a, b]$, then the Taylor series expansion
of $f(x)$ about c:

$$f(c) + f'(c)(x-c) + \frac{f^{(2)}(c)}{2!}(x-c)^2 + \dots$$

or the Taylor series as

$$\sum_{k=0}^{\infty} \frac{1}{k!} f^{(k)}(c)(x-c)^k.$$

In general, the accuracy of a Taylor approximation is always an issue. In general, its accuracy increases with the number of terms used in the series. The remainder, or error term, after n terms is defined by

$$R_n = |f^{(n-1)}(c)|_{max} \frac{|x-c|^{n+1}}{(n+1)!},$$

where the subscript max refers to the maximum magnitude of the derivatives in the interval from $x = x_0$ to $x = b$.

Example 14. Taylor Series for the Function, $f(x) = e^x - x$ Expanded about $x = 2$.

> $evalf(\text{Taylor Approximation}(\exp(x) - x, x = 2, degree = 1..3))$
$-7.389056101 + 6.389056099\,x, -7.389056101 + 6.389056099\,x + 3.694528050\,(x-2.)^2,$
$-7.389056101 + 6.389056099\,x + 3.694528050\,(x-2.)^2 + 1.231509350\,(x-2.)^3$

> $\text{Taylor Approximation}(\exp(x) - x, x = 2, output = plot, degree = 4)$

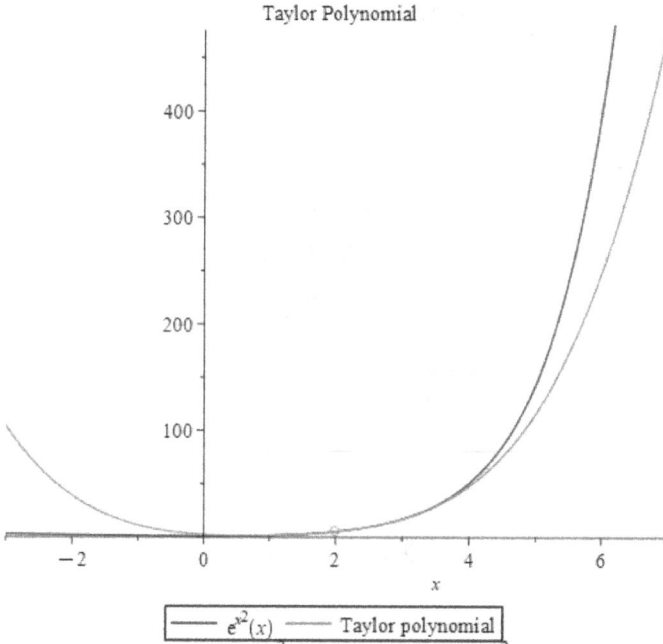

Taylor Polynomial

$e^{x^2}(x)$ ——— Taylor polynomial

At $x = 2$, for the function $e^{x^2}(x) = e^x - x$, a graph of $e^{x^2}(x)$ and the
approximating Taylor polynomial(s) of degree 4.

FIGURE 1.11
Taylor expansion of $f(x) = e^x - x$ about $x = 2$.

From Figure 1.11, we clearly see the closeness of the approximation near $x = 2$.
Perhaps you have encountered a function such as $f(x) = e^{x^2}$.
Integrating on a machine, we get an *erf* function.

$$> f1 := \exp(x^2);$$

$$f1 := e^{x^2}$$

$$> int(f1, x);$$

$$\frac{\sqrt{\text{perfi}(x)}}{2}$$

To most students, this is not helpful. Let's try a Taylor approximation about $x = 0$.

$$t1 := evalf\left(TaylorApproximation\left(\exp\left(x^2\right), x = 0, degree = 10\right)\right)$$

$$1 := 1. + x^2 + 0.5000000000x^4 + 0.1666666667x^6 + 0.04166666667x^8$$

> $+ 0.008333333333x^{10}$

Now, we can integrate the Taylor polynomial from 0 to 1. The plot is provided in Figure 1.12.

x

> $int(t1, x= 0. . .1);$

1.462530063

<

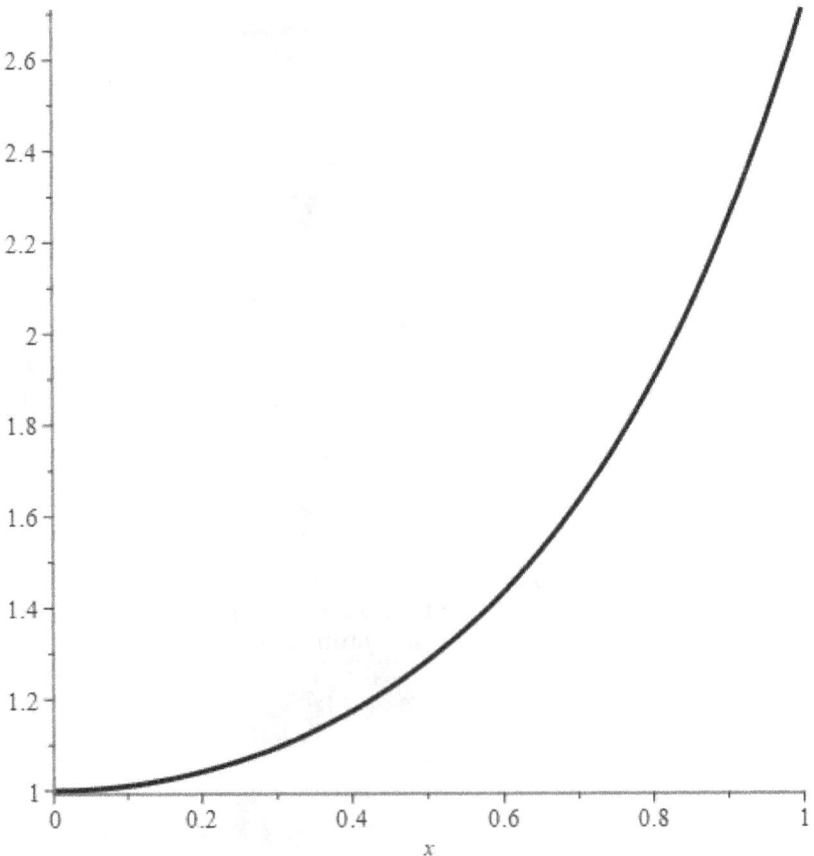

FIGURE 1.12
Plot of e^{x^2} from [0, 1].

Nearly all problems can be approximated using a continuous function, so calculus is extremely important for approximating numerical methods and different types of errors.

Exercises

1. Expand $f(x) = e^x$ around $x = 0$.
2. Expand $f(x) = e^x$ around $x = 1$.
3. Expand $f(x) = e^{2x}$ around $x = 1$.

1.8 Errors

There are different types of errors that we must consider:

Round-off error

Truncation error

Computer arithmetic error

Errors in scientific computing

The errors produced by a computer when performing a real-number calculation are called round-off errors. These occur when the machine uses an approximation to the real numbers, either chopped or rounded off. A simple example is 1/3 * 3, which we know equals 1. However, on a computer or a calculator 1/3 is approximated by 0.333333. . . . When this is multiplied by 3, we get 0.9999999 . . . as our answer is not the exact answer 1.

Truncation errors are attributable to the numerical methods used to solve a problem. We showed in Section 1.7 that the number of terms in a Taylor expansion affects the solution. Therefore, any series introduced as a truncated polynomial to represent a function causes this type error.

Scientific computing errors are obtained by the computer and its accuracy.

The choices if functions and values that are either stable or unstable. Stability and instability related to the equilibrium are explored in Chapter 3 on discrete dynamical systems. We will also see similar behavior in the use of polynomial approximations.

1.9 Algorithm Accuracy

An algorithm is a procedure that has a step-by-step methodology to be performed in a specific order to obtain a desired outcome. The objective is to implement the procedure to solve the problem or approximate a solution to a problem.

As a mathematical modeller, you will face issues. First, which algorithm should you choose to solve or approximate your solution? Second, after choosing your algorithm, what degree of accuracy do you require?

The choices of algorithms are explained in the text within the chapter in which they are contained. Each algorithm is provided with a step-by-step procedure. Some explanations and some geometric interpretations are provided if they help the modeller choose the better algorithm.

Accuracy is determined by many features. An exact answer, if available, is the most appropriate. Often, exact answers cannot be obtained or cannot be obtained in an appropriate amount of time required for a solution.

If f^* is found as an approximate solution, then the absolute error is $|f - f^*|$. In most of our algorithms, there is an input tolerance value where the absolute error must be less than the tolerance value.

Relative error is $|f - f^*|/|f|$, provided $f \neq 0$.

The number of significant digits is related to the relative error by $|f - f^*|/|f| < 5 \times 10^{-t}$, where t is the largest nonnegative integer for which this inequality holds (see Burden and Faires, 2001).

If you are using Maple, we can control the number of digits:

$$>\text{digits}:=$$

This causes the arithmetic operations to be rounded to the number of digits specified.

When you model, you are rarely (if ever) given a stopping criterion or a true value to compare your approximation to. Therefore, we use successive iterations to approximate f and f^* in our stopping criteria. In other words, if a true value is not available, we use the best available approximation of f in computing the error in our approximation to f^*.

Don't "hunt in the dark" and always curtail a procedure after a reasonable number of iterations.

References and Further Readings

Burden, R. and J. D. Faires (2001). *Numerical Analysis*. Cengage Publishing, Belmont, CA.

Burden, R. and J. D. Faires (2003). *Numerical Methods*. Thompson Publishing, Pacific Grove, CA.

Stewart, J. (2018). *Single Variable Calculus, Early Transcendentals*, 8th ed. Cengage Publishing, Belmont, CA.

2

Mathematical Modelling and Introduction to Technology: Perfect Partners

Consider the importance of decision making in such areas as business (B), industry (I), and government (G). BIG decision-making is essential to success at all levels. We do not encourage "shooting from the hip." We recommend good analysis for the decision-maker to examine and question in order to find the best alternative to choose or decision to make. So, why mathematical modelling?

A **mathematical model** is a description of a system using mathematical concepts and language. The process of developing a mathematical model is termed **mathematical modelling**. Mathematical models are used not only in the natural sciences (such as physics, biology, earth science, and meteorology) and engineering disciplines (e.g., computer science, artificial intelligence) but also in the social sciences (such as business, economics, psychology, sociology, and political science); physicists, engineers, statisticians, operations research analysts, and economists use mathematical models most extensively. A model may help explain a system, study the effects of different components, and make *predictions* about behavior.

Mathematical models can take many forms, including but not limited to dynamical systems, statistical models, differential equations, or game-theoretic models. These and other types of models can overlap, with a given model involving a variety of abstract structures. In general, mathematical models may include logical models, as far as logic is taken as a part of mathematics. In many cases, the quality of a scientific field depends on how well the mathematical models developed on the theoretical side agree with results of repeatable experiments. A lack of agreement between theoretical mathematical models and experimental measurements often leads to important advances as better theories are developed.

2.1 Overview and the Process of Mathematical Modelling

Consider that in bridge jumping, a participant attaches one end of a bungee cord to themselves, attaches the other end to a bridge railing, and then drops off the bridge. In this problem, the jumper will be dropping off the

DOI: 10.1201/9781032703671-2

Royal Gorge Bridge, a suspension bridge that is 1053 feet above the floor of the Royal Gorge in Colorado. The jumper will use a 200-foot-long long bungee cord. It would be nice if the jumper has a safe jump, meaning that the jumper does not crash into the floor of the gorge or run into the bridge on the rebound. In this project, you will do some analysis of the fall.

Assume the jumper weighs 160 pounds. The jumper will free-fall until the bungee cord begins to exert a force that acts to restore the cord to its natural (equilibrium) position. In order to determine the spring constant of the bungee cord, you found that that a mass weighing 4 pounds stretches the cord 8 feet. Hopefully, this spring force will help sufficiently slow the descent so that the jumper does not hit the bottom of the gorge. We would like to build a model to test mathematically if the jumper can make a safe jump.

Your team is responsible for collecting, analyzing, and coordinating the use of certain telemetry data from an experimental satellite (being piggy-backed on a rocket) being used by the NASA and the Jet Propulsion Laboratory to collect special space data. The data represent the velocity of the rocket. From the rocket launching through the first 6.5 seconds, these data are transmitted by the satellite back to the ground at varying time intervals. Occasionally, radio interferences cause some of the transmitted data to be garbled and consequently lost. Can we build mathematical models that allow us to replace the missing or garbled data with real numbers to complete our analysis? Mathematical methods in numerical analysis allow us to obtain estimates for the missing data and complete our analysis.

In the sport of bridge jumping, a willing participant attaches one end of bungee cord to himself, attaches the other end to a bridge railing, and then drops off a bridge. Is this a safe sport? Can we describe the typical motion?

You are a new city manager in California. You are worried about your city's water tower and water usage. Data might be sparse or unavailable. We can build a model and use numerical methods to estimate missing data and perform our analysis.

Maybe you have flown lately. Most airplanes are full these days. As a matter of fact, many times, an announcement is made that the plane is overbooked and that the airline is looking for volunteers to take a later flight. Why do airlines overbook? Should they overbook? What impact does this have on the passengers? What impact does it have on the airlines?

These are all events that we can model using mathematics. This textbook will help you understand what a mathematical modeller might do for you as a confident problem solver using the techniques of mathematical modelling. As a decision-maker, understanding the possibilities and asking the key questions will enable better decisions to be made.

2.2 The Modelling Process

In this chapter, we turn our attention to the process of modelling and examine many different scenarios in which mathematical modelling can play a role.

Mathematical modelling requires as much art as it does science. Thus, modelling is more of an *art* than a science. Modellers must be creative—willing to be more artistic or original in their approach to the problem. They must be inquisitive—questioning their assumptions, variables, and hypothesized relationships. Modellers must also think outside the box in order to analyze the models and their results. Modellers must ensure their model and results pass the commonsense test. Science is very important and understanding science enables one to be more creative in viewing and modelling a problem. Creativity is extremely advantageous in problem-solving with mathematical modelling.

To gain insight we should consider one framework that will enable the modeler to address the largest number of problems. The key is that there is something *changing for which we want to know the effects*. We call this the **system** under analysis. The real-world system can be very complicated or very simplistic. This requires a process that allows for both types of real-world systems to be modelled within the same process.

Consider striking a golf ball with a golf club from a tee. Our first inclination is to use the equations about distance and velocity that we used in our high school mathematics class. These equations are very simplistic and ignore many factors that could impact the fall of the ball such as wind speed, air resistance, mass of the ball, and others. As we add more factors, we can improve the precision of the model. Adding these additional factors makes the model more realistic and more complicated to produce. Understanding this model might be a first start in building a model for such situations or similar situation such as a bungee jumper or bridge swinger. These systems are similar for part of the model: the free-fall portion has similar characteristics.

Figure 2.1 provides a closed-loop process for modelling. Given a real-world situation like the earlier one, we collect data in order to formulate a mathematical model. This mathematical model can be one we derive or select from a collection of already built mathematical models depending on the level of sophistication required. Then we analyze the model that we used and reach mathematical conclusions about it. Next, we interpret the model and either make a predict about what has occurred or offer an explanation as to why something has occurred. Finally, we test our conclusion about the real-world system with new data. We may refine or improve the model to improve its ability to predict or explain the phenomena. We might even reformulate a new mathematical model.

Real World   Math World

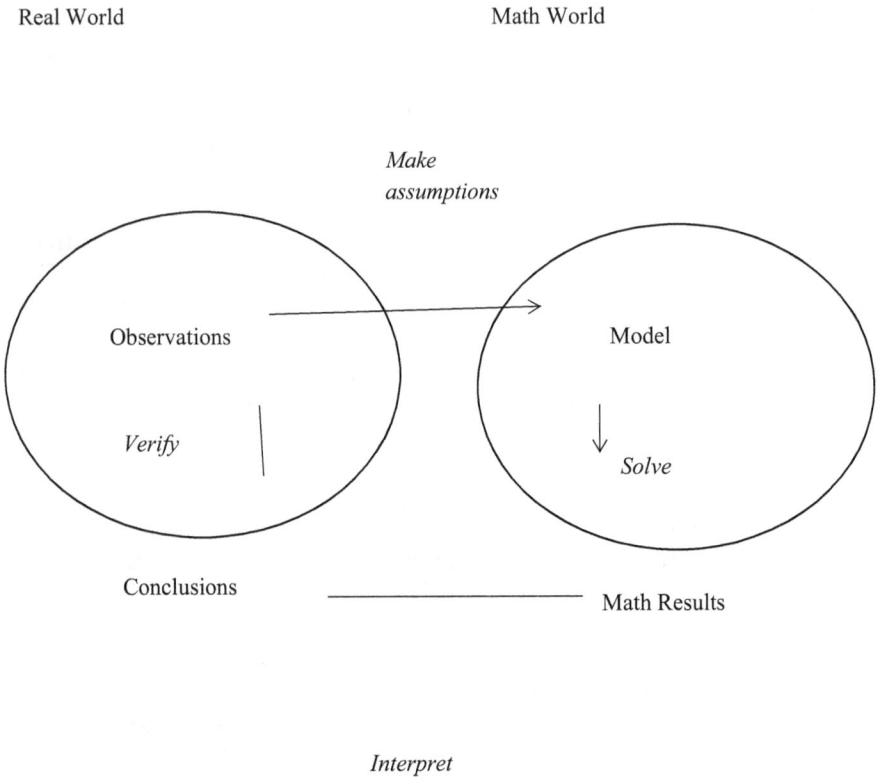

FIGURE 2.1
Modelling real-world systems with mathematics (see Albright, 2010).

Mathematical Modelling

We will build some mathematical models describing change in the real world. We will solve these models and analyze how good our resulting mathematical explanations and predictions are. The solution techniques that we employ in subsequent chapters take advantage of certain characteristics that the various models enjoy. Consequently, after building the models, we will **classify** the models based on their mathematical structure.

When we observe change, we are often interested in understanding why change occurs the way it does, perhaps to analyze the effects of different conditions or perhaps to predict what will happen in the future. Often, a mathematical model can help us understand a behavior better, while allowing us to experiment mathematically with different conditions. For our purposes, we will consider a mathematical model to be a mathematical construct designed to study a particular real-world system or behavior. The model allows us to use mathematical operations to reach mathematical conclusions about the model as illustrated in Figure 2.1.

Models and Real-World Systems

A system is an assemblage of objects joined by some regular interaction or interdependence. Examples include sending a module to Mars, handling the U.S. debt, a fish population living in a lake, a TV satellite orbiting the earth, delivering mail, locations of service facilities—all are examples of a system. The person modelling is interested in understanding not only how a system works but also what interactions cause change and how sensitivity the system is to changes in these inputs. Perhaps the person modelling is also interested in predicting or explaining what changes will occur in the system as well as when these changes might occur.

A possible basic technique used in constructing a mathematical model of some system is a combined mathematical-physical analysis. In this approach, we start with some known physical principles or reasonable assumptions about the system. Then we reason logically to obtain conclusions. Sometimes we have data and let help us come up with a reasonable model. Modellers must be open to many avenues to solve problems.

Figure 2.2 suggests how we can obtain real-world conclusions from a mathematical model. First, observations identify the factors that seem to be involved in the behavior of interest. Often, we cannot consider, or even identify, all the relevant factors, so we make simplifying assumptions excluding some of them. Next, we conjecture tentative relationships among the identified factors we have retained, thereby creating a rough "model" of the behavior. We then apply mathematical reasoning that leads to conclusions about the model. These conclusions apply only to the model and may or may not apply to the actual real-world system in question. Simplifications were made in constructing the model, and the observations on which the model is based invariably contain errors and limitations. Thus, we must carefully account for these anomalies and test the conclusions of the model against real-world observations. If the model is reasonably valid, we can then draw inferences

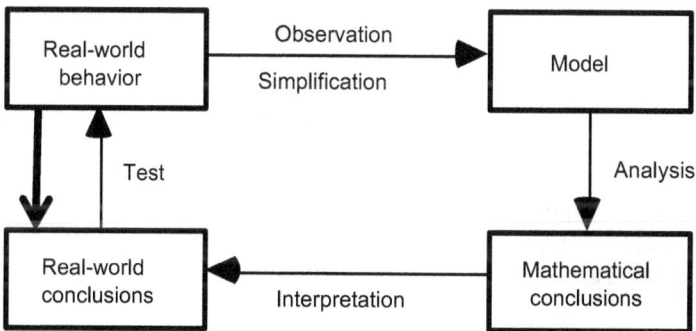

FIGURE 2.2
In reaching conclusions about a real-world behavior, the modelling process is a closed system (adapted from Giordano et al., 2013).

about the real-world behavior from the conclusions drawn from the model. In summary, we have the following procedure for investigating real-world behavior:

Step 1. Observation the system or hypothesis the system (if one does yet exist), identify the key factors involved in real-world behaviors, simplify initially, and refine later as necessary.

Step 2. Conjecture or guess the possible relationships or interrelationships among the factors and variables identified in step 1.

Step 3. Solve the model.

Step 4. Interpret the mathematical conclusions in terms of the real-world system.

Step 5. Test the model conclusions against real-world observations— the commonsense rule.

Step 6. Perform model testing or sensitivity analysis.

There are various kinds of models that we will introduce as well as methods or techniques to solve these models in the subsequent chapters. An efficient process would be to build a library of models and then be able to recognize various real-world situations to which they apply. Another task is to formulate and analyze new models. Still another task is to learn to solve an equation or system in order to find more revealing or useful expressions relating to the variables. Through these activities, we hope to develop a strong sense of the mathematical aspects of the problem, its physical underpinnings, and the powerful interplay between them.

Most models do simplify reality. Generally, models can only approximate real-world behavior. Next, let's summarize a *process* for formulating a model.

Model Construction

Let's focus our attention on the process of model construction. An outline is presented as a procedure to help construct mathematical models. In the next section, we illustrate this procedure with a few examples.

These nine steps are summarized in Figure 2.3 as modified from a six-step approach by Giordano et al. (2013). These steps act as a guide for thinking about the problem and getting started in the modelling process.

Let's discuss each step in more depth.

Step 1. Understand the problem or the question asked.

Identifying the problem to study is usually difficult. In real life, no one walks up to you and hands you an equation to be solved. Usually, it is a comment like "We

Step 1. Understand the problem or the question asked.

Step 2. Make simplifying assumptions. Justify your assumptions.

Step 3. Define all variables and provide units.

Step 4. Construct a model.

Step 5. Solve and interpret the model.

Step 6. Verify the model.

Step 7. Identify the strengths and weaknesses of your model.

Step 8. Sensitivity analysis or model testing of the model. Do the results pass the "common-

sense" test?

Step 9. Implement and maintain the model for future use.

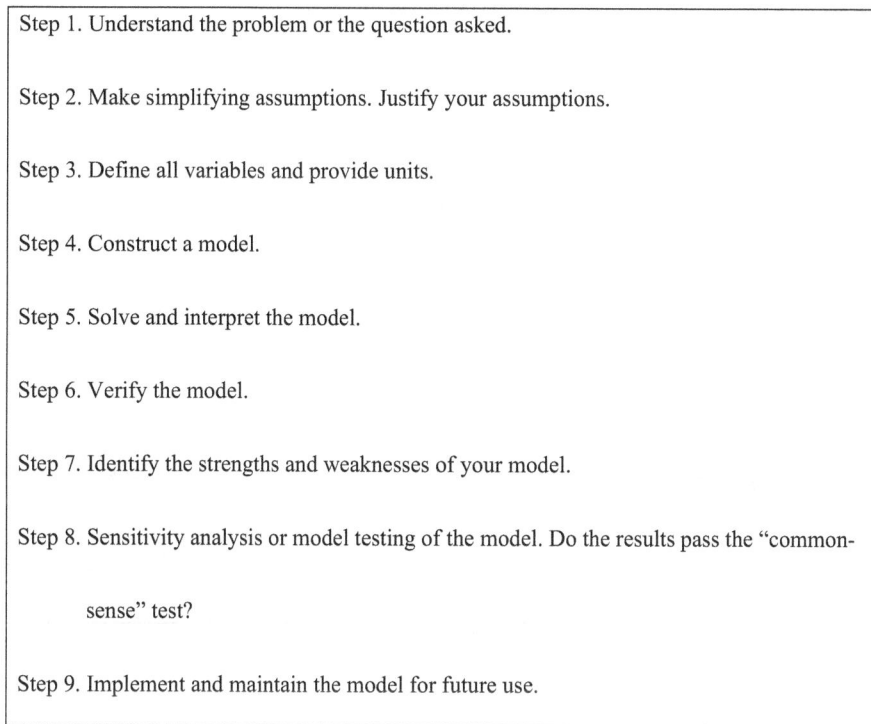

FIGURE 2.3
Mathematical modelling process.

need to make more money" or "We need to improve our efficiency." We need to be precise in our formulation of the mathematics to describe the situation.

Step 2. Make simplifying assumptions.

Start by brainstorming the situation. Make a list of as many factors, or variables, as you can. Realize we usually cannot capture all these factors influencing a problem. The task is simplified by reducing the number of factors under consideration. We do this by making simplifying assumptions about the factors, such as holding certain factors as constants. We might then examine to see if relationships exist between the remaining factors (or variables). Assuming simple relationships might reduce the complexity of the problem. Once you have a shorter list of variables, classify them as independent variables, dependent variables, or neither.

Step 3. Define all variables.

It is critical to define all your variables and provide the mathematical notation to be used for each.

Step 4. Select the modelling approach and formulate the model.

Using the tools in this text and your own creativity, build a model that describes the situation and whose solution helps to answer important questions.

Step 5. Solve and interpret the model.

We take the model we constructed in steps 1–4 and solve it. Often, this model might be too complex or unwieldy, so we cannot solve it or interpret it. If this happens, we return to steps 2–4 and simplify the model further.

Step 6. Verify the model.

Before we use the model, we should test it out. There are several questions we must ask. Does the model directly answer the question, or does the model allow for the answer to the questions to be answered? Is the model usable in a practical sense (can we obtain data to use the model)? Does the model pass the commonsense test?
We like to say that we corroborate the reasonableness of our model rather than verify or validate the model.

Step 7. Strengths and weaknesses.

No model is complete with self-reflection of the modelling process. We need to consider not only what we did right, but we did that might be suspect as well as what we could do better. This reflection also helps in refining models.

Step 8. Sensitivity analysis and model testing.

A modeller wants to know how the inputs affect the ultimate output for any system. Passing the commonsense test is essential. I once had a class model Hooke's law with springs and weights. I asked them all to use their model to see how far the spring would stretch using their weight. They all provided numerical answers, but none said that the spring would break under their weight.

Step 9. Refine, implement, and maintain the model.

A model is pointless if we do not use it. The more user-friendly the model the more it will be used. Sometimes the ease of obtaining data for the model can dictate its success or failure. The model must also remain current. Often this entails updating parameters used in the model.

2.3 Making Assumptions

In its simplest form, we

make assumptions,

do some "math," and

derive and interpret conclusions.

We say that one cannot question the math but can question the assumptions used to get to the model used. Assumptions drive the modelling as well as the analysis. According to Albright, every model is based on some set of assumptions. These can be trivial or more complex depending on what we know or can observe about the problem. It also might be affected by the available data.

That is why we say we can question the assumptions and ensure they are justified. We might also verify a key assumption of "accuracy" in terms of tolerance that is deemed acceptable for our model.

2.4 Illustrated Examples

We now demonstrate the modelling process that was presented in the previous section. Emphasis is placed on identifying the problem and choosing appropriate (usable) variables. We do not build the models as these modelling example are repeated later in the book, and the models are completed and discussed there.

Example 1. Prescribed Drug Dosage

Scenario: Consider a patient who needs to take a newly marketed prescribed drug. To prescribe a safe and effective regimen for treating the disease, one must maintain a blood concentration above some effective level and below an unsafe level.

Understanding the Problem: Our goal is a mathematical model that relates dosage and time between dosages to the level of the drug in the bloodstream. What is the relationship between the amount of drug taken and the amount in the blood after time, t? By answering this question, we are empowered to examine other facets of the problem of taking a prescribed drug.

Assumptions: We should choose or know the disease in question and the type (name) of the drug that is to be taken. We will assume the drug is Rythmol, a drug taken to control the heart rate and is called an antiarrhythmic. We need to know or to find decaying rate of Rythmol in the bloodstream. This might be found from data that has been previously collected. We need to find the safe and unsafe levels of Rythmol based on the drug's "effects" within the body. This will serve as the bounds for our model. Initially, we might assume that the patient's size and weight have no effect on the drug's decay rate. We might assume that all patients are about the same size and weight. All are in good health, and no one takes other drugs that affect the prescribed drug. We assume all internal organs are functioning properly. We might assume that we can model this using a discrete time period even though the absorption rate is a continuous function. These assumptions help simplify the model. We will see this model again in Chapter 3.

Example 2. Determining Heart Weight

Let's assume we are interested in building a model that models heart weight as a function of the a measurement of the heart. We might have access to data that relates, for the following seven mammals, their heart weight in grams and their diameter of the left ventricle of the heart measured in millimeters (mm) shown in Table 2.1. We might use a numerical method to get an interpolating polynomial to find the heart weight of an animal whose heart diameter is 8.5 millimeters.

Problem Identification: Find a relationship between heart weight and the diameter of the left ventricle of the heart.

Assumptions: We will assume that the heart weight is typical for a healthy animal. We further assume that the diameter is measured the same way for each heart. We might assume that all mammals are scale models of other mammals. We assume that all mammals are in good health. We assume that the data was collected the same way for each mammal.

TABLE 2.1

Data for Mammals Heart Sizes

Animal	Heart Weight (g)	Diameter (mm)
Mouse	0.13	0.55
Rat	0.64	1.0
Rabbit	5.8	2.2
Dog	102	4.0
Sheep	210	6.5
Ox	2030	12.0
Horse	3900	16.0

Adapted from special projects in MA 381 at West Point.

Example 3. A Bridge Too Far

Consider an engineering design for a truss bridge as shown in Figure 2.4. Trusses are lightweight structures capable of carrying heavy loads. In civil engineering bridge design, the individual members of the truss are connected with rotatable pin joints that permit forces to be transferred from one member of the truss to another. The accompanying Figure 2.4 shows a truss that is held stationary at the lower left endpoint 1, is permitted to move horizontally at the lower right endpoint 4, and has pin joints at 1, 2, 3, and 4 as shown. A load of 10 kilonewtons (kN) is placed at joint 3, and the forces on the members of the truss have magnitude given by $f_1, f_2, f_3, f_4,$ and f_5 as shown. The stationary support member has both a horizontal force F_1 and a vertical force F_2, but the movable support member has only the vertical force F_3.

If the truss is in static equilibrium, the forces at each joint must add to the zero vector, so the sum of the horizontal and vertical components of the forces at each joint must be zero. This produces the system of linear equations.

Problem Identification: Determine the forces on each joint as a function of the angles between joints.

Assumptions: We assume the bridge is sturdy and will not falter or collapse. We assume that the motion of the bridge due to normal motion is negligible. We assume a design such as the one provided. We might assume that the bridge is not in an earthquake zone. We will see this solution in Chapter 8.

Example 4. Oil Rig and Pumping Station Location

Consider an oil-drilling rig that is 8.5 miles offshore. The drilling rig is to be connected by an underwater pipe to a pumping station. The pumping station is connected by land-based pipe to a refinery, which is 14.7 miles down the shoreline from the drilling rig (see Figure 3.5). The underwater

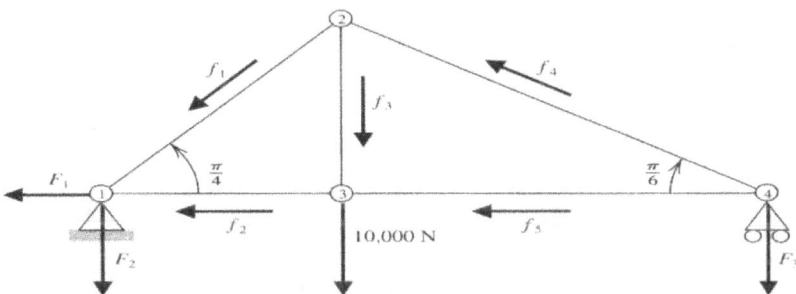

FIGURE 2.4
Bridge truss (from Burden and Faires, 1997, p. 423).

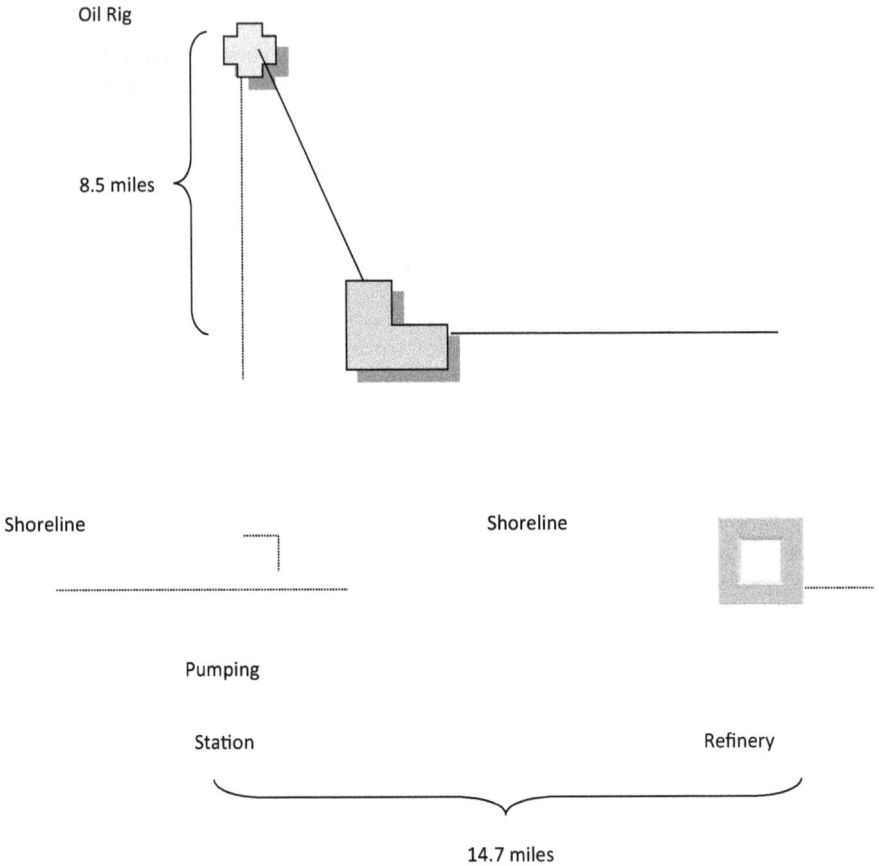

FIGURE 2.5
Oil rig, pumping station, and refinery.

pipe costs $31,575 per mile, and the land-based pipe costs $13,342 per mile. You are to determine where to place the pumping station to minimize the cost of the pipe.

Problem Identification: Build a model to minimize the cost of a pipe from the oil rig to the refinery.

Assumptions: We assume that the cost is based on solid estimates and will not fluctuate during the building of the pipeline. We assume no weather delays or any other natural or unnatural delays to building. All workers are competent in their jobs. The materials are not flawed in any way. We assume that no natural disaster or storms might disrupt work. We assume that, when built, the pipeline works successfully. A schematic is shown in Figure 2.5.

2.5 Technology

In mathematical modelling that we have done, it is impossible to proceed through all the steps without technology. That is why this chapter is called prefect partners. The partnering of technology with modelling is both key and essential to good modelling principles and practices. In this book, we illustrate three different technologies: Excel, Maple, and MATLAB.

Excel

Although Excel might not be the go-to technology for mathematicians or academicians, it is a "go-to" tool for the real world. The more we can empower students in math, science, and engineering to use Excel properly, then the better solutions will be for solving future problems.

Maple

Maple is an excellent technology for mathematics and operations research majors. Its power and graphical interface in two and three dimensions make it an excellent tool. Maple also has an excellent numerical analysis package that uses many, if not all, of the algorithms from Burden and Faires (1997).

R

R is an excellent technology tool for statisticians, engineers, and scientists. R will be illustrated, when appropriate, in this text.

Python

Python is a free downloadable software tool for statisticians, engineers, and scientists. Python will be illustrated, when appropriate, in this text. Python code will also be provided in this text. Jupyter notebook (which is also free) will be our Python interface.

Exercises

1. How would you approach a problem concerning a drug dosage? Do you always assume the doctor is right?
2. In modelling the size of any prehistoric creature, what information would you like to be able to obtain? What additional assumptions might be required?

3. In the oil-rig problem, what other factors might be critical in obtaining a "good" model that predicts reasonably well? What variables could be important that were not considered?

4. For the model in Example 2, are the assumptions about the data reasonable? How would you collect data to build the model. What other variables would you consider?

5. In the bridge example, what is the impact of the location of the bridge?

References and Additional Readings

Albright, B. (2010). *Mathematical Modeling with Excel*. Sudburry, MA: Jones and Bartlett.

Albright, B. and W. Fox (2020). *Mathematical Modeling with Excel*, 2nd ed. Taylor and Francis Publishers, Boca Raton, FL.

Burden, R. and D. Faires (1997). *Numerical Analysis*. Brooks-Cole, Pacific Grove, CA.

COMAP, Modeling Competition Sites Found. www.comap.com/contests

Fox, W. (2018). *Mathematical Modeling for Business Analytics*. Taylor and Francis Publishers, Boca Raton, FL.

Giordano, F., W. Fox and S. Horton (2013). *A First Course in Mathematical Modeling*, 5th ed. Cengage Publishers, Boston, MA.

3

Modelling with Discrete Dynamical Systems and Modelling Systems of Discrete Dynamical Systems

3.1 Introduction Modelling with Discrete Dynamical Systems

Consider car options today. Should we buy or lease a car? How do we go about financing a new car? Once you have looked at the makes and models to determine what type of car you like, it is time to consider the "costs" and finance packages that lure potential buyers into the car dealerships. This process can be modeled as a dynamical system. Payments are made typically at the end of each month. The amount owed is predetermined as we will see later in the chapter.

What if your company is faced with a decision of buying or leasing new computers for the company use for the next short-term horizon? How could this decision be analyzed? We might employ discrete dynamical systems to model and analyze the possible decisions.

A *discrete dynamical system* (DDS) is a system that changes over time. In this case, the time interval is discrete, such as minutes, hours, days, weeks, or years. A DDS is easy to model but maybe difficult or even impossible to solve in closed form. In this chapter, although we present some easy closed-form solutions, we concentrate more on building the models and obtaining numerical and graphical solutions.

Let's define the dynamical system $A(n)$ to be the amount of antibiotic in our blood stream after n time periods. The domain is nonnegative integers representing the time periods from $0, 1, 2, \ldots$ that will be the inputs to the function. Since the domain is discrete, then our function is a discrete function. The range is the values of $A(n)$ determined for each value of the domain. Thus, $A(n)$ also represents the dependent variable. For each input value of the domain from $0, 1, 2, \ldots$, the result is one and only one $A(n)$; thus, $A(n)$ is a function.

There are three components to dynamical systems: an equation for sequence representing $A(n)$, the time period n is well defined and has at least

DOI: 10.1201/9781032703671-3

one starting value. This starting value is called an **initial condition**. For example, if we start with no antibiotic in our system then $A(0) = 0$ milligrams is our initial condition, but if we started after we took an initial 200-milligram tablet, then $A(0) = 200$ milligrams would be our initial condition. An example of a discrete dynamical system with its initial condition would be

$$A(n + 1) = 0.5\,A(n), A(0) = 500.$$

We are interested in modelling discrete *change*. Modelling with DDSs employs a method to explain certain discrete behaviors or make long-term predictions. A powerful paradigm that we use to model with discrete dynamical systems is

future value = present value + change.

The dynamical systems that we will study with this paradigm may differ in appearance and composition, but we will be able to solve a large class of these "seemingly" different dynamical systems with similar methods. In this chapter, we will use iteration and graphical methods to answer questions about the DDSs.

We will use flow diagrams to help us see how the dependent variable changes. These flow diagrams help to see the paradigm and put it into mathematical terms. Let's consider financing a new Ford Mustang. The cost is $25,000, and you can put down $2,000, so you need to finance $23,000. The dealership offers you 2% financing over 72 months. Consider the flow diagram for financing the car below that depicts this situation in Figure 3.1.

We use this flow diagram to help build the discrete dynamical model. Let $A(n)$ equal the amount owed after n months. Notice that the arrow pointing into the circle is the interest to the unpaid balance. This increases your debt. The arrow pointing out of the circle is your monthly payment that decreases your debt. We define the following variables:

$$A(n + 1) = \text{the amount owed in the future}$$
$$A(n) = \text{amount currently owed}$$

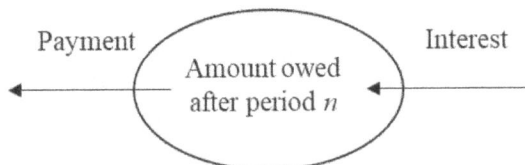

FIGURE 3.1
Flow diagram for financing a car.

Change as depicted in the flow diagram is $i\,A(n) - P$, so the model is

$$A(n + 1) = A(n) + i\,A(n) - P, \text{ where}$$

i is the monthly interest rate and
P is the monthly payment.

We model dynamical systems that have only **constant coefficients**. A dynamical system with constant coefficients may be written in the form

$$A(n + 3) = b_2\,A(n + 2) + b_1\,A(n + 1) + b_0\,A(n)$$

where b_0, b_1, and b_2 are arbitrary constants.

Solutions to DDSs

Although some DDSs have a closed-form analytical solution that we will discuss, our emphasis will be on iterative and graphical solution that we will illustrate in Excel. We will show that DDSs are relatively easy to model. The solutions may always be obtained by iterative and graphical methods. Not all DDSs have closed-form analytical solutions.

Let's begin with a simple DDS model with a closed-form solution.

Theorem 3.1 The solution to a linear discrete dynamical system $a(n + 1) = r\,a(n)$ for $r \ne 0$ is

$$a(k) = b^k\,a(0), \tag{3.1}$$

where $a(0)$ is the initial condition of the system at time period 0 and k is a generic time period.

Technology is an integral part of models with DDSs. Every DDS has a numerical and graphical solution, which can easily be attained with technology.

A Drug Dosage Example

Suppose that a doctor prescribes that their patient takes a pill containing 100 mg of a certain drug every hour. Assume that the drug is immediately ingested into the bloodstream once taken. Also assume that every hour the patient's body eliminates 25% of the drug that is in their bloodstream. Suppose that the patient had 0 mg of the drug in their bloodstream prior to taking the first pill. How much of the drug will be in their bloodstream after 72 hours?

Problem Statement: Determine the relationship between the amount of drug in the bloodstream and time.

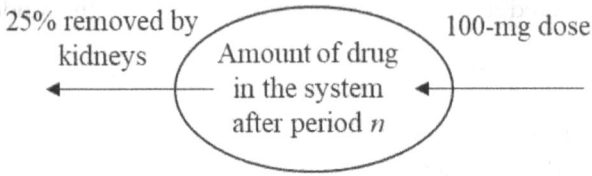

FIGURE 3.2
Change diagram for drugs in system.

Assumptions: The system can be modeled by a DDS. The patient is of average height and health. There are no other drugs being taken that will affect the prescribed drug. We assume that there are no internal or external factors that will affect the drug absorption rate. The patient always takes the prescribed dosage at the correct time. Figure 3.2 shown the change diagram.

Variables: Define $a(n)$ to be the amount of drug in the bloodstream after period n, $n = 0, 1, 2, \ldots$ hours.

Change Diagram: See Figure 3.2.

Model Construction:

Let's define the following variables:

$$a(n + 1) = \text{amount of drug in the system in the future}$$
$$a(n) = \text{amount currently in system}$$

We define change as follows: change = dose − loss in system

$$\text{change} = 100 - 0.25a(n)$$

so, *Future = Present + Change* is

$$a(n + 1) = a(n) - 0.25a(n) + 100$$

or

$$a(n + 1) = 0.75a(n) + 100$$

We note that this is not in the form as Theorem 3.1, so we introduce Theorem 3.2.

Theorem 3.2 The solution to a linear discrete dynamical system $a(n+1) = r\,a(n) + d$ for $r \neq 0$ and $d \neq 0$ is

$$a(k) = b^k C + (d/(1 - r)), \tag{3.2}$$

where $d/(1 - r)$ is the equilibrium value of the system, C is the initial condition of the system at time period 0, and k is a generic time period that provides $a(0) = V$.

We can employ Theorem 3.2 and find

$$A(k) = 0.75kC + d/(1 - r)$$
$$d = 100$$
$$1\text{-}0.75 = 0.25, \text{ so } d/(1 - r) = 400$$

Now,

$$A(0) = 0$$
$$\text{So, } 0 = C(0.75)^0 + 400$$
$$C = -400$$

The model, in general form, is

$$A(k) = 0.75^k(-400) + 400.$$

Let's iterate the system numerically. If we let $a(n)$ (we say "a at n" or "a of n") be the number of milligrams of drug in the bloodstream after n hours, and the initial amount in the bloodstream is 0 mg, then

$$a(0) = 0,$$
$$a(1) = 0.75(0) + 100 = 100,$$
$$a(2) = 0.75(100) + 100 = 175,$$
$$a(3) = 0.75(175) + 100 = 231.25 \text{ mg}.$$

We could write these equations where we do not substitute the numerical values that we calculated:

$$a(0) = 0$$
$$a(1) = 0.75a(0) + 100$$
$$a(2) = 0.75a(1) + 100$$
$$a(3) = 0.75a(2) + 100$$

Here, we see that the amount of drug in the bloodstream is related to the amount of drug in the bloodstream after the previous time period. Specifically, the amount of drug in the bloodstream after any of the first three time periods is 0.75 times the amount of drug in the bloodstream after the previous time period plus an additional 100 milligrams that is injected every hour.

We see that a pattern has developed that describes the amount of drug in the bloodstream. We are now prepared to conjecture (make an educated guess) about the amount of drug in the bloodstream after any hour. Mathematically, we say that the amount of drug in the bloodstream after n hours is

$$a(n + 1) = 0.75\, a(n) + 100.$$

This relationship describes the amount of drug in the bloodstream after n hours. With this, we can see the change that occurs every hour within this "system" (amount of drug in the bloodstream), and the state of the system, after any hour, is dependent on the state of the system after the previous hour. This is a *DSS*.

Analyzing the DDS: We want to find the value of $a(72)$. First, we can apply Theorems 3.1 and 3.2 as before.

$$a_e = 100/(0.25) = 400$$
$$a(k) = c\,(0.75)^k + 400$$

Since $a(0) = 0$,
$$0 = c + 400$$
$$c = -400.$$

The solution is $a(k) = -400(0.75)^k + 400$ as illustrated with Theorem 3.2 earlier.
For $a(72) = -400(0.75)^{72} + 400 = 399.9999996$.

Interpretation of Results: The DDS shows that the drug reaches a value where change stops and, eventually, the concentration in the bloodstream levels at 400 milligrams as shown in Figure 3.3. If 400 milligrams is both a safe and effective dosage level, then this dosage schedule is acceptable. We discuss this concept of change stopping (equilibrium) later in this chapter.

A Simple Mortgage Example

Five years ago, your parents purchased a home by financing $80,000 for 20 years, paying monthly payments of $880.87 with a monthly interest of 1%. They have made 60 payments and wish to know what they actually owe on the house at this time. They can use this information to decide whether they should refinance their house at a lower interest rate for the next 15 or 20 years. The change in the amount owed each period increases by the amount of the interest and decreases by the amount of the payment.

Problem Identification: Build a model that relates the time with the amount owed on a mortgage for a home.

Assumptions: Initial interest was 12%. Payments are made on time each month.

Variables: Let $b(n)$ = amount owed on the home after n months.

Flow Diagram: See Figure 3.4.

	A	B	C	D	E	F	G	H	I	J
	n	*a(n)*								
	0	0								
	1	100								
	2	175								
	3	231.25								
	4	273.438								
	5	305.078								
	6	328.809								
	7	346.606								
	8	359.955								
	9	369.966								
	10	377.475								
	11	383.106								
	12	387.329								
	13	390.497								
	14	392.873								
	15	394.655								
	16	395.991								
	17	396.993								
	18	397.745								
	19	398.309								
	20	398.732								
	21	399.049								
	22	399.286								
	23	399.465								
	24	399.599								
	25	399.699								
	26	399.774								
	27	399.831								

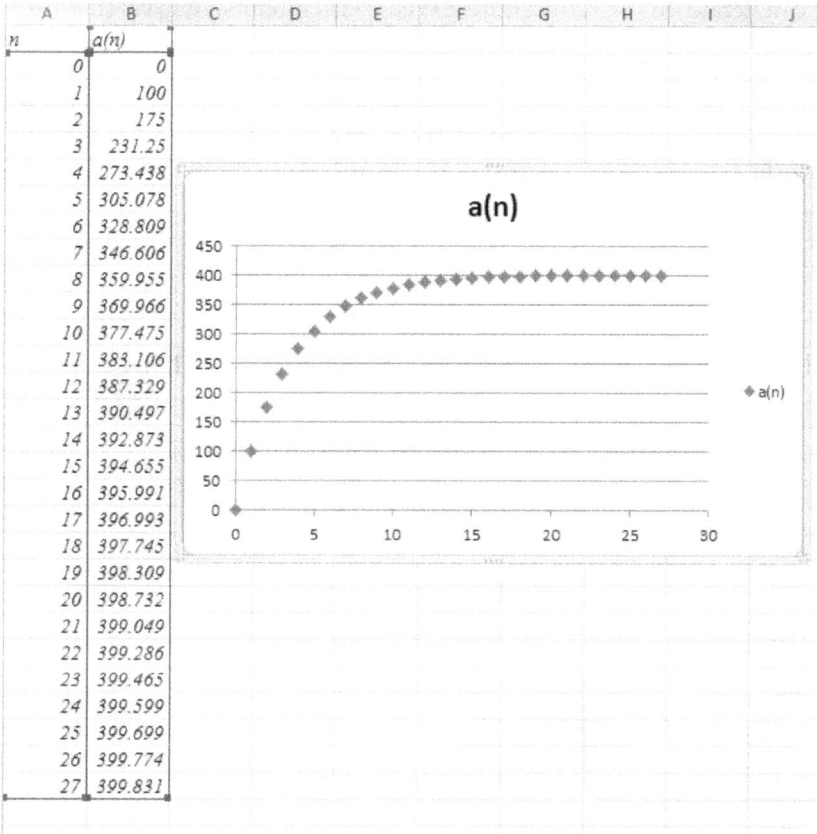

FIGURE 3.3
Behavior of drugs in our systems.

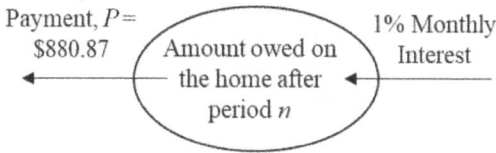

FIGURE 3.4
Flow diagram for mortgage example.

Model Construction:

$$b(n + 1) = b(n) + 0.12/12 \, b(n) - 880.87, \, b(0) = 80{,}000$$
$$b(n + 1) = 1.01 \, b(n) - 880.87, \, b(0) = 80{,}000$$

Model Solution:

$$\text{Mortgage Owed}(n) = -8087(1.01)n + 88{,}087$$

We can iterate this over the entire 20 years (240 months) in Excel (Figure 3.5).

After paying for 60 months, your parents still owe $73,395.37 out of the original $80,000.

In addition, they have paid in a total of $52,852.20 and only $6605 went toward the principal payment of the home. The rest of the money went toward paying only the interest. If the family continues with this loan, then

n	b(n)		int
0	80000	37	76400.67
1	79919.13	38	76283.8
2	79837.45	39	76165.77
3	79754.96	40	76046.56
4	79671.64	41	75926.15
5	79587.48	42	75804.55
6	79502.49	43	75681.72
7	79416.64	44	75557.67
8	79329.94	45	75432.38
9	79242.37	46	75305.83
10	79153.92	47	75178.02
11	79064.59	48	75048.93
12	78974.37	49	74918.55
13	78883.24	50	74786.86
14	78791.2	51	74653.86
15	78698.24	52	74519.53
16	78604.36	53	74383.85
17	78509.53	54	74246.82
18	78413.76	55	74108.42
19	78317.02	56	73968.64
20	78219.32	57	73827.45
21	78120.65	58	73684.86
22	78020.98	59	73540.84
23	77920.32	60	73395.37
24	77818.66		
25	77715.97		
26	77612.26		
27	77507.51		
28	77401.72		
29	77294.87		
30	77186.95		
31	77077.95		
32	76967.85		
33	76856.66		
34	76744.36		
35	76630.93		
36	76516.37		

FIGURE 3.5
Screenshot of mortgage iterations in Excel.

they will make 240 payments of $880.87 or $211,400.80 total in payments. This is $133,400.80 in interest. They have already paid $46,647.20 in interest. They would pay an additional $86,753.60 in interest over the next 15 years. What should they do? Perhaps an alternative scheme is available, such as refinance.

3.2 Equilibrium and Stability Values and Long-Term Behavior

We have previous mention Theorem 3.1 concerning equilibrium values to linear DDSs. Let's further discuss these equilibriums.

Equilibrium Values

Let's go back to our original paradigm,

$$Future = Present + Change$$

When change stops, the change equals zero, and the future equals the present. The value for which this happens, if any, is the equilibrium value. This gives us a context for the concept of the equilibrium value.

Models of the Form $a(n + 1) = r\, a(n) + b$, Where r and b Are Constants

Let's return to our drug dosage problem and consider adding the constant dosage each time period (time periods might be 4 hours). Our model is $a(n + 1) = 0.5\, a(n) + 16$ mg. We will also assume that there is an initial dosage applied prior to beginning the regime. We will let these initial values be as follows:

$$a(0) = 10$$

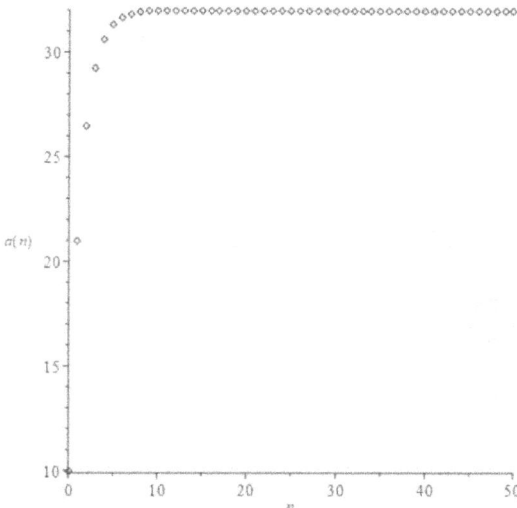

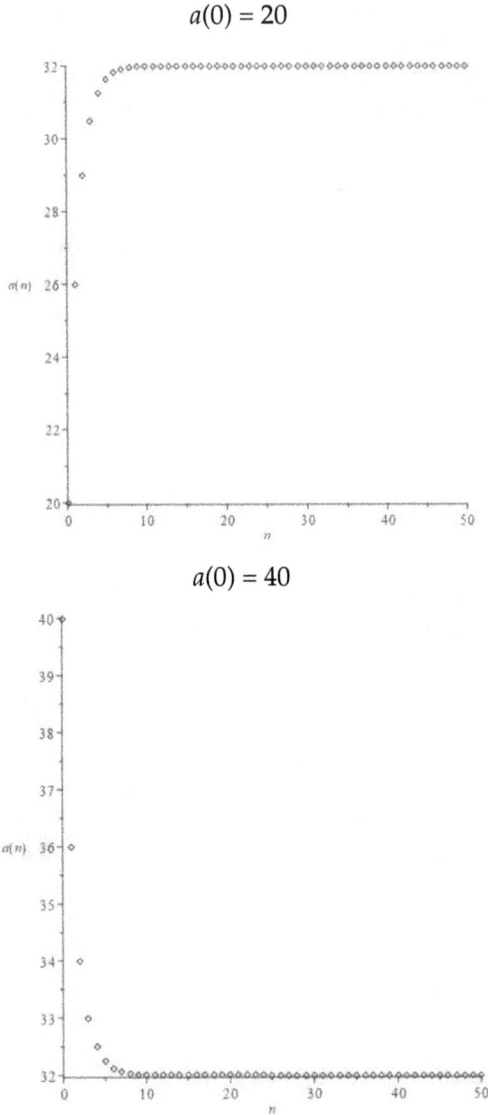

$$a(0) = 20$$

$$a(0) = 40$$

FIGURE 3.6
Plot of drugs in our systems with different initial starting conditions.

Regardless of the starting value, the future terms of $a(n)$ approach 32, as seen in Figure 3.6. Thus, 32 is the equilibrium value. We could have solved this algebraically as well.

$$a(n + 1) = 0.5 \, a(n) + 16$$
$$ev = 0.5ev + 16$$
$$0.5ev = 16$$
$$ev = 32$$

Another method for finding the equilibrium values involves solving the equation $a = ra + b$, and solving for a (where a is ev), we find

$$a = \frac{b}{1-r}, \text{ if } r \neq 1.$$

Using this formula in our previous example, the equilibrium value is

$$a = 16/(1 - 0.5) = 32.$$

Stability and Long-Term Behavior

For a dynamical system, $a(n + 1)$ with a specific initial condition, $a(0) = a_0$, we have shown that we can compute $a(1)$, $a(2)$, and so forth. Often these particular values are not as important as the long-term behavior. By long-term behavior, we refer to what will eventually happen to $a(n)$ for larger values of n. There are many types of long-term behavior that can occur with DDS, we will only discuss a few here.

If the $a(n)$ values for a DDS eventually get close to the equilibrium value, ev, no matter the initial condition, then the equilibrium value is called a **stable equilibrium value** or **an attracting fixed point**.

Often, we characterize the long-term behavior of the system in terms of its stability. If a DDS has an equilibrium value and if the DDS tends to the equilibrium value from starting values near the equilibrium value, then the DDS is said to be stable.

Thus, for the dynamical system $a(n + 1) = r\, a(n) + b$, where $b \neq 0$:

If $r \neq 1$, an equilibrium exists at $a = b/(1 - r)$.

If $r = 1$, no equilibrium value exists.

Relationship to Analytical Solutions

If a discrete dynamical system has an ev value, we can use the ev value to find the analytical solution, see Table 3.1.

TABLE 3.1

Stability for Linear Functions

Value of r	DDS Form	Equilibrium	Stability of Solution	Long-Term Behavior
$r = 0$	$a(n + 1) = b$	b	Stable	Stable equilibrium
$r = 1$	$a(n+1) = a(n) + b$	*None*	Unstable	
$r < 0$	$a(n + 1) = r * a(n) + b$	$b/(1 - r)$	Depends on $|r|$	Oscillations
$|r| < 1$	$a(n + 1) = r * a(n) + b$	$b/(1 - r)$	Stable	Approaches $b/(1 - r)$
$|r| > 1$	$a(n + 1) = r * a(n) + b$	$b/(1 - r)$	Unstable	Unbounded

Note. DDS = discrete dynamic system.

Recall the mortgage example from Section 3.1,

$$B(n + 1) = 1.00541667\, B(n) - 639.34,\ B(0) = 73{,}395$$

The *ev* value is found as 118,031.9274.

The analytical solution may be found using the following form:

$$B(k) = (1.00541667^{k})C + D,\ \text{where } D \text{ is the } ev.$$
$$B(k) = (1.00541667^{k})C + 118{,}031.9274,\ B(0) = 73{,}395$$

Since $B(0) = 73{,}395 = 1.00541667^{0}(C) + 118{,}031.9274$

$$C = -44{,}636.92736$$

Thus,

$$B(k) = -\ 44{,}636.92737\ (1.00541667^{k}) + 118{,}031.9274$$

Let's assume we did not know the payment was \$639.34 a month. We could use the analytical solutions to help find the payment.

$$B(k) = (1.00541667^{k})C + D$$

We build a system of two equations and two unknowns.

$$B(K) = (1.00541667^{k})C + D$$
$$B(0) = 73395 = C + D$$
$$B(180) = 0 = 1.00541667^{180}\,C + D$$
$$C = -44638.70,\ D = 118033.7$$
$$B(K) = -44638.70(1.00541667)^{k} + 118033.7$$

D represents the equilibrium value and we accepted some round-off error. From our model form

$$B(n + 1) = 1.00541667B(n) - P,\ \text{we can find } P.$$

Solving analytically for the equilibrium value,

$$X - 1.00541667X = -P$$
$$X = P/0.00541667$$
$$X \text{ is } 118{,}033.70,\ \text{so}$$

$$118{,}033.70 = P/0.00541667$$
$$P = 639.34.$$

Example 1. Growth of a Bacteria Population

We often model population growth by assuming that the change in population is directly proportional to the current size of the given population. This produces a simple first-order DDS similar to those seen earlier. It might appear reasonable at first examination, but the long-term behavior of growth without bound is disturbing. Why would growth without bound of a yeast culture in a jar (or controlled space) be alarming?

There are certain factors that affect population growth. Things include resources (food, oxygen, space, etc.) These resources can support some maximum population. As this number is approached, the change (or growth rate) should decrease, and the population should never exceed its resource-supported amount.

Problem Identification: Predict the growth of yeast in a controlled environment as a function of the resources available and the current population.

Assumptions and Variables: We assume that the population size is best described by the weight of the biomass of the culture. We define $y(n)$ as the population size of the yeast culture after period n. There exists a maximum carrying capacity, M, that is sustainable by the resources available. The yeast culture is growing under the conditions established.

Model:

$$y(n + 1) = y(n) + k\,y(n)\,(M - y(n)), \text{ where}$$
$$y(n) \text{ is the population size after period } n,$$
$$n \text{ is the time period measured in hours,}$$
$$k \text{ is the constant of proportionality, and}$$
$$M \text{ is the carrying capacity of our system.}$$

We have data shown in Table 3.2 for the growth of bacteria in a Petri dish. The variable, $y(n)$, is the number of bacteria at the end of period n.

It is often convenient to think about the way the variables change between time periods. we compute $\Delta y(n) = y(n + 1) - y(n)$. The values are provided in Table 3.3 and used to find the proportionality constant.

We find the constant slope is 0.0008. We also find the *ev* of 621.

TABLE 3.2

Bacteria Growth in a Petri Dish

n	0	1	2	3	4	5	6	7	8	9
y	10.3	17.2	27	45.3	80.2	125.3	176.2	256.6	330.8	390.4
n	10	11	12	13	14	15	16	17	18	19
y	440	520.4	560.4	600.5	610.8	614.5	618.3	619.5	620	621

TABLE 3.3

Bacteria Growth in a Petri Dish Solution Values

n	0	1	2	3	4	5	6	7	8
$y(n)$	10.3	17.2	27	45.3	80.2	125.3	176.2	256.6	330.8
$\Delta y(n)$	6.9	9.8	18.3	34.9	45.1	50.9	79.4	74.2	59.6
N	9	10	11	12	13	14	15	16	17
$y(n)$	390.4	440	520.4	560.4	600.5	614.5	618.3	619.5	620
$\Delta y(n)$	46.4	79.4	40	40.1	14	3.9	1.2	0.5	1

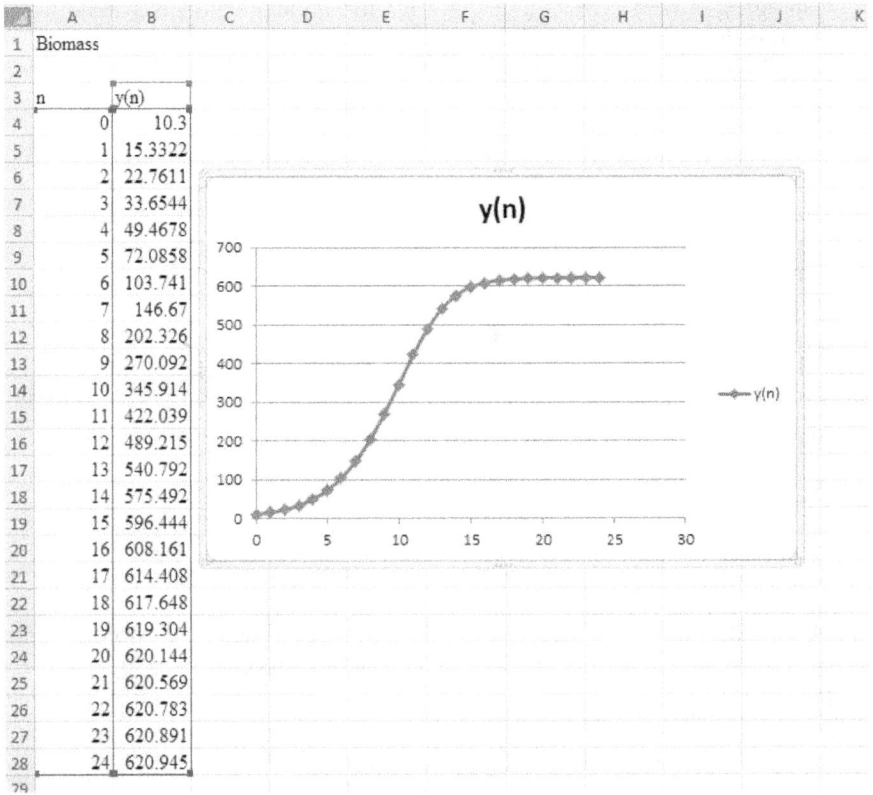

FIGURE 3.7

Screenshot of DDS for bacteria growth in a petri dish.

In our experiment, we first plot $y(n)$ versus n and find a stable equilibrium value of approximately 621. Next, we plot $y(n + 1) - y(n)$ versus $y(n)$ $(621 - y(n))$ to find the slope, k, is approximately 0.0008. With $k = 0.0008$ and the carrying capacity in biomass is 621. This model is

$$y(n + 1) = y(n) + 0.0008y(n)(621 - y(n)).$$

n	M		rate =	0.00009
0	2			
1	2.89964			
2	4.203721			
3	6.093805			
4	8.832676			
5	12.80036			
6	18.54577			
7	26.86042			
8	38.88267			
9	56.2438			
10	81.26881			
11	117.2454			
12	168.7686			
13	242.151			
14	345.8416			
15	490.7058			
16	689.8521			
17	957.4549			
18	1305.805			
19	1739.956			
20	2250.466			
21	2807.361			
22	3361.359			
23	3857.084			
24	4253.833			
25	4539.5			
26	4727.639			
27	4843.525			
28	4911.735			
29	4950.753			
30	4972.696			
31	4984.916			
32	4991.683			
33	4995.42			
34	4997.479			
35	4998.613			
36	4999.237			
37	4999.58			
38	4999.769			

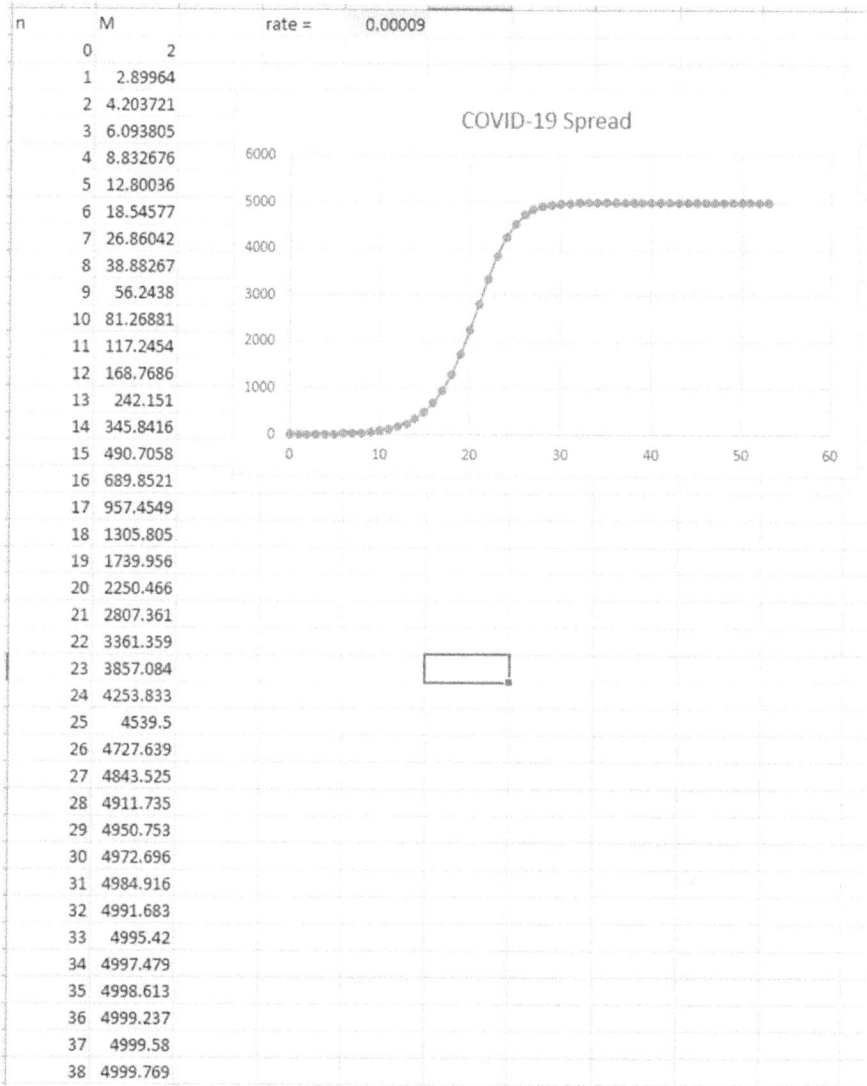

FIGURE 3.8
Screenshot of plot and iteration for the spread of COVID-19.

Again, this is nonlinear because of the $y^2(n)$ term. There is no closed form analytical solution for this equation, however we may obtain a solution through iteration and graphing the iterated values in Excel.

The model shows stability in that the population (biomass) of the yeast culture approaches 621 as n gets large. Thus, the population is eventually stable at approximately 621 units as shown in Figure 3.7.

Example 2. Spread of a Contagious Disease

There are 5000 students in college dormitories, and some students have been diagnosed with COVID-19, a highly contagious disease. The health center wants to build a model to determine how fast the disease will spread.

Problem Identification: Predict the number of students affected with COVID-19 as a function of time.

Assumptions and Variables: Let $m(n)$ be the number of students affected with COVID-19 after n days. We assume all students are susceptible to the disease. The possible interactions of infected and susceptible students are proportional to their product (as an interaction term).

Model:

$$m(n + 1) - m(n) = k\, m(n)\, (5000 - m(n))\ \text{or}$$
$$m(n + 1) = m(n) + k\, m(n)\, (5000 - m(n))$$

Two students returned from spring break with COVID-19, so $m(0) = 2$. The rate of spreading per day is characterized by $k = 0.00090$. It is assumed that there is no vaccine that can be introduced to slow the spread.

Interpretation: The results show that most students will be affected within 2 weeks. Since only about 10% will be affected within 1 week, every effort must be made to get the vaccination at the school and get the students vaccinated within 1 week. This is illustrated graphically in Figure 3.8.

3.3 Using Python for a Drug Problem

Consider the prescribed drug problem where the body eliminates 50% of the drug every 8 hours and every 8 hours we take 16 milligrams of the prescribed drug. Let $D(n)$ equal the amount of the drug in our system after n periods, where $n = 0, 1, 2, \ldots$ and is measured in 8-hour periods. Assume that $D(0) = 0$. The plot is seen in Figure 3.9

Python Code:

```
# Program to print the drug DDS up to n_terms

# Recursive function
def recursive_drug(n):
    if n <= 1:
```

```
        return n
   else:
      return(recursive_drug(n-1) -.5* recursive_drug(n-1)+16)

n_terms = 20

# check if the number of terms is valid
if n_terms <= 0:
   print("Invalid input ! Please input a positive value")
else:
   print("Drug over time")

for i in range(n_terms):
print(recursive_drug(i))
import pylab as pl
for i in range(n_terms):
y1 = [0,1,2,3,4,5,6,7,8,9,10,11,12,13,14,15,16,17,18,19,20]
 y2 = [0,1,16.5,24.25,30.06,31.03,31.52,31.76,31.88,
31.94,31.97,31.98, 31.99, 32,32,32,32,32,32,32,32]
 x1 = range(len(y1))
x2 = range(len(y1))
pl.plot(x2, y2,'bo')
pl.show()
```

Output

```
Drug over time
0
1
16.5
24.25
28.125
30.0625
31.03125
31.515625
31.7578125
31.87890625
31.939453125
31.9697265625
31.98486328125
31.992431640625
31.9962158203125
31.99810791015625
31.999053955078125
31.999526977539062
31.99976348876953
31.999881744384766
```

Plot: See Figure 3.9.

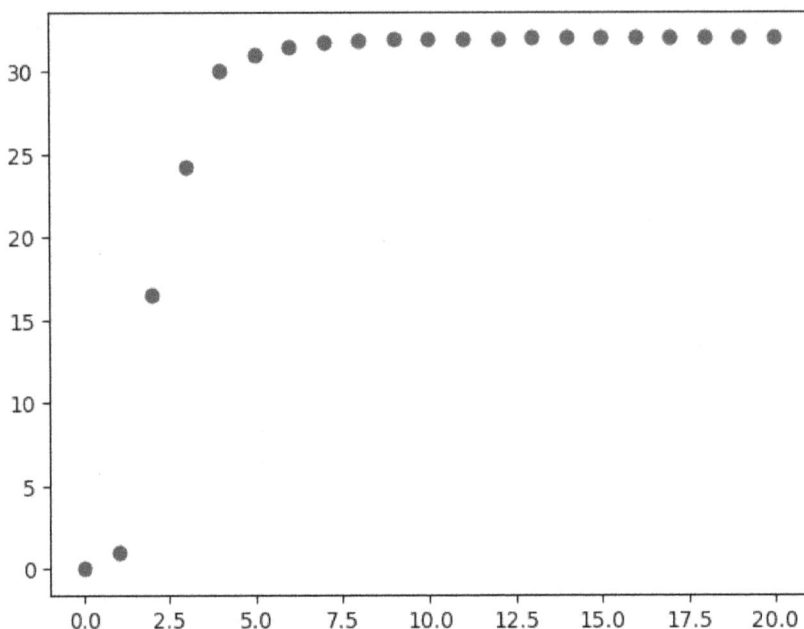

FIGURE 3.9
Drug problem with a plot from Python.

3.4 Introduction to Systems of DDSs

In the previous sections, we reviewed linear and nonlinear DDS models. Now, we extend the discussion to systems of systems, but we still use our paradigm:

Future = Present + Change.

Consider wanting to retire on a lake that you stock with bass and trout for endless fishing. Will it be endless? Can the species co-exist in your lake? How often do you need to restock the lake? This is an example of a competitive hunter model where both species compete for the same resources.

Let's define a system of DDS:

$$A(n) = f(A(n), B(n))$$
$$B(n) = g(A(n), B(n)).$$

As before, simple linear systems of DDS have closed-form analytical solutions; however, most systems do not. We will analyze all those DDS through iteration and graphs.

For selected set of initial conditions, we build numerical solutions to get a sense of long-term behavior for the system. For the systems that we will study, we will find their equilibrium values. We then explore starting values near the equilibrium values to see if by starting close to an equilibrium value, the system will

a. remain close,
b. approach the equilibrium value, or
c. not remain close.

What happens near these values gives great insight concerning the long-term behavior of the system. We can study the resulting pattern of the numerical solutions and the resulting plots.

Simple Linear Systems and Analytical Solutions

Let's consider school vouchers. There are both students in the public school (PS) and the private magnet (PM) school. Suppose a pre-survey of families used as historical records determined that 75% of students at the PM school remain while 25% preferred to transfer. We found the 65% of the PS students want to remain but 35% preferred to transfer. Let's build a model to determine the long-term behavior of these students based on this historical data. The change diagram is shown in Figure 3.10.

Problem Identification: Determine the school voucher over time.

Assumptions and Variables: Let n represent the number of student months. We define

$PS(n)$ = the number of students in public school at the end of n months
$PM(n)$ = the number of students in the magnet school at the end of n months

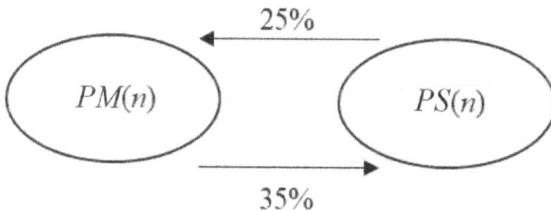

FIGURE 3.10
Change diagram for the school vouchers.

We assume that no other incentives are given to the students for either staying or moving.

The Model:

To build the dynamical system, mathematically, this is written as

$$PM(n + 1) = 0.75\ PM\ (n) + 0.35\ PS(n)$$
$$PS(n + 1) = 0.25\ PM(n) + 0.65\ PS(n).$$

There are initially 1500 students in the magnet school and 2000 students in the public school. We seek to find the long-term behavior of this system.

We rewrite the model as a system of DDS:

$$PM(n + 1) = 0.75\ PM\ (n) + 0.35\ PS(n)$$
$$PS(n + 1) = 0.25\ PM(n) + 0.65\ PS(n)$$

$PM(0) = 1500$ and $PS(0) = 1000$, respectively.

This is a simple linear system. We may use the initial conditions to iterate this system of discrete dynamical systems as shown in Figure 3.11. We see

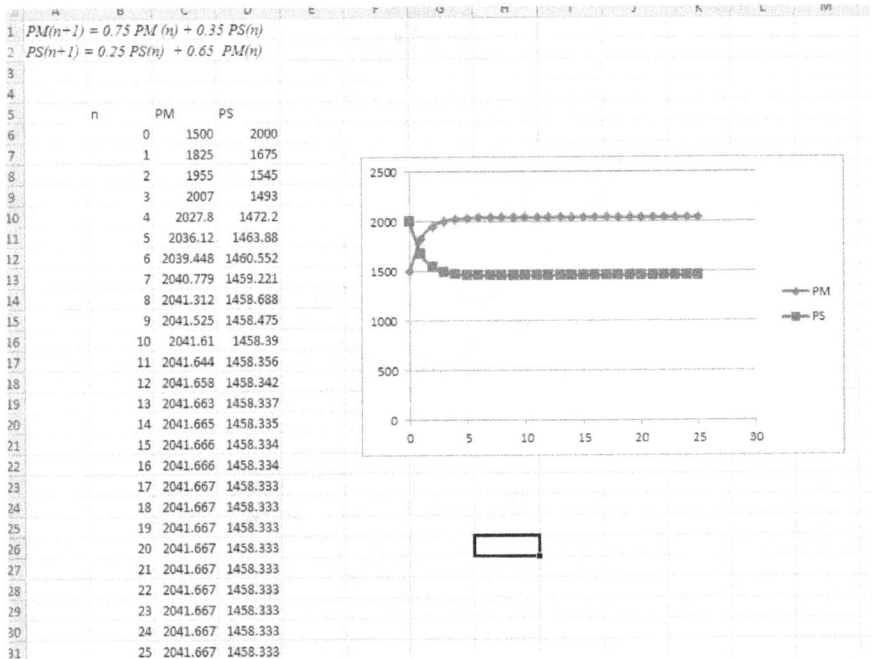

FIGURE 3.11
Screenshot iterative solution and plot for voucher students.

from the solution plot that there are stable equilibriums at about 2042 and 1458 in the magnet and public schools, respectively.

Analytical Solutions

Analytical solutions assume knowledge of linear algebras through eigenvalues and eigenvectors.

We rewrite the DDS in matrix form:

$$X(n + 1) = MX(n), X(0) = B$$

$$\begin{bmatrix} PM(n+1) \\ PS(n+1) \end{bmatrix} = \begin{bmatrix} 0.75 & 0.35 \\ 0.25 & 0.65 \end{bmatrix} \begin{bmatrix} PM(n) \\ PS(n) \end{bmatrix}, \begin{bmatrix} PM(0) \\ PS(0) \end{bmatrix} = \begin{bmatrix} 1500 \\ 2000 \end{bmatrix}$$

We define the analytical solution with two distinct real eigenvalues and eigenvectors. The general form of the solution is

$$X(k) = 1_1^k c_1 V_1 + \lambda_2^k c_2 V_2, \text{ where}$$

λ_1 and l_2 are the two distinct eigenvalues,

V_1 and V_2 are the corresponding eigenvectors, and

c_1 and c_2 are the constants.

The characteristic polynomial is $\lambda^2 - 1.4\lambda + 0.40 = 0$. This provides two eigenvalues: 1, 0.4. The corresponding eigenvalues can be found easily as vectors for $\lambda = 1$ of [0.35, 0.25] and for $\lambda = 0.40$ as [1, −1].

The general solution is

$$X(k) = c_1 \left(1^k\right) \begin{bmatrix} 0.35 \\ 0.25 \end{bmatrix} + c_2 \left(.4^k\right) \begin{bmatrix} 1 \\ -1 \end{bmatrix}.$$

With our initial conditions we find c_1 and c_2 to be 416.6666 and −4.16666, so

$$X(k) = 5833.333\left(1^k\right) \begin{bmatrix} .35 \\ .25 \end{bmatrix} - 541.6666\left(.4^k\right) \begin{bmatrix} 1 \\ -1 \end{bmatrix}.$$

When $k = 10$, we find 2041.609 and 1458.31 for students, respectively. We see this again in Figure 3.12.

Analytically, we can solve for the equilibrium values. We let $X = D(n)$ and $Y = M(n)$. From the DDS, we obtain the equations

$$X = 0.75X + 0.35Y$$
$$Y = 0.25X + 0.65Y,$$

and both equations reduce to $X = 0.35/0.25Y$. There are two unknowns, so we need a second equation.

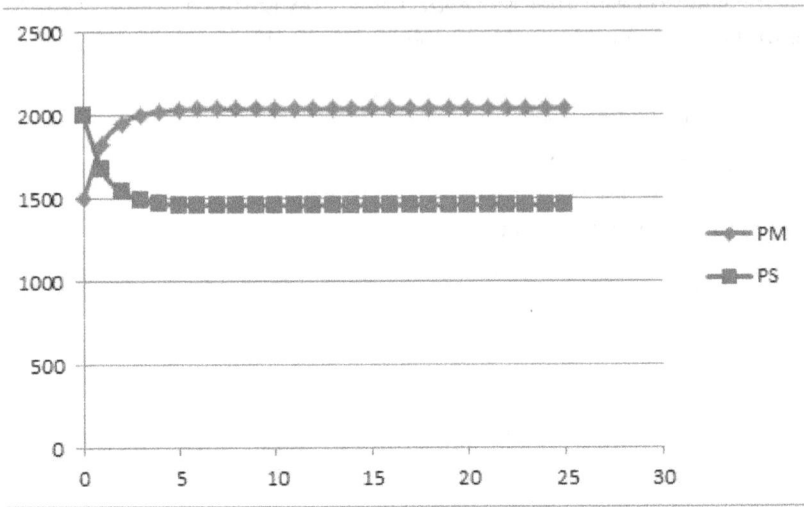

FIGURE 3.12
Plot of DDS for student's relocation example.

From the initial conditions, we know that $X + Y = 3500$. We can use the equations

$$X + Y = 3500 \text{ and } X = 0.35/0.25Y \text{ to find the equilibrium values:}$$
$$X = 2041.6667 \text{ and } Y = 1458.3333.$$

We previously iterated the solution and now we start with initial conditions near f those equilibrium values and we find the sequences tend toward those values. We conclude the system has *stable* equilibrium values.

You should go back and change the initial conditions and see what behavior follows.

Interpretation: The long-term behavior shows that eventually (without other influences) that of the 3500 students that about 1458 remain in public school and 2042 go to the magnet school. We might want to try to attract students with advertising and perhaps add incentives.

Iteration and Graphical Solution

Example 3. Competitive Hunter Models

Competitive hunter models involve species vying for the same resources (such as food or living space) within the habitat. The effect of the presence of a second species diminishes the growth rate of the first species. We now consider a specific example concerning trout and bass in a small pond. Hugh Ketum owns a small pond that he uses to stock fish and eventually allows fishing. He has decided to stock both bass and trout. The fish and game warden, after inspecting his pond for environmental conditions, tells Hugh that he has a solid pond for growth of his fish. In isolation, bass grow at a rate of 20% and trout at a rate of 30%. The warden tells Hugh that the interaction for the food affects trout more than bass. They estimate the interaction affecting bass is 0.0010 bass * trout and for trout is 0.0020 bass * trout. Assume no changes in their habitant occur.

Model:

Let's define the following variables:

$B(n)$ = the number of bass in the pond after period n.

$T(n)$ = the number of trout in the pond after period n.

$B(n) * T(n)$ = interaction of the two species.

$B(n + 1) = 1.20B(n) - 0.0010B(n) * T(n)$

$T(n + 1) = 1.30T(n) - 0.0020B(n) * T(n)$

The equilibrium values can be found by allowing $X = B(n)$ and $Y = T(n)$ and solving for X and Y:

$$X = 1.2X - 0.001X * Y$$
$$Y = 1.3Y - 0.0020X * Y.$$

We rewrite these equations as

$$0.2X - 0.001X * Y = 0$$
$$0.3Y - 0.002X * Y = 0.$$

We can rewrite the two equations to obtain

$$X (0.2 - 0.001Y) = 0$$
$$Y (0.3 - 0.002X) = 0.$$

Solving we find $X = 0$ or $Y = 2000$ and $Y = 0$ or $X = 1500$.

We want to know the long-term behavior of the system and the stability of the equilibrium points.

Hugh initially considers 151 bass and 199 trout for his pond. The solution is left to the student as an exercise. From Hugh's initial conditions, bass will grow without bound and trout will eventually die out.

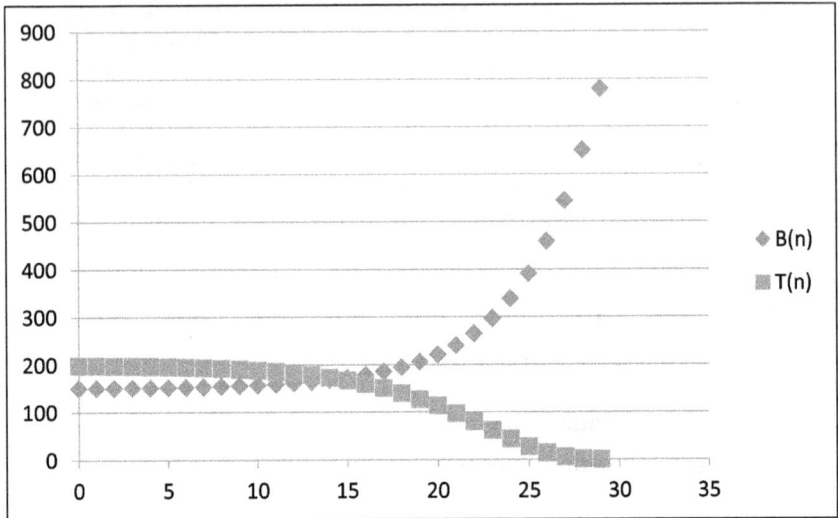

FIGURE 3.13
Bass and trout over time.

We iterated the system and obtained the plot, Figure 3.13, of bass and trout over time. The trout die out at about period 29.

This is certainly not what Hugh had in mind so he must find ways to improve the environment for the fish that alter the parameters from the model.

Example 4. Fast-Food Tendencies

Consider that your student union center desires to have three fast-food chains available to students serving: burgers, tacos, and pizza. These chains run a survey of students and find the following information concerning lunch: 75% of those who ate burgers will eat burgers again at the next lunch, 5% will eat tacos next, and 20% will eat pizza next. Of those who ate tacos last, 20% will eat burgers next, 60% will stay will tacos, and 35% will eat pizza next. Of those who ate pizza, 40% will eat burgers next, 20 % tacos, and 40% pizza again.

We formulate the problem as follows:

Let n represent the nth day's lunch, and so we define the following variables:

$B(n)$ = the number of burger eaters in the nth lunch
$T(n)$ = the number of taco eaters in the nth lunch
$P(n)$ = the number of pizza eaters in the nth lunch

Formulating the system, we have the following dynamical system:

$$B(n + 1) = 0.75B(n) + 0.20T(n) + 0.40P(n)$$
$$T(n + 1) = 0.05B(n) + 0.60T(n) + 0.20P(n)$$
$$P(n + 1) = 0.20B(n) + 0.20T(n) + 0.40P(n).$$

Analytically, we let $X = B(n)$, $Y = T(n)$, and $Z = P(n)$ so that

$$X = 0.75X + 0.2Y + 0.4Z$$
$$Y = 0.05X + 0.6Y + 0.2Z$$
$$Z = 0.2X + 0.2Y + 0.4Z.$$

These equations reduce to

$$X = 20/9Z$$
$$Y = 7/9Z$$
$$Z = Z.$$

Since we have 14,000 students, then we assume that $X + Y + Z = 14,000$. We substitute and solve for Z first.

$$4Z = 14,000$$
$$Z = 3500$$
$$X = 20/9Z = 20/9(3500) = 7777.77$$
$$Y = 7/9Z = 7/9(3500) = 2722.222$$

Suppose the campus has 14,000 students that eat lunch. The graphical results also show that an equilibrium value is reached at a value of about 7778 burger eaters, 2722 taco eaters, and 3500 pizza eaters. This allows the fast-food establishments to plan for a projected future. We see this in the iterated table in Figure 3.14. By varying the initial conditions for 14,000 students, we find that these values stable equilibrium values.

n	B(n)	T(n)	P(n)
0	14000	0	0
1	10500	700	2800
2	9135	1505	3360
3	8496.25	2031.75	3472
4	8167.338	2338.263	3494.4
5	7990.916	2510.204	3498.88
6	7894.78	2605.444	3499.776
7	7842.084	2657.961	3499.955
8	7813.137	2686.872	3499.991
9	7797.224	2702.778	3499.998
10	7788.473	2711.528	3500
11	7783.66	2716.34	3500
12	7781.013	2718.987	3500
13	7779.557	2720.443	3500
14	7778.756	2721.244	3500
15	7778.316	2721.684	3500
16	7778.074	2721.926	3500
17	7777.941	2722.059	3500
18	7777.867	2722.133	3500
19	7777.827	2722.173	3500
20	7777.805	2722.195	3500

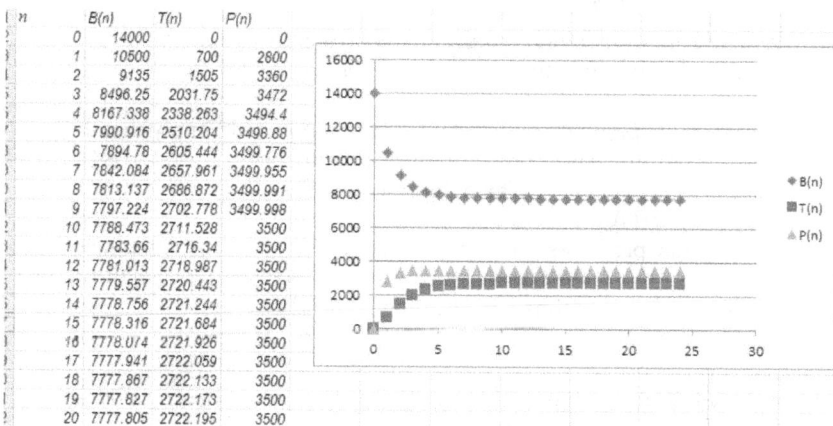

FIGURE 3.14
Screenshot Excel iterated solution and plot for fast food on campus.

3.5 Modelling of Predator–Prey, SIR, and Military Models

Example 5. A Predator–Prey Model: Foxes and Rabbits

In the study of the dynamics of a single population, we typically take into consideration such factors as the "natural" growth rate and the "carrying capacity" of the environment. Mathematical ecology requires the study of populations that interact, thereby affecting each other's growth rates. In this module, we study a very special case of such an interaction in which there are exactly two species and the predator eats the prey. Such pairs exist throughout nature such as lions and gazelles, birds and insects, pandas and bamboo plants, and Venus fly traps and flies.

To keep our model simple, we will make some assumptions that would be unrealistic in most of these predator–prey situations. Specifically, we will assume that

- the predator species is totally dependent on a single prey species as its only food supply,
- the prey species has an unlimited food supply, and
- there exist no other threats to the prey other than the specific predator.

In this modelling process, we will use the Lotka–Volterra model for predator–prey example. Students can read more about the Lotka–Volterra models in the sources in the suggested readings. Here, we simply present the model that we use.

We repeat our two key assumptions:

- The predator species is totally dependent on the prey species as its only food supply.
- The prey species has an unlimited food supply and no threat to its growth other than the specific predator.

If there were no predators, the second assumption would imply that the prey species grows exponentially without bound; that is, if $x = x(n)$ is the size of the prey population after a discrete time period n, then we would have $x(n + 1) = ax(n)$.

But there *are* predators, which must account for a negative component in the prey growth rate. Suppose we write $y = y(n)$ for the size of the predator population at time t. Here are the crucial assumptions for completing the model:

- The rate at which predators encounter prey is jointly proportional to the sizes of the two populations
- A fixed proportion of encounters lead to the death of the prey

These assumptions lead to the conclusion that the negative component of the prey growth rate is proportional to the product xy of the population sizes; that is,

$$x(n+1) = x(n) + ax(n) - bx(n)y(n).$$

Now we consider the predator population. If there were no food supply, the population would die out at a rate proportional to its size; that is, we would find $y(n+1) = -cy(n)$.

We assume that is the simple case that the "natural growth rate" is a composite of birth and death rates, both presumably proportional to population size. In the absence of food, there is no energy supply to support the birth rate.) But there is a food supply: the prey. And what's bad for hares is good for lynx. That is, the energy to support growth of the predator population is proportional to deaths of prey, so

$$y(n+1) = y(n) - cy(n) + px(n)y(n).$$

This discussion leads to the discrete version of the Lotka–Volterra predator–prey model:

$$x(n+1) = (1+a)x(n) - bx(n)y(n)$$
$$y(n+1) = (-c)y(n) + px(n)y(n)$$
$$n = 0,1,2,...$$

where a, b, c, and p are positive constants.

The Lotka–Volterra model consists of a system of linked dynamical systems equations that cannot be separated from each other and that cannot be solved in closed form. Nevertheless, they can be solved numerically and graphed in order to obtain insights into the scenario being studied.

Let us return to our fox-and-rabbit scenario. Let's assume this discrete model is as explained earlier. Furthermore, data investigation yields the following estimates for the parameters that we require: $\{a, b, c, p\} = \{0.04, 0.0004, 0.09, 0.001\}$. Let's further assume that initially there are 600 rabbits and 125 foxes.

We iterate and plot the results for rabbits and foxes versus time and then plot rabbits versus foxes shown in Figures 3.15 and 3.16.

If we ran this model for many more iterations, we would find the plot of foxes-versus-rabbits spiral in a similar fashion as above. We conclude that the model appears reasonable. We could find the equilibrium values for the system. There are a set of feasible equilibrium points for rabbits and foxes at (0, 0) and at (2725, 960). The orbits of the spiral indicate that the system is moving away from both (0, 0) and (2725, 960), so we conclude the system is not stable.

In predator–prey models, it is often that "managers" of the ecological system must intervene in some way to keep both species flourishing.

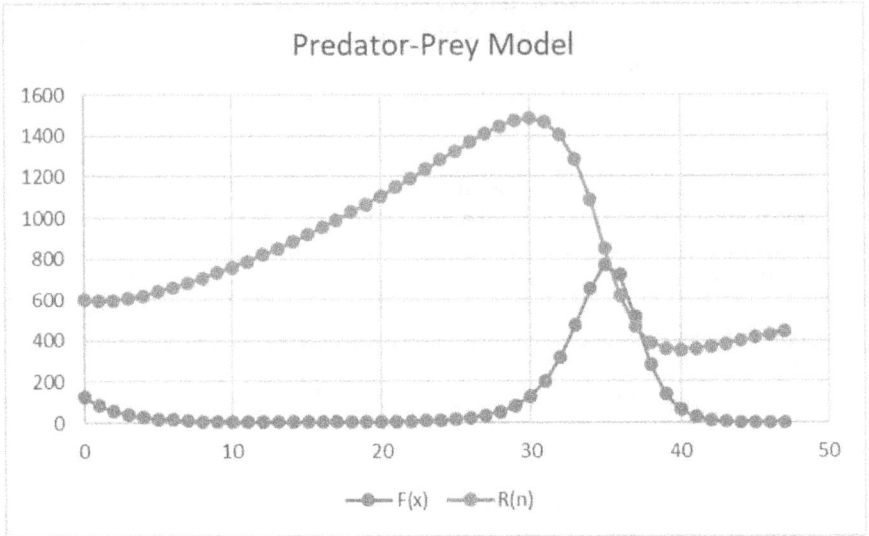

FIGURE 3.15
Foxes and rabbits over time.

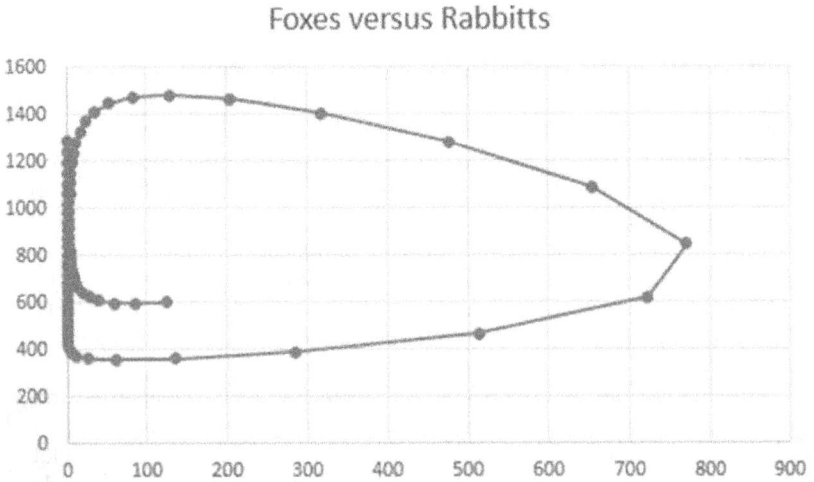

FIGURE 3.16
Foxes versus rabbits in a spiral motion.

Example 6. Discrete SIR Model of Epidemics

Consider a disease that is spreading throughout the Unites States such as the new flu. The Centers for Disease Control is interesting in know and experimenting with a model for this new disease prior to it actually becoming an "real" epidemic. Let us consider the population being divided into three categories: susceptible, infected, and removed. We make the following assumptions for our model:

- No one enters or leaves the community, and there is no contact outside the community.
- Each person is either susceptible, S (able to catch this new flu); infected, I (currently has the flu and can spread the flu); or removed, R (already had the flu and will not get it again, including those who died).
- Initially, every person is either S or I.
- Once someone gets the flu this year, they cannot get it again.
- The average length of the disease is two weeks, over which the person is deemed infected and can spread the disease.
- Our time period for the model will be per week.

The model we will consider is the SIR model (Allman and Rhodes, 2004). Let's assume the following definition for our variables:

$S(n)$ = number in the population susceptible after period n
$I(n)$ = number infected after period n
$R(n)$ = number removed after period n

Let's start our modelling process with $R(n)$. Our assumption for the length of time someone has the flu is 2 weeks. Thus, half the infected people will be removed each week:

$$R(n + 1) = R(n) + 0.5I(n).$$

The value, 0.5, is called the removal rate per week. It represents the proportion of the infected persons who are removed from infection each week. If real data are available, then we could do "data analysis" in order to obtain the removal rate.

$I(n)$ will have terms that both increase and decrease its amount over time. It is decreased by the number that are removed each week, $0.5 * I(n)$. It is increased by the numbers of susceptible that come into contact with an infected person and catch the disease, $aS(n)I(n)$. We define the rate, a, as the rate at which the disease is spread, or the transmission coefficient. We realize this is a probabilistic coefficient. We will assume, initially, that this rate is a constant value that can be found from initial conditions.

Let's illustrate as follows: Assume we have a population of 1000 students in the dorms. Our nurse found 3 students reporting to the infirmary initially. The next week, 5 students came into the infirmary with flu-like symptoms. $I(0) = 3$, $S(0) = 997$. In week 1, the number of newly infected is 30.

$$5 = a\,I(n)S(n) = a(3) * (\,995)$$
$$a = 0.00167$$

Let's now consider $S(n)$. This number is decreased only by the number that becomes infected. We may use the same rate, a, as before to obtain the model:

$$S(n + 1) = S(n) - aS(n)I(n).$$

Our coupled SIR model is

$$R(n+1) = R(n) + 0.5I(n)$$
$$I(n+1) = I(n) - 0.5I(n) + 0.00167\,I(n)S(n)$$
$$S(n+1) = S(n) - 0.00167\,S(n)I(n)$$
$$I(0) = 3, S(0) = 997, R(0) = 0.$$

The SIR model can be solved iteratively and viewed graphically. Let's iterate the solution and obtain the graph shown in Figure 3.17 to observe the behavior to obtain some insights.

The worse of the flu epidemic occurs around week 8, at the maximum of the infected graph. The maximum number is slightly larger than 400, from the table it is 427. After 25 weeks, slightly more than 9 persons never get the flu. You will be asked to check for sensitivity to the coefficient in the exercise set.

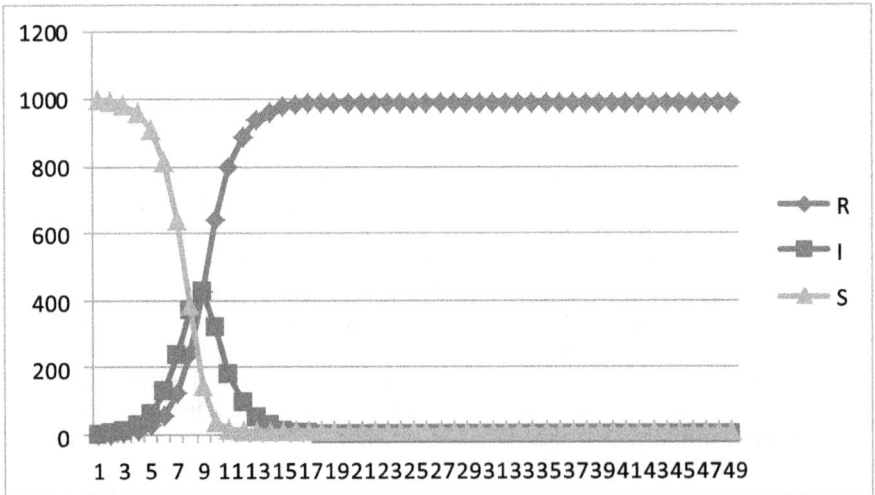

FIGURE 3.17
Plot of SIR model over time.

3.6 Technology Examples for DDSs

Using DDS is an interesting and productive approach. The use of a computer technology is essential to the methods described in this article. It allows for interactions between instructors and students. It provides a means to use technology in a nonstandard way. It provides another way to educate students concerning squares and cubes using discrete mathematics. In this section, we present technology and examples of solving linear and nonlinear discrete dynamical systems.

Excel for Linear and Nonlinear DDS

Let's consider a discrete dynamical system such as $a(n + 1) = 0.5a(n)$, with $a(0) = 100$.

Steps to iterate and graph in Excel are as follows:

Step 1. Open a new worksheet and name is DDS or some appropriate name.

Step 2. Label the following columns as n and $a(n)$ in cell *a1* and *b1*.

Step 3. In cells *a2* and *b2*, input the initial condition by putting in 0 in cell *a2* and 100 in cell *b2*.

Step 4. In cell *a3*, type = 1 +cell *a2*

Step 5. In cell *b3*, type = 0.5 * cell *b2*

Step 6. Highlight cells *a3* and *b3* and drag the curser down to fill in cells as far as desired or needed, in this case to about *a4:b16*.

Step 7. Highlight cells *a1:b16*, INSERT scatterplot to obtain the graph.

Step 8. Interpret the results.

A screenshot of the model and results is provided in Figure 3.18, where we see that our DDS tends to zero over time.

In Figure 3.19, we show the appropriate formulas used.

Maple for Linear and Nonlinear DDS

In Maple, DDS are referred to as recursion equations. One might obtain closed-form solutions, if they exist. One might also iterate and graph the behavior of the recursion equation. We will use both commands and libraries from Maple such as with(plots), and we will add some commands to our Maple toolbox, **rsolve** and **seq**.

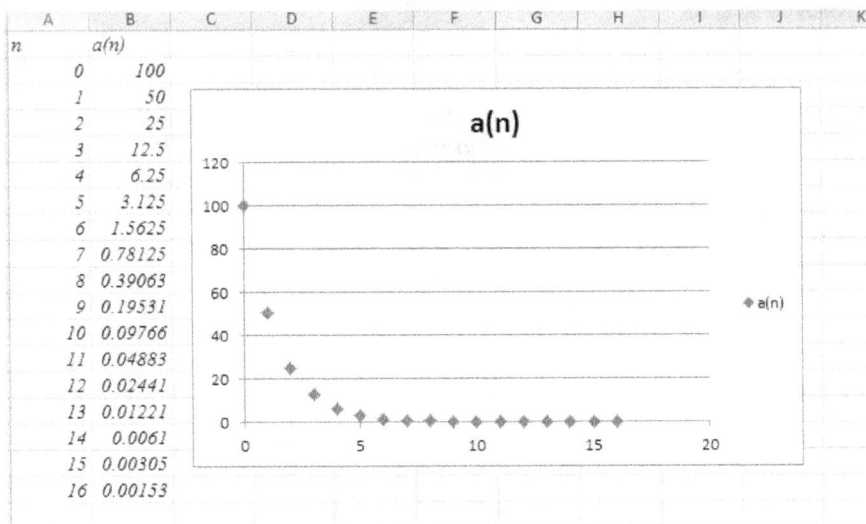

	A	B	C	D	E	F	G	H	I	J	K
n		*a(n)*									
	0	100									
	1	50									
	2	25									
	3	12.5									
	4	6.25									
	5	3.125									
	6	1.5625									
	7	0.78125									
	8	0.39063									
	9	0.19531									
	10	0.09766									
	11	0.04883									
	12	0.02441									
	13	0.01221									
	14	0.0061									
	15	0.00305									
	16	0.00153									

FIGURE 3.18
Screenshot of Excel's solution.

A	D
n	*a(n)*
0	100
=1+A2	=0.5*B2
=1+A3	=0.5*B3
=1+A4	=0.5*B4
=1+A5	=0.5*B5
=1+A6	=0.5*B6
=1+A7	=0.5*B7
=1+A8	=0.5*B8
=1+A9	=0.5*B9
=1+A10	=0.5*B10
=1+A11	=0.5*B11
=1+A12	=0.5*B12
=1+A13	=0.5*B13
=1+A14	=0.5*B14
=1+A15	=0.5*B15
=1+A16	=0.5*B16
=1+A17	=0.5*B17

FIGURE 3.19
Screenshot of Excel formulas used in our example.

```
rsolve- recurrence equation solver
Calling Sequence
      rsolve(eqns, fcns)
      rsolve(eqns, fcns, 'genfunc'(z))
      rsolve(eqns, fcns, 'makeproc')
Parameters
      eqns - single equation or a set of equations
      fcns - function name or set of function names
      z - name, the generating function variable
seq- create a sequence
Calling Sequence
      seq(f, i = m..n)
      seq(f, i = x)
Parameters
      f - any expression
      i - name
      m, n - numerical values
      x - expression
```

Many of the models that we will solve have closed-form solutions so that we can use the command rsolve to obtain the closed solution, and then we can use the sequence command (seq) to obtain the numerical values in the solution. Many dynamical systems do not have closed form analytical solutions so we cannot use the rsolve and seq commands to obtain solutions. When this occurs, we will write a small program using PROC to obtain the numerical solutions. To plot the solution to the dynamical systems, we will use plot commands to plot the sequential data pairs. We will illustrate all these commands in our example.

Example 7. Solve the DDS $a(n + 1) = 0.75a(n) + 100$, $a(0) = 0$

Determine the value of $a(72)$.

Let's illustrate the iterative technique for analyzing a DDS in Maple. Figure 3.20 shows the graphical representation of the solution.

We type the following commands:

```
> restart;
> drug:=rsolve({a(n+1)=.75*a(n)+100,a(0)=0}, a(n));
```

$$drug = 400 - 400\left(\frac{3}{4}\right)^n$$

```
> L:=limit(drug,n=infinity);
L = 400
> with(plots):
> pointplot({seq([i,-400*(3/4)^i+400],i=0..48)});
> drug_table:=seq(-400.0*(0.75)^i+400.,i=0..48);
```

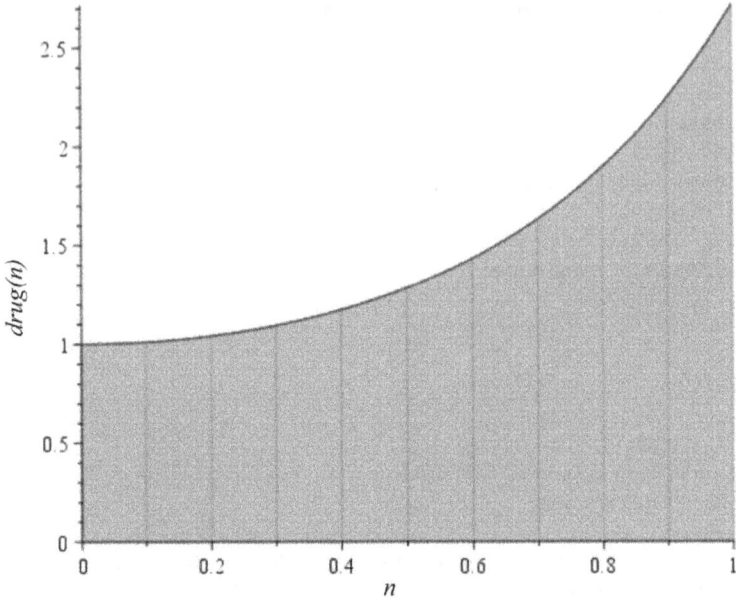

An approximation of $\int_0^1 f(x)\,dx$ using Simpson's rule, where $f(x) = e^{x^2}$ and the partition is uniform. The approximate value of the integral is 1.462653625. Number of subintervals used: 10.

FIGURE 3.20
Behavior of drugs in our systems.

```
drug_table := 0., 100.000, 175.00000, 231.2500000,
273.4375000, 305.0781250, 328.8085938, 346.6064453,
359.9548340, 369.9661255, 377.4745941, 383.1059456,
387.3294592, 390.4970944, 392.8728208, 394.6546156,
395.9909617, 396.9932213, 397.7449160, 398.3086870,
398.7315152, 399.0486364, 399.2864773, 399.4648580,
399.5986435, 399.6989826, 399.7742370, 399.8306777,
399.8730083, 399.9047562, 399.9285672, 399.9464254,
399.9598190, 399.9698643, 399.9773982, 399.9830487,
399.9872865, 399.9904649, 399.9928487, 399.9946365,
399.9959774, 399.9969830, 399.9977373, 399.9983030,
399.9987272, 399.9990454, 399.9992841, 399.9994630,
399.9995973
```

Interpretation of Results: The DDS shows that the drug reaches a value where change stops and eventually the concentration in the bloodstream levels at 400 milligrams seen in both Figure 3.20 and the numerical output. If 400 milligrams is both a safe and effective dosage level, then this dosage schedule is acceptable.

Using Maple for a System of DDS

Again, we use systems of DDS with rsolve, numerical, and plotting for the problem:

$$PS(n) = 0.65\ PS(n-1) + 0.25\ PM(n-1)$$
$$PM(n) = 0.35\ PS(n-1) + 0.75\ PM(n-1)$$
$$PS(0) = 2000$$
$$PM(0) = 1500$$

```
> dds := rsolve({PS(n) = .65*PS(n-1)+.25*PM(n-1), PM(n) = .35*
PS(n-1)+.75*PM(n-1), PS(0) = 2000, PM(0) = 1500}, {PM, PS});
```

$$dds := \left\{ PM(n) = -\frac{1625}{3}\left(\frac{2}{5}\right)^n + \frac{6125}{3},\ PS(n) = \frac{1625}{3}\left(\frac{2}{5}\right)^n + \frac{4375}{3} \right\}$$

```
> plot({-(1625/3)*(2/5)^n+6125/3, (1625/3)*(2/5)^n+4375/3},n=0
..15,thickness=3, title=`Student Vouchers`);
```

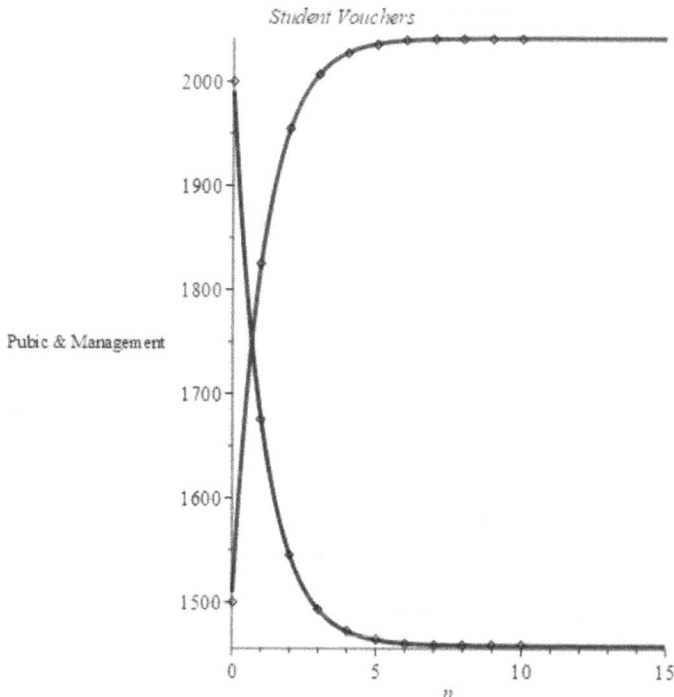

Student Vouchers

```
> public:=n-> if n=0 then 2000 else .65*public(n-
1)+.25*magnent(n-1) end if;
```

$$public := n \rightarrow \textbf{if } n = 0 \textbf{ then } 2000 \textbf{ else}$$
$$0.65\, public(n-1) + 0.25\, magnent(n-1)$$
$$\textbf{end if}$$

```
> magnent:=n-> if n=0 then 1500 else 0.35*public(n-
1)+.75*magnent(n-1) end if;
```

$$magnent := n \rightarrow \textbf{if } n = 0 \textbf{ then } 1500 \textbf{ else}$$
$$0.35\, public(n-1) + 0.75\, magnent(n-1)$$
$$\textbf{end if}$$

```
> seq([public(n),magnent(n)],n=0..10);
[2000, 1500], [1675.00, 1825.00], [1545.0000, 1955.0000],
[1493.000000, 2007.000000], [1472.200000, 2027.800000]
[1463.880000, 2036.120000], [1460.552000, 2039.448000]
[1459.220800, 2040.779200], [1458.688320, 2041.311680]
[1458.475328, 2041.524672], [1458.390131, 2041.609869]
> u:=seq(public(n),n=0..10);
u := 2000. 1675.00, 1545.0000, 1493.000000, 1472.200000,
1463.880000, 1460.552000, 1459.220800, 1458.688320,
1458.475328, 1458.390131
> w:=seq(magnent(n),n=0..10);
w :=1500, 1825.00, 1955.0000, 2007.000000, 2027.800000, 2036.120000,
2039.448000, 2040.779200, 2041.311680, 2041.524672, 2041.609869
> with(plots):
> a:=plot({-(1625/3)*(2/5)^n+6125/3, (1625/3)*(2/5)^n+4375/3},
n=0..15,thickness=3, title=`Student Vouchers`):
> b:=pointplot({seq([n,public(n)],n=0..10)}):
> c:=pointplot({seq([n,magnent(n)],n=0..10)}):
> display(a,b,c);
```

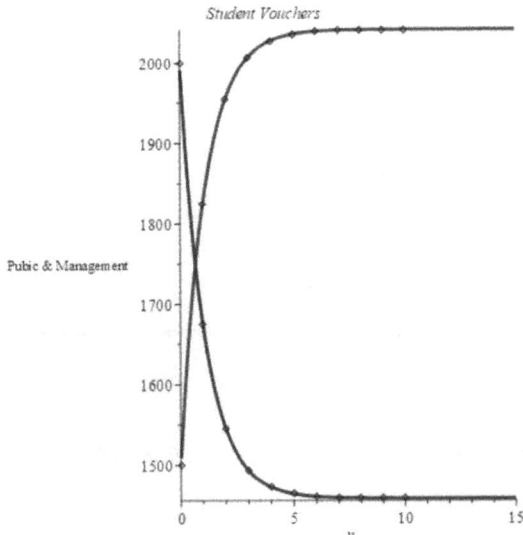

```
> pointplot({seq([public(n),magnent(n)],n=0..20)});
```

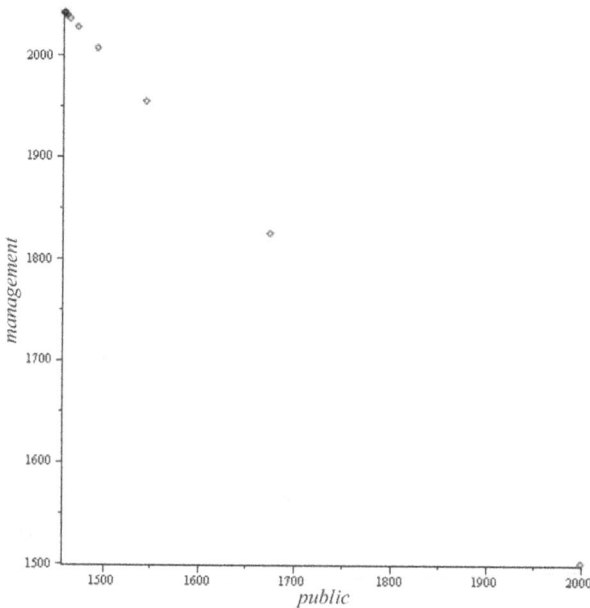

R for Linear and Nonlinear DDS

Example 8. Population Dynamics Using R

Given the DDS, $N[t + 1] = \lambda N[t]$, where λ is $(1 + r)$.

We open R studio, and we are going to use the **for** loop to address this problem. We type the following commands:

```
> generations <- 10
> N <- numeric(generations)
> lambda <- 2.1
> N [1] <- 3
> for (t in 1: (generations-1)) { N [t+1] <- lambda* N [t]}
> N
 [1] 3.0000 6.3000 13.2300 27.7830 58.3443 122.5230
257.2984
 [8] 540.3266 1134.6858 2382.8401
> plot(0:(generations-1),N, type="o",xlab="Time", ylab="Pop
Size")
```

We see unbounded growth in the plot shown in Figure 3.21.

Example 9. Repeat Example 7 $a(n + 1) = 0.75a(n) + 100$, $a(0) = 0$ Using R

We type the following commands:

```
> gener<-20
> D<-numeric(gener)
```

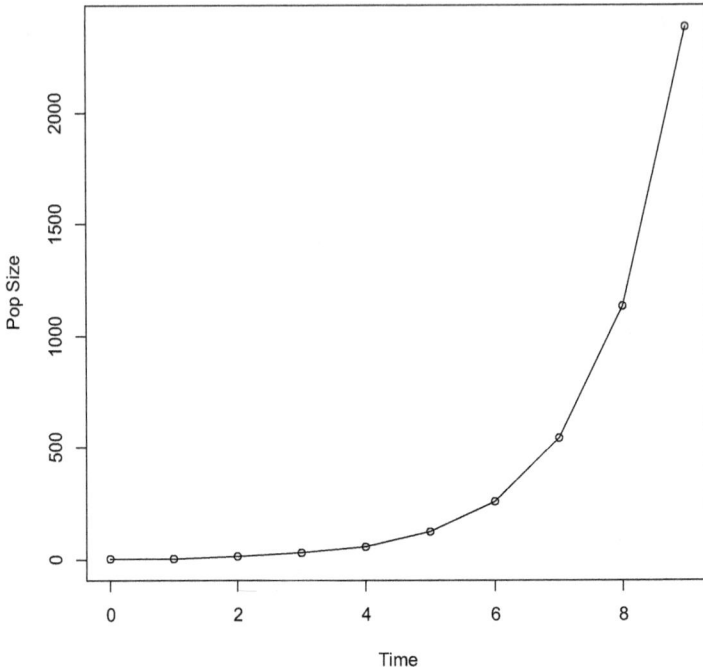

FIGURE 3.21
Screenshot from R.

```
> lam<-0.75
> D[1]<-0
> for (t in 1:(gener-1)) {D[t+1]<-lam*D[t]+100}
> D
 [1]  0.0000 100.0000 175.0000 231.2500 273.4375 305.0781
328.8086 346.6064
 [9] 359.9548 369.9661 377.4746 383.1059 387.3295
390.4971 392.8728 394.6546
[17] 395.9910 396.9932 397.7449 398.3087
> plot(0:(gener-1),D,type="o",xlab="Time",ylab="D
rug_in_Sys")
```

In R, we can see that the drug becomes stable at approximately 400 units as shown in Figure 3.22.

We now present a drug dosage model analytically.

Logistics Growth

Let's modify this model to a nonlinear model using a logistics growth DDS in R.

Given the DDS, $N[t+1] = N[t] + r\lambda N[t] (1 - N[t]/K)$, we have the following R script using the **function** command.

DDSL <- function(K,r,N0,generations)

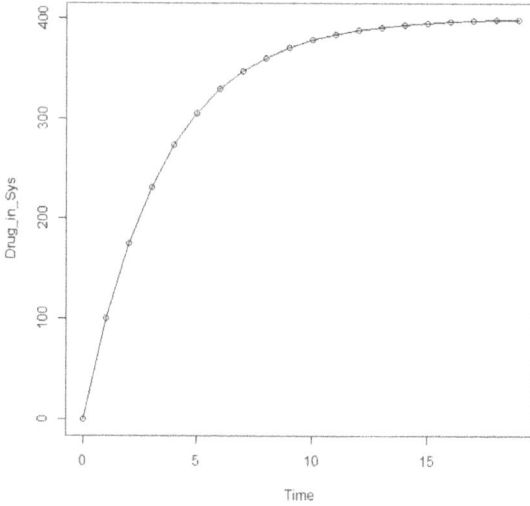

FIGURE 3.22
Screenshot from R solution graph for $a(n + 1) = 0.75a(n) + 100$, $a(0) = 0$.

```
+ {N <- c(N0,numeric(generations-1))
+ for ( t in 1:(generations-1)) N [t+1] <- {N [t] + r*N [t]*
(1- (N [t]/K))}
+ return(N)}
> Output <-DDSL(K=1000,r=1.5,N0=10,generations=30)
> generations <-30
> plot(0:(generations-1),Output, type='o', xlab="time",ylab="P
opulation")
```

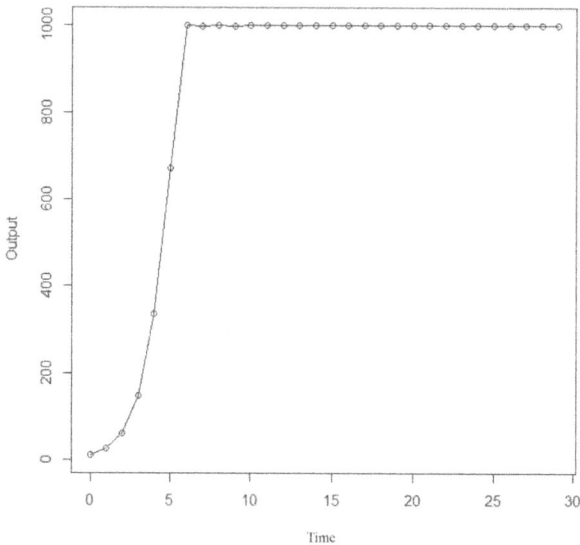

The plot shows the exponential growth at the beginning and then level off to the carrying capacity, $K = 1000$.

Next, we present a system of DDS using R.

Systems of DDS using a multiple **for** loop command.

- nn <-11
- a[1] <-100
- b[1]<-150
- for (t in 1: (nn-1)) {
- for (tt in 1: (nn-1)){

```
>plot(a, type="l",col="green")
>par(new=TRUE)
> plot( b, type="l",col="red", axes=FALSE)

> for (t in 1:(nn-1)) {
+ for (tt in 1: (nn-1)) {
+ a[t+1] <- .6*a[t]+.3*b[t]
+ b[tt+1] <- .4 *a[tt]+.7*b[tt]
+ }
+ }
> a
 [1] 100.0000 105.0000 106.5000 106.9500 107.0850 107.1255
107.1376 107.1413
 [9] 107.1424 107.1427 107.1428
> b
 [1] 150.0000 145.0000 143.5000 143.0500 142.9150 142.8745
142.8623 142.8587
 [9] 142.8576 142.8573 142.8572
```

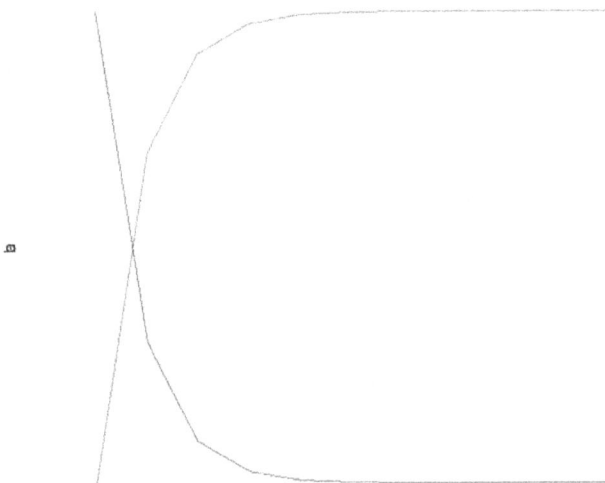

Steady-state probabilities

```
> a[1] <- 1
> b[1] <-0
> for (t in 1:(nn-1)) {
    + for (tt in 1: (nn-1)) {
    + a[t+1]<-.6*a[t]+.3*b[t]
    + b[tt+1]<-.4*a[tt]+.7*b[tt]
    + }
    + }
> a
 [1] 1.0000000 0.6000000 0.4800000 0.4440000 0.4332000
0.4299600 0.4289880
 [8] 0.4286964 0.4286089 0.4285827 0.4285748
> b
 [1] 0.0000000 0.4000000 0.5200000 0.5560000 0.5668000
0.5700400 0.5710120
 [8] 0.5713036 0.5713911 0.5714173 0.5714252
```

0.42857 and 0.57143, respectively.
    The matrix

$$\begin{bmatrix} 0.42857 & 0.42857 \\ 0.57143 & 0.57143 \end{bmatrix} \text{ times } \begin{bmatrix} 100 \\ 150 \end{bmatrix} = \begin{bmatrix} 107.1428 \\ 142.8572 \end{bmatrix} \text{ as before.}$$

Getting multiple plots: Here are some suggested R commands:

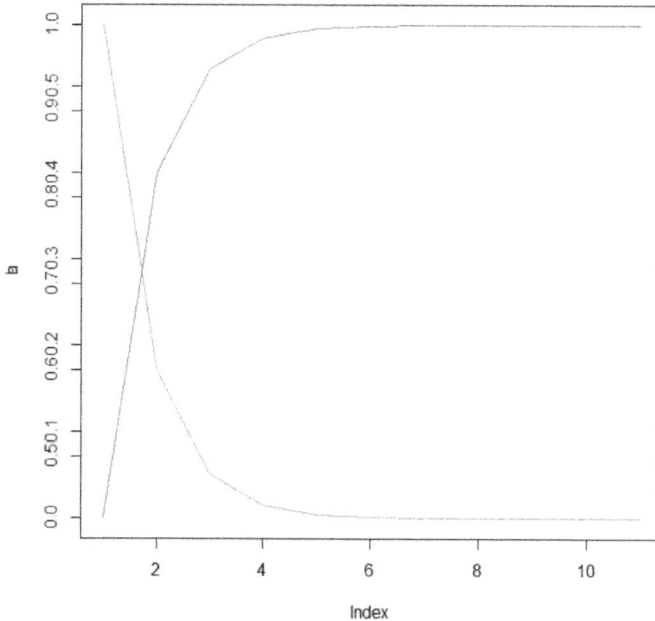

```
plot.new()
plot.window(xlim=range(x1),ylim=range(y1))
lines(x1,y1)
axis(1); axis(2); box()
plot.window(xlim=range(x2),ylim=range(y2))
lines(x2,y2)
axis(1); axis(2); box()
## Using the `deSolve` package
library(deSolve)
## Time
t <- seq(0, 100, 1)
## Initial population
N0 <- 10
## Parameter values
params <- list(r=0.1, K=1000)
## The logistic equation
fn <- function(t, N, params) with(params, list(r * N * (1 - N
/ K)))
## Solving and plotting the solution numerically
out <- ode(N0, t, fn, params)
plot(out, lwd=2, main="Logistic equation\nr=0.1, K=1000,
N0=10")
## Plotting the analytical solution
with(params, lines(t, K * N0 * exp(r * t) / (K + N0 * (exp(r *
t) - 1)), col=2, lwd=2))
```

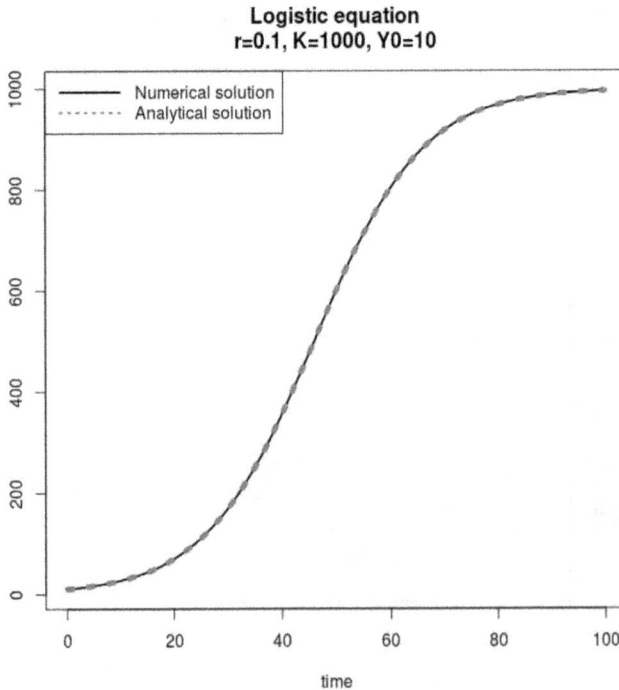

Logistic equation
r=0.1, K=1000, Y0=10

Exercises

Consider the model $a(n + 1) = r\, a(n)\, (1 - a(n))$. Let $a(0) = 0.2$. Determine the numerical and graphical solution for the following values of r. Find the pattern in the solution.

1. $r = 2$
2. $r = 3$
3. $r = 3.6$
4. $r = 3.7$

For Exercises 5–8, find the equilibrium value by iteration and determine if it is stable or unstable.

5. $a(n + 1) = 1.7a(n) - 0.14a(n)^2$
6. $a(n + 1) = 0.8a(n) + 0.1a(n)^2$
7. $a(n + 1) = 0.2a(n) - 0.2a(n)^3$
8. $a(n + 1) = 0.1a(n)^2 + 0.9a(n) - 0.2$
9. Consider spreading a rumor through a company of 1000 employees all working in the same building. We assume that the spread of a rumor is similar to the spread of a contagious disease in that the number of people hearing the rumor each day is proportional to the product of the number hearing the rumor and the number who have not heard the rumor. This is given by the formula:

$$r(n + 1) = r(n) + 1000kr(n) - kr(n)^2,$$

where k is the parameter that depends on how fast the rumor spreads. Assume $k = 0.001$ and further assume that four people initially know the rumor. How soon will everyone know the rumor?

10. Determine the equilibrium values of the bass and trout model presented in Section 3.2. Can these levels ever be achieved and maintained? Explain.
11. Test the fast-food models with different starting conditions summing to 14,000 students. What happens? Obtain a graphical output and analyze the graph in terms of long-term behavior.
12. Find the equilibrium values for the predator–prey model presented in Section 3.5.
13. In the predator–prey model, presented in Section 3.5, determine the outcomes with the following sets of parameters:
    a. Initial foxes are 200, and initial rabbits are 400.

  b. Initial foxes are 2000, and initial rabbits are 10,000.

  c. The birth rate of the rabbits increases to 0.1.

14. In the SIR model, presented in Section 3.5, determine the outcome with the following changed parameters:

  a. The flu lasts 1 week.

  b. Initially 5 are sick and 10 the next week.

  c. The flu lasts 4 weeks.

  d. There are 4000 students in the dorm, and 5 are initially infected and 30 more the next week.

Projects

1. Consider the contagious disease as the Ebola virus. Look on the internet and find out some information about how deadly this virus actually is. Now consider an animal research laboratory in Reston, Virginia, a suburb of Washington, D.C., with a population of 856,900 people. A monkey with the Ebola virus has escaped its captivity and infected one employee (unknown at the time) during its escape. This employee reports to University Hospital later with Ebola symptoms. The Infectious Disease Center (IDC) in Atlanta gets a call and begins to model the spread of the disease. Build a model for the IDC with the following growth rates to determine the number infected after 2 weeks:

  a. $k = 0.00025$

  b. $k = 0.000025$

  c. $k = 0.00005$

  d. $k = 0.000009$

  e. List some ways of controlling the spread of the virus.

2. Consider the spread of a rumor concerning termination among 1000 employees of a major company. Assume that the spreading of a rumor is similar to the spread of contagious disease in that the number hearing the rumor each day is proportional to the product of those who have heard the rumor and those who have not heard the rumor. Build a model for the company with the following rumor growth rates to determine the number having heard the rumor after 1 week:

  a. $k = 0.25$

  b. $k = 0.025$

  c. $k = 0.0025$

  d. $k = 0.00025$

  e. List some ways of controlling the spread of the rumor.

3. Lions and spotted hyena: Predict the number of lions and spotted hyena in the same environment at a function of time.

Assumptions:

The variables: $L(n)$ = number of lions at the end of period n
$H(n)$ = number of hyenas at the end of period n

Assume the following model:

$$L(n + 1) = 1.2L(n) - 0.001L(n)H(n)$$
$$H(n + 1) = 1.3H(n) - 0.002H(n)L(n).$$

a. Find the equilibrium values of the system.
b. Iterate the system from the following initial conditions and determine what happens to the lions and the spotted hyenas in the long term.

Lions	Spotted Hyena
150	200
151	199
149	210
10	10
100	100

4. It is getting close to election day. The influence of the new Independent Party is of concern to the Republicans and Democrats. Assume that in the next election that 75% of those who vote Republican vote Republican again, 5% vote Democratic, and 20% vote Independent. Of those that voted Democratic before, 20% vote Republican, 60% vote Democratic, and 20% vote Independent. Of those that voted Independent, 40% vote Republican, 20% vote Democratic, and 40% vote Independent.

a. Formulate and write the system of discrete dynamical systems that models this situation.
b. Assume that there are 399,998 voters initially in the system, how many will vote Republican, Democratic, and Independent in the long run? (Hint: You can break down the 399,998 voters in any manner that you desire as initial conditions.)
c. (New Scenario) In addition to the preceding, the community is growing (18-year-olds + new people – deaths – losses to the community, etc.). Republicans predict a gain of 2000 voters between elections. Democrats estimate a gain of 2000 voters between

elections. The Independents estimate a gain of 1000 voters between elections. If this rate of growth continues, what will be the long-term distribution of the voters?

5. You are a new high school graduate on your way back from graduation practice. In your mind, you could not be happier. You are thinking about your plans after graduation. You are going to start a new job in two weeks that will allow you to work full-time during the summer and part-time when college starts in the fall. This job will allow you to save $260 a month. You are considering buying a car with this money. Your mind is a thousand miles away when all of a sudden, you see something in the road ahead of you. Several deer have run out in front of your car. You try everything you can to avoid them, but you still hit one. At this point, you are not too upset because you know that you have insurance on your car and you assume that everything will be taken care of with no problems. However, when you begin talking to your insurance agent, you run into some problems, and things are not as simple as you thought they would be.

You plan on having your car repaired at the local body shop. The owner informs you that it will take two to three weeks to repair your car. You are not able to drive the car, so you have to get a rental car. Furthermore, when the insurance adjuster comes out to talk to you, he informs you that your policy does not cover a rental car and that you have a $300 deductible on your collision coverage. Therefore, you are going to have to pay for the rental car and the deductible. Suddenly you realize you have a problem: You don't have the money to pay for these expenses. Because you are only 17, you cannot borrow money from a financial institution. A friend tells you that he will loan you the money at 10% interest, compounded monthly. As a result, you gather the following information:

- You will need a rental car for 20 days at $25.50 per day plus taxes (8.25%).
- You need to pay the $300 deductible.
  a. How much money do you need to borrow?
  b. What will your monthly payments be if you repay your friend in 3 monthly payments or 4 monthly payments? (Set up a DDS for the amount that you owe your friend and figure out the amount for each payment.)
  c. How much interest will you pay if you make 3 payments? What about 4 payments?
  d. Is there a difference in the amount of interest that you will pay? Will it be easy to repay the loan in 3 payments? Or 4 payments?

Which is best and why? (Think about your income and the amount of each payment. Consider the amount of interest you will pay.)

Three or four months later, you are planning on buying a used car with the money you will save from your job. Between the body shop owner and the insurance settlement, you will receive $5850 for your wrecked car. The used car you want to purchase will cost a total of $11,000. Your friend agrees to loan you $5150 for 3 years at 10% annual interest, compounded monthly. He says your payments should be $175 each month.

e. Is this figure $175 a month correct? (Use a discrete dynamical system to model your new loan and figure the amount of your monthly payment.

f. What should your monthly payment be? Using iteration what should you last payment be?

g. How much interest will you pay?

h. If you decide to make payments of $260 per month, how many payments should you make? How much will the last payment be?

i. How much interest will you pay?

j. Is there much difference in the amount of interest that you will pay at $175 per month and $260 per month? Which plan is best for you and why?

6. Pollution in the Great Lakes

Scenario: Let $e(n)$ and $o(n)$ be the total amount of pollution in Lake Erie and Lake Ontario, respectively, in year n. It has been determined that each year, the percentage of water replaced in Lakes Erie and Ontario is approximately 36% and 11%, respectively. Since most of the water flowing from Lake

Ontario is from Lake Erie, this means that each year, 36% of the water in Lake Erie flows into Lake Ontario and is replaced by rain and water flowing from other sources. Also, each year 11% of Lake Ontario's water flows out and is replaced by the water flowing in from Lake Erie. Assuming the concentration of pollution in each lake is constant throughout the lake, then 36% of the pollution in Lake Erie is removed each year. Each year, 11% of the pollution in Lake Ontario is removed, but the pollution removed from Lake Erie is added to Lake Ontario. Furthermore, assume that 2 tons of pollutants are added directly to Lake Erie each year and 12 tons of pollutants are added directly to Lake Ontario each year.

    A. Make a flow diagram that summarizes the scenario.

    b. Develop a dynamical system for $e(n)$ and $o(n)$.

    c. Assume there are now 10 tons and 30 tons of pollution in Lakes Erie and Ontario, respectively. Iterate your system to find the pollution levels twenty years from now.

    d. Analytically find the equilibrium point for Lake Erie and Lake Ontario.

    e. Suppose the Environmental Protection Agency determines that an equilibrium level of a total of 5 tons of pollutants in Lake Erie and a total of 15 tons in Lake Ontario would be acceptable. In order to achieve these equilibrium levels, what restrictions should be placed upon the total amount of additional pollutants that are dumped directly into Lake Erie and Lake Ontario?

References and Suggested Future Readings

Albright, B. (2010). *Mathematical Modeling with Excel*. Jones and Bartlett, Sudberry, MA.

Alfred, U. (1967). Sums of Squares of Consecutive Odd Integers. *Mathematics Magazine*, 40(4): 194–199.

Allman, E. and J. Rhodes (2004). *Mathematical Models in Biology: An Introduction*. Cambridge University Press, Cambridge.

Arney, D., F. Giordano and J. Robertson (2002). *Mathematical Modeling with Discrete Dynamical Systems*. McGraw Hill, Boston, MA.

Fox, W. P. (2010). Discrete Combat Models: Investigating the Solutions to Discrete Forms of Lanchester's Combat Models. *International Journal of Operations Research and Information Systems (IJORIS)*, 1(1): 16–34.

Fox, W. P. (2012a). *Mathematical Modeling with Maple*. Cengage Publishing, Boston, MA.

Fox, W. P. (2012b). Discrete Combat Models: Investigating the Solutions to Discrete Forms of Lanchester's Combat Models. In *Innovations in Information Systems for Business Functionality and Operations Management*. IGI Global & SAGE Publishers, pp. 106–122.

Fox, W. P. (2012c). Mathematical Modeling of the Analytical Hierarchy Process Using Discrete Dynamical Systems in Decision Analysis. *Computers in Education Journal*, 3(3): 27–34.

Fox, W. P. and R. Burks (2021). *Advanced Mathematical Modeling*. Taylor and Francis, CRC Press, Boca Raton, FL.

Fox, W. P. and R. Burks (2022). *Mathematical Modeling Under Change, Uncertainty, and Machine Learning*. Taylor and Francis, CRC, Boca Raton, FL.

Fox, W. P. and P. J. Driscoll (2011). Modeling with Dynamical Systems for Decision Making and Analysis. *Computers in Education Journal (COED)*, 2(1): 19–25.

Giordano, F. R., W. P. Fox and S. Horton (2014). *A First Course in Mathematical Modeling*, 5th ed. Cengage Publishing, Boston, MA.

Leyendekker, J. and A. Shannon (2015). The Odd-Number Sequence: Squares and Sums. *International Journal of Mathematical Education in Science and Technology*, 46(8): 1222–1228.

Sandefur, J. (1990). *Discrete Dynamical System: Theory and Applications*. Oxford University Press, New York.

Sandefur, J. (2002). *Elementary Mathematical Model: A Dynamic Approach*, 1st ed. Brooks-Cole Publishers, Belmont, CA.

Sandefur, J. (2003). *Elementary Mathematical Modeling: A Dynamic Approach*. Thomson Publishing, Belmont, CA.

4

Numerical Solutions to Equations in One Variable

4.1 Introduction and Scenario

In this chapter, we illustrate the methods of numerical solutions for equations of one variable. We accomplish this through applied projects that require mathematical modelling and the use of previous mathematical skills developed in other courses (such as calculus). In our teaching of mathematics courses, we have found that the use of projects and applications are tremendous motivators for students. It helps them focus on why they are learning the techniques being presented. The students pay closer attention to the comparisons and contrasts of techniques as they consider their use within the project. In our project, we desire to illustrate the techniques found for numerical solutions to equations with one variable. We illustrate bisection, fixed-point iteration, and Newton's method. In each method, we need to consider the error analysis from the technique. You should be aware of the precision of any numerical computing device that is used. Precision is defined as the largest number of decimal digits that is calculated. For whatever technology that you are using, you should find its precision.

Your analysis group has been assigned to design spherical ships for the coast guard. Archimedes' principle states that when a solid of density ρ_s is placed in a liquid of density ρ_l, where $\rho_s < \rho_l$, the solid displaces an amount of liquid whose weight equals that of the solid. A sphere of radius r sinks to a height h in liquid as shown in Figure 4.1.

We need to find the roots where $f(x) = 0$.

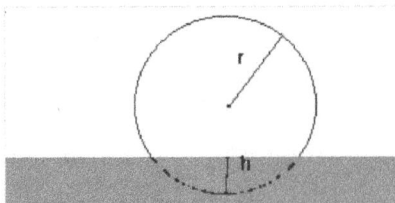

FIGURE 4.1
Archimedes' principal.

                    DOI: 10.1201/9781032703671-4

4.2 Archimedes' Design of Ships

Because of Archimedes' principle (Equation 4.1),

$$\frac{4}{3}\pi r^3 \rho_s = \frac{1}{3}\pi h^2 (3r - h)\rho_1\, 4/3. \tag{4.1}$$

We need to find the height, h, if $r = 1$ ft, $\rho_1 = 62.5$ lb/ft^3 (density of water), and $\rho_s = 0.61$ lb/ft^3. Letting h be x and substituting we obtain the function in Equation 4.2, shown in figure 4.2:

$$f(x) = 196.35x^2 - 65.449x^3 - 2.5551. \tag{4.2}$$

Assume all quantities are accurate to five significant figures. We will illustrate several of the root finding techniques as we develop our solution to finding the height. We want to develop more rigorously the concept and then compare and contrast procedures of numerical root finding methods.

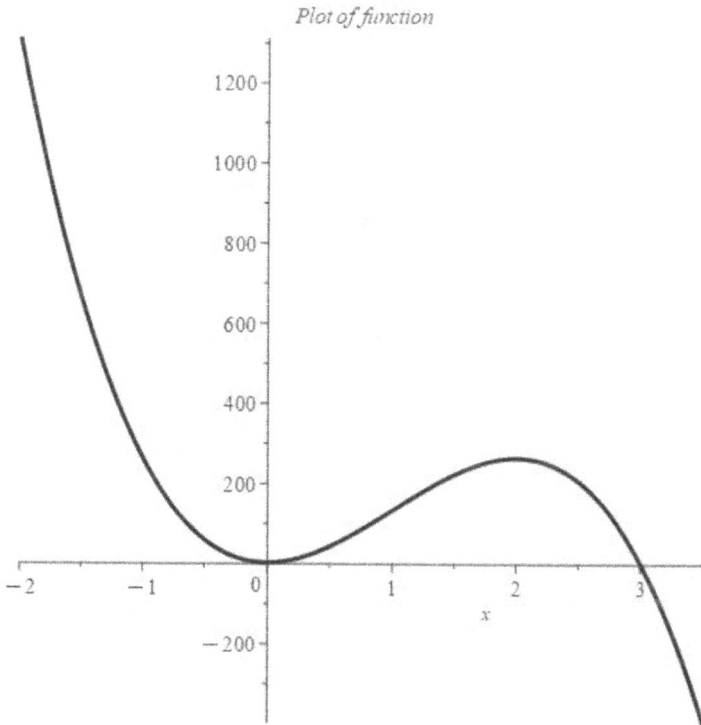

Plot of function

FIGURE 4.2
Plot of $f(x)=196.35x^2 - 65.449x^3 - 2.5551$.

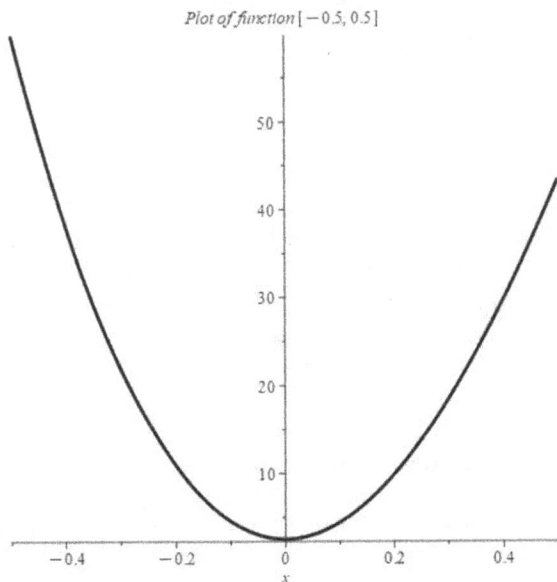

FIGURE 4.3
Plot of $f(x)$ from [−0.5, 0.5].

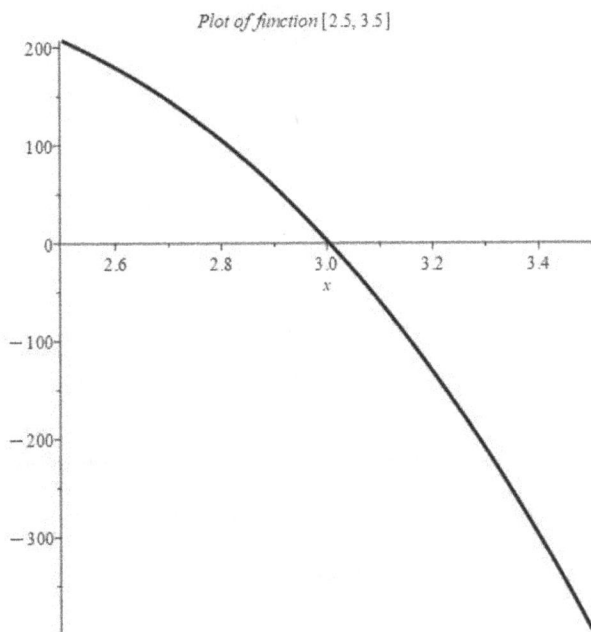

FIGURE 4.4
Plot of $f(x)$ from [2.5, 3.5].

We begin by plotting the function (Equation 4.2) in different regions to find intervals for all the possible roots shown in Figures 4.3 and 4.4.

We now can numerically search for roots in the following intervals found from the graphs:

a. $[-0.2, 0]$
b. $[0, 0.2]$
c. $[2.8, 3.2]$

We have not made a predetermined choice on which method is preferred. We will illustrate several root-finding methods and ask you to determine which might be more appropriate to use.

We will begin with the bisection method.

4.3 Bisection Method

The bisection method is used to find the root of $f(x) = 0$. Bisection is useful when the function is continuous but may not be differentiable. We developed a pseudo-code for our bisection method that can be implemented on any technology from graphing calculator to computers. The following is a summary of the bisection algorithm.

Bisection Algorithm

To find a solution to $f(x) = 0$ given the continuous function f on the interval $[a, b]$, where $f(a)$ and $f(b)$ have opposite signs.

```
INPUTS:
Endpoints: a , b ;
Tolerance: TOL ;
Maximum number of iterations, N₀
OUTPUTS
   Approximate solution p or an error message if a solution is
not found in N₀ iterations.
STEP 1. Set i=1; Let FA=f( a ) .
STEP 2. While i≤ N₀ do Steps 3-6.
Step 3. Set p=a+ ( b - a ) /2; compute p i . FP=f(p sub i)
Step 4. If F P = 0 or (b-a)/2<TOL, then OUTPUT (p); Procedure
complete. STOP ; otherwise
Step 5. Set i=i+1.
Step 6. If FA·FP>0 then set a = p ; (Compute aᵢ , bᵢ ) . F A =
F P else set b=p.
STEP 7. OUTPUT ('Method Failed after N₀ iterations') STOP
```

Execution on the computer of the bisection algorithm:

INPUTS >alg(031); This is the Bisection Method. Input the function F(x)
in terms of x

$f(x) = 196.35x^2 - 65.449x^3 - 2.5551$ from Burden and Faires (1997).

```
Input
 endpoints A < B separated by blank
 > 0   0.2
Input tolerance > .00001
 Input maximum number of iterations - no decimal point > 50
```

For the INPUT for the bisection method, we rearrange our equation to the form $f(x) = 0$ and input $F(x)$. Then we insert our interval with left endpoint A and the right endpoint B. In our routine, we will stop when the absolute error bound (one half of the interval) is less than the tolerance. Since we are seeking five significant digits, we will simply make the tolerance 0.00001 for five decimal places. The maximum number of iterations is just to make sure that our routine will stop if we make an error on input ($F(x)$ or A or B) or in our estimate of the number of iterations needed to attain our desired accuracy.

```
OUTPUT
[-0.05, 0.5]
[-0.05, 0.2250000000]
 [0.08750000000, 0.2250000000]
 [0.08750000000, 0.1562500000]
 [0.08750000000, 0.1218750000]
 [0.1046875000, 0.1218750000]
 [0.1132812500, 0.1218750000]
 [0.1132812500, 0.1175781250]
 [0.1154296875, 0.1175781250]
 [0.1154296875, 0.1165039062]
 [0.1159667968, 0.1165039062]
p=0.1162353515
F(p) - 0.00506
```

MAPLE
```
MAPLE currently has a package:
with(Student[NumericalAnalysis]) that may be used to obtain
solutions.
Bisection(f1, x = [2.5, 3.1], tolerance = 10^(-4), output =
sequence);
          [2.5, 3.1],
[2.800000000, 3.1],
[2.950000000, 3.1],
[2.950000000, 3.025000000],
[2.987500000, 3.025000000],
```

```
[2.987500000, 3.006250000],
 [2.987500000, 2.996875000],
 [2.992187500, 2.996875000],
[2.994531250, 2.996875000],
[2.994531250, 2.995703125],
[2.995117187, 2.995703125]
One root is between [2.995117187, 2.995703125].
Bisection(f1, x = [2.8, 3.2], tolerance = 10^(-4), output =
sequence);
 [2.8, 3.2],
[2.8, 3.000000000],
[2.900000000, 3.000000000],
[2.950000000, 3.000000000],
[2.975000000, 3.000000000],
[2.987500000, 3.000000000],
[2.993750000, 3.000000000],
[2.993750000, 2.996875000],
[2.995312500, 2.996875000],
[2.995312500, 2.996093750],
[2.995312500, 2.995703125]
Maple has a Bisection call-out in its Numerical Analysis
package.
Bisection(f1, x = [-0.5, 0.], tolerance = 10^(-4), output =
sequence);
 [-0.5, 0.], [-0.2500000000000000000, 0.],

[-0.1250000000000000000, 0.],
[-0.1250000000000000000, -0.06250000000000000000],
[-0.1250000000000000000, -0.09375000000000000000],
[-0.1250000000000000000, -0.1093750000000000000],
[-0.1171875000000000000, -0.1093750000000000000],
[-0.1132812500000000000, -0.1093750000000000000],
[-0.1132812500000000000, -0.1113281250000000000],
[-0.1123046875000000000, -0.1113281250000000000],
[-0.1123046875000000000, -0.1118164062500000000]
```

We have found the three roots by the method of bisection. The roots via the Computer Algebra System (CAS) Maple are $x = -0.11200562$, 2.99569702, and 0.11635132. Let's compare these results to the roots found by some of the other root finding methods. Using our TI-83 Plus program, we find the roots as $-.1120056152$, 2.995697021, and 0.1163513184.

4.4 Fixed-Point Algorithm

The fixed-point algorithm requires a choice of $G(x)$. Choosing the function is critical and without prior knowledge of the roots, it is possible to not choose a robust enough $G(x)$ to find all your roots. Solving for $x = G(x)$, we start with $196.35 \cdot x^2 - 65.449 \cdot x^3 - 2.5551 = 0$.

$$\text{So, } 196.35 \cdot x^2 = 65.449 \cdot x^3 + 2.5551.$$

$x^2 = (65.449 \cdot x^3 + 2.5551)/196.35$, and $x = ((65.449 \cdot x^3 + 2.5551)/196.35)^{\wedge}(1/2)$.

Thus, we begin by choosing $G(x) = ((2.5551 + 65.449x^3)/196.35)^{1/2}$.

Fixed-Point Algorithm

```
To find a solution to p=g( p ) given an initial approximation
p 0 :
INPUT initial approximation p₀ ;
Tolerance, TOL ;
maximum number of iterations N₀
OUTPUT approximate solution p or message of failure
Step 1. Set i=1.
Step 2. While i≤N ₀ do Steps 3-6.
Step 3. Set p=g( p₀ ) . Compute pᵢ .
Step 4. If | p - p= | <TOL then OUTPUT p (Procedure is
complete.) STOP
Step 5. Set i=i+1.
Step 6. Set p₀ =p.(Update p₀ . )
Step 7. OUTPUT (''Method Failed after N₀ iterations') STOP
```

Using the preceding algorithm we obtain, INPUT:
   This is the fixed-point method using the input the function $G(x)$ in terms

of x, $G(x) = \sqrt{\dfrac{\left(2.5552296.53 + 65.449x^3\right)}{196.35}}$.

   For example:

```
Input initial approximation > 1.5
Input tolerance > .0005
Input maximum number of iterations - no decimal point > 25
```

For the INPUT for the fixed-point iteration, we have to rewrite the equation $f(x) = 0$ to $x = G(x)$, as we did earlier, and input $G(x)$. Next, we choose a good initial guess: From the graph 2–2, we pick 0.1, which is halfway between our two bisection endpoints. The error tolerance is the same as with the bisection method. However, the stopping criterion uses an estimate of the absolute error as the difference between the current iteration and the previous iteration. Again, the maximum number of iterations is set to stop the routine in the case that we pick a bad $G(x)$ and the fixed-point iteration never converges to the solution.

Using the Numerical Analysis package in MAPLE we used Fixed Point Iteration:

$$g := sqrt((2.5551 + 65.449*x\wedge3)/196.35);$$

FixedPointIteration(fixedpointiterator = g, x = 1.0, tolerance = $10\wedge(-4)$, output = sequence);

```
1.0, 0.5885076273, 0.2845231688, 0.1438422012, 0.1183428550,
0.1164707847, 0.1163599512, 0.1163534967
```

It has converged in eight iterations.

This root agrees to only two decimal places for the roots found by bisection. Additionally, we were not able to find the other roots. From any starting value, the algorithm converged to approximately $x = 0.11635306$. In summary, two different starting points converged to 0.11635306538.

4.5 Newton's Method

Now, let us try one the student's favorite methods, Newton's method. To use Newton's method, the function must be differentiable over the search region. We used a typical Newton's Method algorithm.

Newton's Algorithm

```
To find a solution to f(x) =0 given the differentiable
function f and an initial approximation p 0 :
 INPUT initial approximation p0 ; tolerance, TOL ; maximum
number of iterations, N0
OUTPUT approximate solution p or message of failure
Step 1. Set i = 1.
Step 2. While i ≤ N0 do
Steps 3-6. Step 3. Set p = p 0 - f( p 0 )/f' ( p 0 ) . Compute
p i .
Step 4. If | p - p 0 | <TOL then OUTPUT p (Procedure is
complete.) STOP
Step 5. Set i = i + 1.
Step 6. Set p 0 = p . (Update p 0 . )
 Step 7. OUTPUT (''Method Failed after N 0 iterations') STOP
Note: in the above f' = df/dx.
```

Example 1. Newton's Method Roots for $196.35 \cdot x^2 - 65.449 \cdot x^3 - 2.5551$

```
INPUT: f( x ) =196.35· x 2 -65.449· x 3 -2.5551
f ' (x ) = d/d x( 196.35· x 2 -65.449· x 3 -2.5551 ) =
392.7 x - 196.35 x 2
 T O L = .00001 p 0 = 0.1 N 0 = 10
```

Newton's method is a special case of fixed-point iteration, except that we input $F(x)$ from $f(x) = 0$, and after we compute $f'(x)$, we input that. If we cannot compute $f'(x)$, then we must try some other technique. The initial guess and tolerance are the same as for fixed-point iteration. The initial guess for Newton's method needs to be very good and away from points where $f'(x) = 0$. So the maximum number of iterations is very important as a default stopping criterion. OUTPUT

Using Maple's numerical analysis package:

```
Newton(f1, x = 1, output = sequence);
 1., 0.3463512144, 0.1838058612, 0.1277838269, 0.1168315448,
0.1163540171, 0.1163530980
Newton(f1, x = 3, output = sequence);
3., 2.995707977, 2.995695635
Newton(f1, x = 1, tolerance = 10^(-4), output = sequence);
 1., 0.3463512144, 0.1838058612, 0.1277838269, 0.1168315448,
0.1163540171, 0.1163530980
Newton(f1, x = 3, tolerance = 10^(-4), output = sequence);
3., 2.995707977, 2.995695635
```

Finally,

Newton(f1, x = –1, tolerance = 10^(–4), output = sequence);

```
-1., -0.55989267409901077503, -0.30943508100343872947,
-0.17983720444419085833, -0.12558643471647619372,
-0.11277777145308352332, -0.11200569859458019363,
-0.11200289516897509591
```

Newton's method converged to all three roots, but we had to choose "good" starting values. You should think about why this happened.

A summary of our root finding techniques for the root we were looking for is displayed in the following table.

Method	Root	# of Iterations	# of Functions Evaluations
Bisection	0.11635132	15	16
Fixed Point	0.11635371	5	6
Newton–Raphson	0.116350982	3	4

4.6 Secant Method

Secant Method Definition

The approximation, p_{n+1}, for $n > 1$, to a root of $f(x) = 0$ is computed from the approximation p_n and p_{n-1} using the equation:

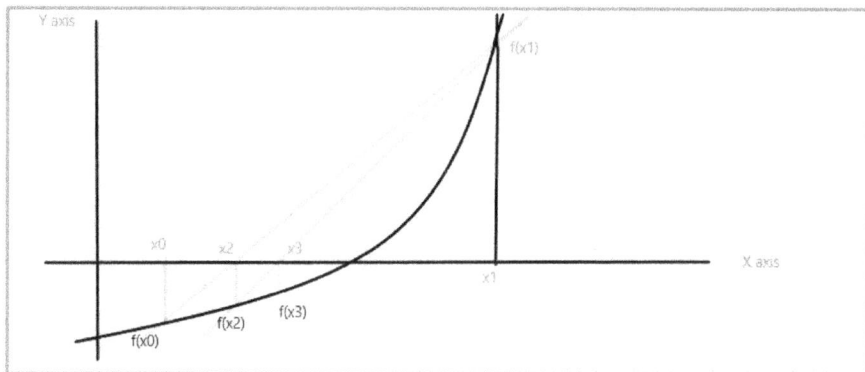

FIGURE 4.5
Graphical depiction of the secant method.

$$p_{n+1} = p_o - \frac{f(p_n)(p_n - p_{n-1})}{f(p_n) - f(p_{n-1})}.$$

The secant method usually converges faster than the bisection method.

Again, in the Maple Numerical Analysis package, there is a secant method command.

Archimedes' Example with the Secant Method

Using the same problem as before, we show how to use the secant method searching for the two roots. One root is between 2.7 and 3.2, and the other root is between 0 and 0.2.

```
Secant(f1, x = [2.7, 3.2], tolerance = 10^(-4), output =
sequence);
 2.7, 3.2, 2.953643824, 2.990388380, 2.995848590, 2.995695090
Secant(f1, x = [0, 0.2], tolerance = 10^(-4), output = sequence);
0., 0.2, 0.0697123543, 0.1027619726, 0.1198261734,
0.1161538853,
 0.1163503482, 0.1163531002
```

Now, let's more closely examine the secant method.

Example 2. Buying a Car Using the Secant Method

Now, let's look at buying a new car where we combine our methods to ensure accurate and efficient solutions. Consider a car dealer who is offering a new 2024 car for $18,000. He also offered to sell the same car for payments of $375 per month for 5 years. What monthly interest rate is the dealer charging? To begin, we will use the formula for the present value A of an annuity consisting of n equal payments of size R with interest rate i per time period.

$$A = (R/i)[1 - (1 + i)^{-n}]$$

If we replace i with x, we can show that the following equation is valid: $f(x) = 48 \cdot x \cdot (1 + x)^{60} - (1 + x)^{60+1} = 0$.

Now, we examine the graph of our functions and then perform a few iterations of bisection to get a good initial guess.

If we could find the derivative of $f(x)$, we would now use Newton's method. However, since we may not remember the product rule and know that a 60-degree polynomial can get messy, we will use the secant method.

Secant Method Algorithm

To find a solution to $f(x) = 0$ given the differentiable function f and an initial approximations p_0 and p_1:

```
INPUT initial approximations p0 , p1 ;
tolerance T O L ; maximum number of iterations N₀
OUTPUT approximate solution p or message of failure
Step 1. Set i=2; q₀ =f(p₀); q₁ = f ( p₁ ) .
Step 2. While i ≤ N 0 do Steps 3-6.
Step 3. Set p = p₁ - q₁ ⊠ p₁ - p ₀ q₁ - q₀ . Compute pᵢ .
Step 4. If | p - p₁ | < T O L then OUTPUT p (Procedure is
complete.)
 STOP Step 5. Set i = i + 1. Step 6.
Set p₀ = p₁ . (Update p₀ , q₀ , q₁ , p₁ )      q ₀ = q₁
p₁ = p               q ₁ = f ( p )
Step 7. OUTPUT (''Method Failed after N 0 iterations')
STOP
```

Using the output=plot from Maple's Numerical Analysis package, we see Figure 4.6.

The secant method is also a special case of fixed-point iteration. The secant method approximates the first derivative with two good guesses. So, for INPUT, after we insert $f(x)$, we use two best initial guesses we can find. In our case, they are the last iterations of the bisection method. The tolerance and maximum iterations are the same as for fixed-point iteration. Similar to Newton's method, the initial guesses must be very good and avoid all places where $f'(x) = 0$. Thus, the maximum number of iterations is a very important default stopping criterion.

```
Secant(sec1, x = [0.007, 0.05], tolerance = 10^(-4), output =
sequence);
0.007, 0.05, 0.0070144149837644415741, 0.0070285318572268873650,
 0.0077040828198610209499, 0.0076202739067732916043,
```

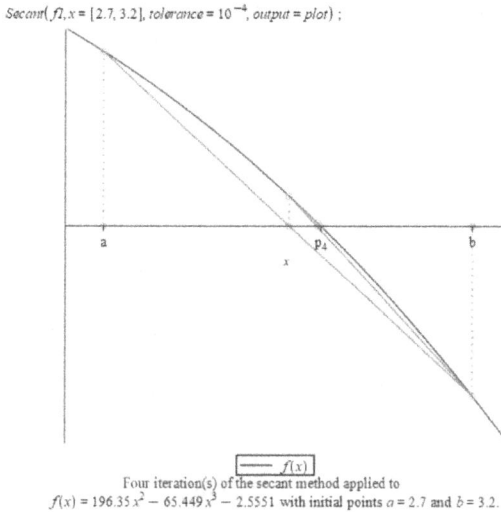

Secant(f1, x = [2.7, 3.2], tolerance = 10^{-4}, output = plot);

Four iteration(s) of the secant method applied to
$f(x) = 196.35\,x^2 - 65.449\,x^3 - 2.5551$ with initial points $a = 2.7$ and $b = 3.2$.

FIGURE 4.6
Secant method.

```
0.0076284963325304915341, 0.0076286029616969787321
```

To highlight the importance of good initial guesses, if we used $p_0 = 0.007$ and $p_1 = 0.05$, then we converged in eight iterations as opposed to two previously found with a root of $p = 0.0074726922$.

The monthly interest rate is 0.007628605, which is usually given as a 9.15% annual interest rate.

4.7 Root Find as a DDS

Many of the root finding methods that we discuss are discrete dynamical systems. Newton's method, fixed-point iteration, and the secant method are, in fact, discrete dynamical systems that we studied in Chapter 3.

Newton's method as a DDS:

$$x(n+1) = x(n) - \frac{f(x(n))}{f'(x(n))}.$$

Secant method:

$$x(n+1) = x(n) - \frac{f(x(n))(x(n+1) - x(n))}{f(x(n+1)) - f(x(n))}.$$

So if we have points (data) and do not know the derivative of the function f, then the secant method is obviously better than Newton's method.

Example of Newton's Method Using Excel

Newton's method is really a discrete dynamical system (Chapter 3).
    Using Future = Present + Change, we have

$$x_{n+1} = x_n - f(x_n)/f'(x_n)$$

We could use Excel to iterate within an acceptable tolerance.
    Consider finding the roots to $f(x) = x^2 - 2x - 3$. Now we could factor this into $f(x) = (x - 3)(x + 1)$, so the roots are 3 and −1.

Our root is found numerically as 3. We might say this is within our acceptable tolerance. Not all functions are this easy, however.

Newton's Method

$x(n + 1) = x(n) - f(xn)/f'(xn)$

n	x	$f(x)$	$f'(x)$	New x
1	2	−3	2	3.5
2	3.5	2.25	5	3.05
3	3.05	0.2025	4.1	3.00061
4	3.00061	0.00244	4.00122	3
5	3	3.7E-07	4	3
6	3	9.8E-15	4	3

Example of the Secant Method

Compute two iterations for the function $f(x) = x^3 - 5x + 1 = 0$ using the secant method in which the real roots of the equation $f(x)$ lie in the interval $(0, 1)$.

Solution:

```
Using the given data, we have,
x₀ = 0, x₁ = 1, and
f(x₀) = 1, f(x₁) = -3
Using the secant method formula, we can write
x₂ = x₁ - [(x₀ - x₁) / (f(x₀) - f(x₁))]f(x₁)
Now, substitute the known values in the formula,
= 1 - [(0 - 1) / ((1-(-3))]](-3)
= 0.25.
Therefore, f(x₂) = - 0.234375
```

```
Performing the second approximation,
x₃ = x₂ - [( x₁ - x₂) / (f(x₁) - f(x₂))]f(x₂)
=(- 0.234375) - [(1 - 0.25)/(-3 - (- 0.234375))]
(- 0.234375)
= 0.186441
Hence, f(x₃) = 0.074276
```

Using Excel to iterate, we obtain the following:

x_0	x_1	$f(x_0)$	$f(x_1)$	$((x_0-x_1)/$ $f(x_0)-f(x_1))*f(x_1)$	x_2	$f(x_2)$
0	1	1	-3	0.75	0.25	-0.234375
1	0.25	-3	-0.234375	0.0635593	0.186440678	0.0742773
0.25	0.1864407	-0.234375	0.0742773	-0.015296	0.201736256	-0.000471
0.1864407	0.2017363	0.0742773	-0.000471	9.64E-05	0.201639853	-8.64E-07
0.2017363	0.2016399	-0.000471	-8.64E-07	1.772E-07	0.201639676	1.035E-11
0.2016399	0.2016397	-8.64E-07	1.035E-11	-2.12E-12	0.201639676	0
0.2016397	0.2016397	1.035E-11	0	0	0.201639676	0

Our estimated root, between [0, 1] is 0.2016397.

Fixed-Point Iteration

Here we need the function, $f(x)$, and another selected function, $g(x)$ such that $|g'(x)| < 1$, on an interval where there is a change in sign.

Example 3. Fixed-Point iteration to find roots

Find the approximate root of the equation $2x^3 - 2x - 5 = 0$ up to 4 decimal places on the interval [0, 2].

Solution:

Given $f(x) = 2x^3 - 2x - 5 = 0$
    As per the algorithm, we find the value of x_0, for which we have to find a and b such that $f(a) < 0$ and $f(b) > 0$.

```
Now, f(0) = - 5
f(1) = - 5
f(2) = 7
Thus, we can use a = 1 and b = 2
Therefore, x₀ = (1 + 2)/2 = 1.5
Now, we shall find g(x) such that |g'(x)| < 1 at x = x₀
2x³ - 2x - 5 = 0
⇒ x = [(2x + 5)/2]¹ᐟ³
g(x) = [(2x + 5)/2]¹ᐟ³ which satisfies |g'(x)| < 1 at x =
1.5
```

```
Now, applying the iterative method x_n = g(x_{n-1}) for n = 1,
2, 3, 4, 5, …
For n = 1; x_1 = g(x_0) = [{2(1.5) + 5}/2]^{1/3} = 1.5874
For n = 2; x_2 = g(x_1) = [{2(1.5874) + 5}/2]^{1/3} = 1.5989
For n = 3; x_3 = g(x_2) = [{2(1.5989) + 5}/2]^{1/3} = 1.60037
For n = 4; x_4 = g(x_3) = [{2(1.60037) + 5}/2]^{1/3} = 1.60057
For n = 5; x_5 = g(x_4) = [{2(1.60057) + 5}/2]^{1/3} = 1.60059
```

In Excel, we start at X(0) = 1.5 and iterate $((2 * x(0) + 5)/2)^{(1/3)}$.

iteration	x_0	$g(x)$
0	1.5	1.5874011
1	1.5874011	1.5988796
2	1.5988796	1.6003749
3	1.6003749	1.6005694
4	1.6005694	1.6005948
5	1.6005948	1.6005981
6	1.6005981	1.6005985
7	1.6005985	1.6005985
8	1.6005985	1.6005985

Example 4

Given our $f(x) = x^3 - 5x + 1$ and $g(x) = (5x - 1)^{(1/3)}$, we are unable to find the root between [0, 1] as we did in the other methods. However, you can find the root between [2, 3]. You will be asked to do this as an exercise.

Root Finding with Python

Secant method for Archimedes' problem

```python
# Defining Function
def f(x):
return 196.35· x² -65.449· x³ -2.5551
# Implementing Secant Method
def secant(x0,x1,e,N):
    print('\n\n*** SECANT METHOD IMPLEMENTATION ***')
    step = 1
    condition = True
    while condition:
      if f(x0) == f(x1):
      print('Divide by zero error!')
      break
      x2 = x0 - (x1-x0)*f(x0)/( f(x1) - f(x0) )
      print('Iteration-%d, x2 = %0.6f and f(x2) = %0.6f' %
(step, x2, f(x2)))
      x0 = x1
```

```
      x1 = x2
      step = step + 1
      if step > N:
         print('Not Convergent!')
         break
      condition = abs(f(x2)) > e
   print('\n Required root is: %0.8f' % x2)
# Input Section
x0 = input('Enter First Guess: ')
x1 = input('Enter Second Guess: ')
e = input('Tolerable Error: ')
N = input('Maximum Step: ')
# Converting x0 and e to float
x0 = float(x0)
x1 = float(x1)
e = float(e)
# Converting N to integer
N = int(N)
#Note: You can combine above three section like this
# x0 = float(input('Enter First Guess: '))
# x1 = float(input('Enter Second Guess: '))
# e = float(input('Tolerable Error: '))
# N = int(input('Maximum Step: '))
# Starting Secant Method
secant(x0,x1,e,N)
Output
Enter First Guess: 2.5
Enter Second Guess: 3.2
Tolerable Error: 0.0001
Maximum Step: 50
*** SECANT METHOD IMPLEMENTATION ***
Iteration-1, x2 = 1.581228 and f(x2) = 292.028062
Iteration-2, x2 = 1.270160 and f(x2) = 117.870837
Iteration-3, x2 = 1.059628 and f(x2) = 21.561917
Iteration-4, x2 = 1.012493 and f(x2) = 2.384679
Iteration-5, x2 = 1.006632 and f(x2) = 0.060990
Iteration-6, x2 = 1.006478 and f(x2) = 0.000182
Iteration-7, x2 = 1.006478 and f(x2) = 0.000000
 Required root is: 1.00647787
Enter First Guess: 2
Enter Second Guess: 3
Tolerable Error: 0.000001
Maximum Step: 10
*** SECANT METHOD IMPLEMENTATION ***
Iteration-1, x2 = 2.785714 and f(x2) = -1.310860
Iteration-2, x2 = 2.850875 and f(x2) = -0.083923
Iteration-3, x2 = 2.855332 and f(x2) = 0.002635
Iteration-4, x2 = 2.855196 and f(x2) = -0.000005
Iteration-5, x2 = 2.855197 and f(x2) = -0.000000
 Required root is: 2.85519654
```

We have not found the correct root between 2.5 and 3.2.

Bisection Method

Code:

```
Defining Function
def f(x):
    return 196.35· x ² -65.449· x ³ -2.5551
# Implementing Bisection Method
def bisection(x0,x1,e):
    step =1
    print('\n\n*** BISECTION METHOD IMPLEMENTATION ***')
    condition =True
    while condition:
      x2 =(x0 + x1)/2
      print('Iteration-%d, x2 = %0.6f and f(x2) = %0.6f'%(step,
x2, f(x2)))
      if f(x0)* f(x2)<0:
          x1 = x2
      else:
          x0 = x2
      step = step +1
      condition =abs(f(x2))> e
    print('\nRequired Root is : %0.8f'% x2)
# Input Section
x0 =input('First Guess: ')
x1 =input('Second Guess: ')
e =input('Tolerable Error: ')
# Converting input to float
x0 =float(x0)
x1 =float(x1)
e =float(e)
#Note: You can combine above two section like this
# x0 = float(input('First Guess: '))
# x1 = float(input('Second Guess: '))
# e = float(input('Tolerable Error: '))
Output
First Guess: 2.5
Second Guess: 3.2
Tolerable Error: 0.0001
*** BISECTION METHOD IMPLEMENTATION ***
Iteration-1, x2 = 2.850000 and f(x2) = 77.210693
Iteration-2, x2 = 3.025000 and f(x2) = -17.500130
Iteration-3, x2 = 2.937500 and f(x2) = 32.767867
Iteration-4, x2 = 2.981250 and f(x2) = 8.378457
Iteration-5, x2 = 3.003125 and f(x2) = -4.372634
Iteration-6, x2 = 2.992188 and f(x2) = 2.049705
Iteration-7, x2 = 2.997656 and f(x2) = -1.149734
Iteration-8, x2 = 2.994922 and f(x2) = 0.452915
```

```
Iteration-9,  x2 = 2.996289 and f(x2) = -0.347677
Iteration-10, x2 = 2.995605 and f(x2) = 0.052802
Iteration-11, x2 = 2.995947 and f(x2) = -0.147392
Iteration-12, x2 = 2.995776 and f(x2) = -0.047283
Iteration-13, x2 = 2.995691 and f(x2) = 0.002762
Iteration-14, x2 = 2.995734 and f(x2) = -0.022260
Iteration-15, x2 = 2.995712 and f(x2) = -0.009749
Iteration-16, x2 = 2.995702 and f(x2) = -0.003493
Iteration-17, x2 = 2.995696 and f(x2) = -0.000366
Iteration-18, x2 = 2.995694 and f(x2) = 0.001198
Iteration-19, x2 = 2.995695 and f(x2) = 0.000416
Iteration-20, x2 = 2.995696 and f(x2) = 0.000025
Required Root is : 2.99569559
```

We have found the correct root.

Newton's Method in Python

Code:

```
import numpy as np
f = lambda x: 196.35*x**2-65.449*x**3-2.5551
f_prime = lambda x: 392.7*x-196.342*x**2
xo=3
newton_raphson = xo – (f(xo))/(f_prime(xo))
print("newton_raphson =", newton_raphson)
```

The key is the value of x0 given. Here with xo= 3 we get our root. Our root by Newton-Raphson is 2.9957076495217136.

Exercises

1. Do three iterations by hand to the root(s) for $f(x) = x^3 - 2$ using
   a. the bisection method.
   b. Newton's method.
   c. the secant method.
2. Find the root(s) accurate to a tolerance of 10^{-4} using technology for each method, bisection, Newton's, and secant, over the specified interval:
   a. $f(x) = (x + 1) - e^{-x^2}, 0 \le x \le 1$
   b. $f(x) = \ln x + x, 0.1 \le x \le 1$

c. $f(x) = x^3 - 2x - 1, -1 \le x \le 0$

d. $x^3 - 2x - 1, 1 \le x \le 2$

e. $f(x) = 20x - e^{-x^2} + 2, x > 0$

f. $f(x) = 10\cos(x) + e^x + 4, x > 0$

Projects

1. Archimedes

   Using the description and the figure from our Archimedes' illustrative example, complete the following requirements:

   a. Verify (re-derive) the Archimedes' equation by a multiple integration technique of your choice.

   b. Bracket all regions in which you expect to find roots. This should be a combination of graphics, function evaluations, and citing the appropriate theorem(s). Discuss which roots are physically relevant.

   c. Solve for the physically relevant root using all four methods available to you (bisection, fixed point, Newton's, secant). Stopping criteria should be chosen to satisfy required tolerances. The output for each method should include the starting values used, the solution and the number of iterations required for convergence reported in a tabular form. All answers should be accurate to 10^{-10}.

   d. Compare and contrast the four methods, including rates of convergence, order, function evaluations per iteration (efficiency), ease of programming, and information required to implement.

   e. Include the following in your project:
      i. Copies of graphs and calculations used in choosing the initial root guess.
      ii. Preliminary calculation of the number of iterations you expect for the convergence of the bisection method and a comparison to the actual number of iterations. Also discuss how you chose the initial start-up seeds for the algorithm.
      iii. Various different forms of $G(x)$ are considered for the fixed-point method, including calculations of the bound on $|G'(x)|$ and the number of iterations expected to converge.
      iv. Discuss how close is "close enough" for the Newton–Raphson method; that is, what happens when you vary the location of the initial seed or root guess?

2. Supply and Demand

   I. The precision of a numerical computing device is defined as the largest such that is calculated. Find the precision of your mathematical software.

   II. SITUATION: In order to support a specific mission, the company must maintain the ability to resupply mission-support contractors

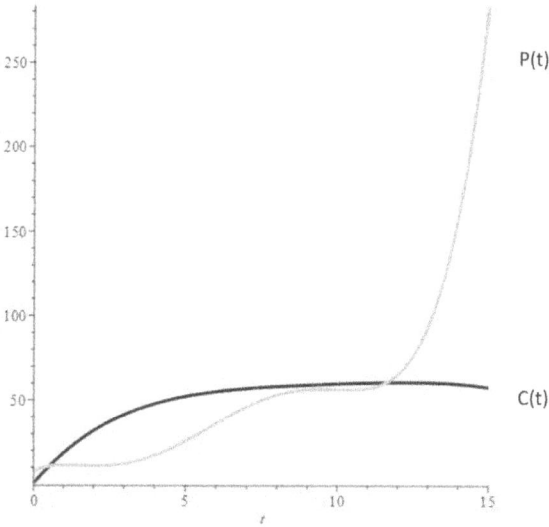

FIGURE 4.7
Plot of supply and demand.

when they are committed to build in third-world countries. Resupply planning involves not only maintaining the capability to produce items of equipment but also stockpiling items to replace initial construction losses. In planning for resupply of a certain item, one important date in the planning sequence is the date that the production rate equals and, from that time on, surpasses the consumption rate of an item. (See Figure 4.7.) This date is called the date to production equality, or D to P. Fluctuations in production and consumption rates sometimes create false D to Ps that must be disregarded by the analyst.

Given a production rate function, $P(t)$ and a consumption rate function, $C(t)$, where t is measured in months as

$$P(t) = 0.007240t^4 + 1.984t^3 - 6.540t^2 + 8.240t + 8.080$$

$$C(t) = -0.0049t^4 + 0.193t^3 - 2.920t^2 + 20.72t + 0.700.$$

Your team is required to graphically and analytically determine the D to P for the system.

a. Graph and label the curves on an appropriate domain and label the curves. Label the intersections of the curve and determine which are appropriate for analysis.

b. Bracket your regions to find the roots and discuss why those regions are selected.

   c. Solve for the relevent roots using the four methods (bisection, fixed point, Newton's, secant). Use a tolerance of 10^{-5}. Output should include starting values used, the solution with number of iterations required for convergence.

   d. Compare and contrast all four methods

3. Machine Replacement

   Consider a machine that is t years old. This machine earns revenue at a rate of e^{-t} dollars per year. After t years of use, the machine can be sold for parts and earns $1/t + 1$ dollars.

   a. Model the revenue function for this machine as a function of time.

   b. Knowing calculus, set up the function to maximize the revenue.

   c. Use several numerical methods to find the number of years that the machine needs to be kept in order to maximize the revenue and compare and contrast your results. Let Newton's method be one of your choices. Explain the values the Newton's method is moving toward from the following starting values: Use the following starting points: 0.00, 0.35, 0.50, 0.57, 0.65, 3.62, 3.63, 3.65, 3.80, 4.32, 4.33, 5.0, 8.0.

4. Pacific Oil Spill

   An oil spill has fouled 200 miles of the Pacific shoreline. Your oil company has been given 14 days by the Environmental Protection Agency to clean up the shoreline, after which a fine will be levied in the amount of $10,000/day. The local cleanup crew can scrub 5 miles of beach per week at a cost of $500/day. Additional crews can be brought in at a cost of $18,000 plus $800/day for each crew. You have been asked by your company to minimize the costs for this cleanup. Formulate this problem and solve for the value that minimizes cost using a numerical method of your choice.

5. Buying a Truck

   Consider a truck dealer who is offering a new 2024 truck for $25,000. He also offered to sell the same car for payments of $475 per month for 5 years. What monthly interest rate is the dealer charging?

References and Further Readings

Burden, R. and D. Faires (1997). *Numerical Analysis*. Brooks-Cole, Pacific Grove, CA.

Fox, W. (2018). *Mathematical Modeling for Business Analytics*. Taylor and Francis Publishers, Boca Raton, FL.

Giordano, F., W. Fox and S. Horton (2013). *A First Course in Mathematical Modeling*, 5th ed. Cengage Publishers, Boston, MA.

5

Interpolation and Polynomial Approximation

5.1 Introduction

In this chapter, we use two modelling scenarios to illustrate both the power and the limitation of interpolating and approximating with polynomials. Often engineers, scientists, and operation research analysts must develop functions over the domain given specific data. The results might be to find values at points in the domain but not specifically given, differentiate or integrate the approximating function, or develop a smooth or continuous curve representing the variables of a problem.

Interpolation refers to determining a function that can be used to exactly represent intermediate values for data in the domain.

Our first scenario deals with rocket telemetry data. Several entries are garbled, and we need to approximate them to determine information about the rocket. In the second scenario, a doctor needs to prescribe a radiation dosage for his patients in the treatment of cancerous tumors. The amount of radiation prescribed is critical not only to kill the cancerous cells but to keep neighbor cells alive, if possible.

In this chapter, we describe Lagrange polynomials, Neville's method, divided differences, and cubic splines.

5.2 Methods

Lagrange Polynomials

In general, the requirement that an $(n - 1)$–degree polynomial passes through n distinct data points yields a system of n linear algebraic equations in n unknowns. It is important to realize that large systems of equations can be difficult to solve with great accuracy, and small round-off errors in computer arithmetic can cause large oscillations to occur due to the presence of the higher order terms.

Using a quadratic polynomial $ax^2 + bx + c$ to interpolate requires us to solve for the coefficients a, b, and c.

The process of solving for a, b, and c by inverting matrix A and multiplying it by vector B is computed by MAPLE using the Linear Algebra package and the ReducedRowEchelonForm command. Often in using high-order polynomials oscillations in the curve between data pairs and near end points occurs. Therefore, although we can build $(n - 1)$–order polynomials, we might want to consider capturing the trend with lower order polynomials. We illustrate with the following examples.

Example 1. An $(n - 1)$–Degree Polynomial

The Maple Linear Algebra package is capable of in-depth matrix manipulations. Specifically, Maple quickly solves the type of problem discussed earlier with one command. Reviewing that earlier example:

$$\begin{bmatrix} 1 & 1 & 1 \\ 1 & 2 & 4 \\ 1 & 3 & 9 \end{bmatrix} \bullet \begin{bmatrix} a \\ b \\ c \end{bmatrix} = \begin{bmatrix} 5 \\ 8 \\ 25 \end{bmatrix}$$

> *with (LinearAlgebra) :*
> $B := \langle\langle 1,1,1,5\rangle|\langle 1,2,4,8\rangle|\langle 1,3,9,25\rangle\rangle;$

$$B := \begin{bmatrix} 1 & 1 & 1 \\ 1 & 2 & 3 \\ 1 & 4 & 9 \\ 5 & 8 & 25 \end{bmatrix}$$

> $B1 := Transpose(B);$

$$B1 := \begin{bmatrix} 1 & 1 & 1 & 5 \\ 1 & 2 & 4 & 8 \\ 1 & 3 & 9 & 25 \end{bmatrix}$$

> *ReducedRowEchelonForm(B1);*

$$\begin{bmatrix} 1 & 0 & 0 & 16 \\ 0 & 1 & 0 & -18 \\ 0 & 0 & 1 & 7 \end{bmatrix}$$

This provides the coefficients for the solution of $P_2(x) = 16 - 18x + 7x^2$. The graphical presentation demonstrates the fit of the solution, presented in Figure 5.1.

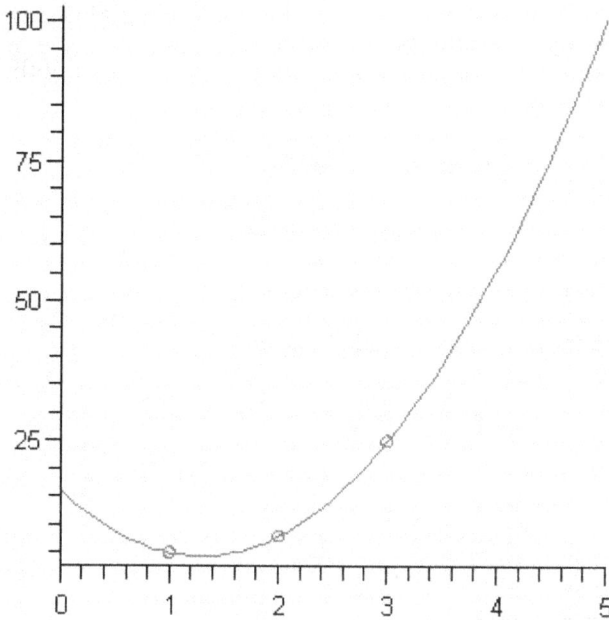

FIGURE 5.1
The quadratic curve passes through the points.

$$> x := [1,2,3]:$$
$$y := [5,8,25]:$$
$$xy := \{seq(x[i], y[i]\ ], i = 1..3)\}:$$
$$with(plots):$$
$$plot1 := plot(xy, style = point, symbol = circle):$$
$$plot2 := plot(16 - 18 * z + 7 * z \wedge 2, z = 0..5):$$
$$display(\{plot1, plot2\});$$

Problems with Higher Order Polynomials

There are problems that might exist in higher order polynomial fitting. As we said earlier, sometimes in fitting the curve, the data have oscillations and snaking near the endpoints of the domain. This oscillating and snaking behavior makes the polynomial fit ill fated for interpolation near the end-points and makes predictions outside the end points pointless. We provide an illustrative example in which we fit a complete polynomial. Note that we have lost the trend of the data (although we fit each datum exactly), and at the endpoints, we have oscillations. Therefore, we suggest a close examination of all higher order polynomials before accepting any.

Here are the data and the scatterplot suggesting a smooth curve with concave-up trends.

```
> Xvalues:=[.55,1.2,2,4,6.5,12,16];Yvalues:=[.13,.64,5.8,102,
210,2030,3900];
```

$$Xvalues := [0.55, 1.2, 2, 4, 6.5, 12, 16]$$

$$Yvalues := [0.13, 0.64, 5.8, 102, 210, 2030, 3900]$$

```
> pointplot(zip((x,y)->[x,y],Xvalues,Yvalues));
```

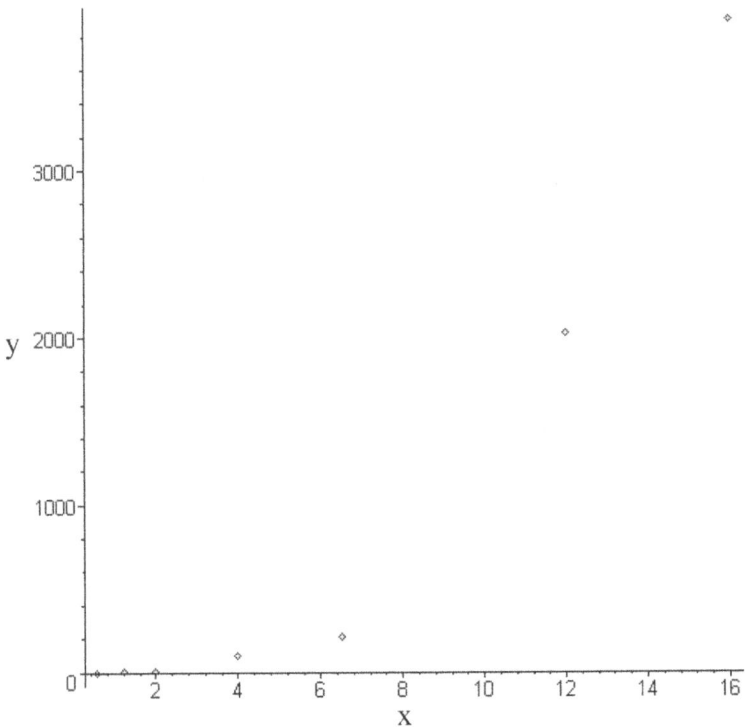

Next, we fit a sixth-order polynomial ($n = 7$) and then plot the polynomial. We will use the *interp* command to obtain an ($n - 1$)–order polynomial.

```
> FI:=interp([.55,1.2,2,4,6.5,12,16],[.13,.64,5.8,102,210,2030,
3900], z);
```

$$FI := -0.01383726235\, z^6 + 0.5084246673\, z^5 - 18.09506969 - 6.437923862\, z^4$$
$$+ 64.31279044\, z + 34.85731000\, z^3 - 73.99155349\, z^2$$

```
>plot(-.1383726235e-1*z^6+.5084246673*z^5-18.09506969-6.437923862*
z^4+64.31279044*z+34.85731000*z^3-73.99155349*z^2,z=0..16);
```

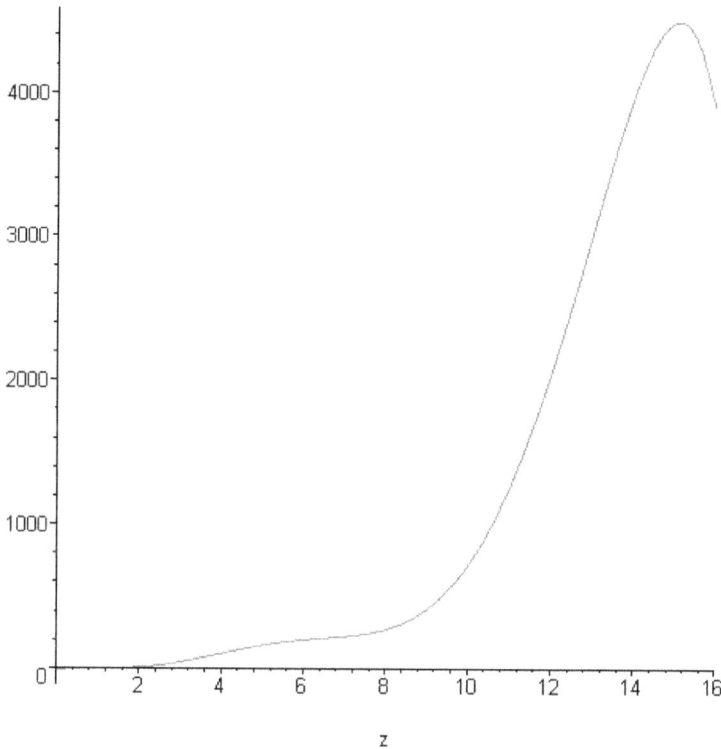

FIGURE 5.2
Higher order fit.

Again note the oscillation at the end points in Figure 5.2. Thus, although the higher order $(n - 1)$–order polynomial gives a perfect fit through the data points, its oscillations make interpolation and prediction less accurate.

Example 2. Fitting a Fifth-Order Polynomial Using Least-Squares

Given a set of data points (see Table 5.1), an analyst decides to attempt to fit a curve to the data using a high-order polynomial. We show the Maple commands required, and Figure 5.3 displays the polynomial curve superimposed on the data.

TABLE 5.1

Data for Example 2

x	1	2	3	4	5	6
y	305	266	135	−16	125	1230

```
> with(stats)with(plots) :
> xdata := [1,2,3,4,5,6];
  xdata := [1,2,3,4,5,6]
> ydata := [305,266,135,-16,125,1230];
  ydata := [305,266,135,-16,125,1230]
```

$$> xyfit := fit\left[leastsquare\left[[x,y], y = a*x \wedge 5 + b*x \wedge 4 + c\right.\right.$$
$$x \wedge 3 + d*x \wedge 2 + e*x + g, \{a,b,c,d,e,g\}\right]\right]([xdata,$$
$$ydata]);$$

$$xyfit := y = x^5 - 5x^4 - 3x^3 + 7x^2 + 5x + 300$$

```
> f := unapply(rhs(xyfit),x);
```
$$f := x \rightarrow x^5 - 5x^4 - 3x^3 + 7x^2 + 5x + 300$$

```
> xy := {seq([xdata[i], ydata[i]], i = 1..6)};
```
$$xy := \{[1,305],[2,266],[3,135],[4,-16],[5,125],[6,1230]\}$$

```
> c1 := pointplot({seq([xdata[i], ydata[i]], i = 1..6)},
    thickness = 3):
  c2 := plot(f(x), x = 0..6):
  display({c1,c2});
```

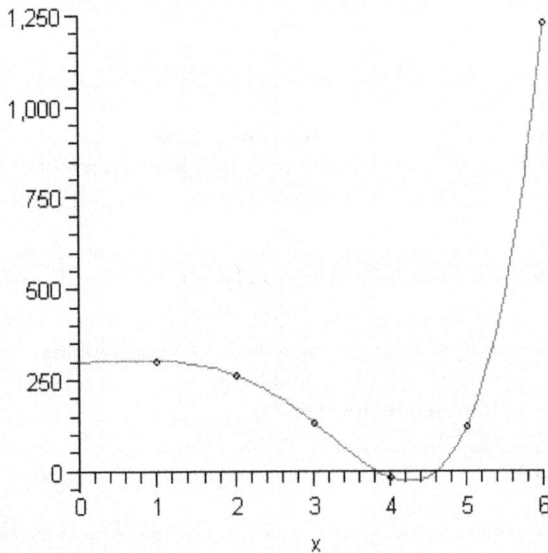

FIGURE 5.3a
The plot for Example 2.

Do we have to use a fifth-order polynomial? The answer lies in the need and use of the model as well as how well the trend is captured regardless of the perfect fit.

5.3 Lagrange Polynomials

So, now let's define an nth-degree interpolating Lagrange polynomial:

$$P_n(x) = f(x_0)L_{n,0}(x) + \ldots + f(x_n)L_{n,n}(x) = \sum_{k=0}^{n} f(x_k)L_{n,k}(x)$$

where

$$L_{n,k}(x) = \frac{(x-x_0)(x-x_1)\ldots(x-x_{k-1})(x-x_k))\ldots(x-x_n)}{(x_k-x_0)(x_k-x_1)\ldots(x_k-x_{k-1})(x_k-x_k))\ldots(x_k-x_n)}$$

for each $k = 0, 1, \ldots, n$.

Assume that we have x values 2, 2.5, and 5 and that we want to find an interpolating polynomial for $f(x) = 1/x$.

We need to apply the $L_{n,k}(x)$ formula:

$$L_0(x) = \frac{(x-2.5)(x-4)}{(2-2.5)(2-4)} = (x-6.5)x + 10$$

$$L_1(x) = \frac{(x-2)(x-4)}{(2.5-2.)(2.5-4)} = \frac{(-4x+24)x-32}{3}$$

$$L_2(x)\frac{(x-2)(x-2.5)}{(4-2.)(4-2.5)} = \frac{(x-4.5)x+5}{3}.$$

Since $f(x_0) = f(2) = \frac{1}{2} = 0.5$, $f(x_1) = f(2.5) = 0.4$, and $f(x_2) = f(4) = 0.25$.
Our second-order polynomial is

$$p(x) = \sum_{k=0}^{2} f(x_k)L_k(x).$$

By substituting

$$= 0.5((x-6.5)x+10) + 0.4((-4x+24)x-32)/3 + 0.25((x-4.5)x+5)/$$
$$3 = 0.05\ x^2 - 0.425x + 1.15$$

To find $f(3)$, we substitute 3 for x.

$$P(3) = 0.325$$

Since we know $f(x) = 1/x$, then $f(3)$ is exactly $1/3$.
The exact error is $|1/3 - 0.325| = .008333333\ldots$

```
Remainder Term(p2);
  / (x - 1.) (x - 2.5) (x - 4.)\
 |- --------------------------| &where {1. <= xi and xi <= 4.}
 |            4               |
 \           xi               /
```

> *RemainderTerm* $(p2)$

$$\left(-\frac{(x-1.)(x-2.5)(x-4.)}{\xi^4} \right) \& \text{ where } \{1. \le \xi \le 4.\}$$

5.4 Divided Differences

Another method that we could use is divided difference tables.

Divided Difference Tables

We offer a detailed discussion of how to answer these questions using divided difference tables as a qualitative method to access low order polynomial trends in the data.

Let's assume we have a polynomial in the form

$$y = ax^2 + bx + c.$$

Now,

$$\frac{dy}{dx} = 2ax + b,$$

$$\frac{d^2y}{dx^2} = 2a,$$

$$\frac{d^3y}{dx^3} = 0$$

Thus, the third derivative and all other higher order derivatives are zero. Now, consider a data set that comes directly from this polynomial such as the following data in Table 5.2.

Recall the definition of the derivative is

$$\frac{dy}{dx} = \lim_{\Delta x \to 0} \frac{\Delta y}{\Delta x}.$$

Now, a slope may not be a very good approximation to dy/dx, but if we take the difference between successive functional values, we can gain insight into what the derivative is doing. Our goal is to find the successive functional differences that might approximate zero. Let's illustrate as shown in Table 5.3.

Our divided difference polynomial is

$$P(x) = 6 + 5(x - 1) + 1(x - 1)(x - 2).$$

For $x = 3.5$, the answer is 22.25.

Using Maple:

> $xy := \big[[1,6],[2,11],[3,18],[4,27],[5,38]\big]$

$$xy := \big[[1,6],[2,11],[3,18],[4,27],[5,38]\big]$$

> $New3 := polynomialInterpolation(xy, independentvar = x, method = newton):$

> $LL5 := expand(Interpolant(New3))$

$$LL5 := x^2 + 2x + 3$$

> $subs(x = 3.5, LL5)$

$$22.25$$

TABLE 5.2

Data for Divided Differences

x	1	2	3	4	5
y	6	11	18	27	38

TABLE 5.3

Divided Difference Table

x	y	Δ^1	Δ^2	Δ^3	Δ^4
1	6				
		5			
2	11		1		
		7		0	0
3	18		1		
		9		0	
4	27		1		
		11			
5	38				

TABLE 5.4

Generic Divided Difference Table Formulas

X	Y	Δ^1	Δ^2
X1	Y1		
		$\dfrac{(Y2-Y1)}{(X2-X1)}$	
X2	Y2		$\dfrac{\dfrac{(Y3-Y2)}{(X3-X2)}-\dfrac{(Y2-Y1)}{(X2-X1)}}{(X3-X1)}$
		$\dfrac{(Y3-Y2)}{(X3-X2)}$	
X3	Y3		

In the Tables 5.3 and 5.4, the Δ^1s are all positive, indicating an increasing function. The Δ^2s are all positive indicating concave up. The Δ^3s are all 0, indicating the third derivative is approximately 0. This corresponds to what we saw earlier when we took successive derivatives of the function: $y = ax^2 + bx + c$.

The divided difference table is composed of two columns of our original data and successive columns of divided differences that qualitatively (not numerically) give information about the derivatives.

We have created a Maple program that takes the input x and y data and produces a divided difference table. The modeller must interpret the divided difference table. This divided difference program with example is provided on the website.

Example 3. Using Divided Differences

```
>with (stats):with(LinearAlgebra):
>xdata:=[0,2,4,6,8]:
>ydata:=[0,4,16,36,64]:
>ddproc(xdata,ydata);
```

$$\begin{bmatrix} 0 & 0 & 0 & 0 & 0 \\ 2 & 4 & 2 & 0 & 0 \\ 4 & 16 & 6 & 1 & 0 \\ 6 & 36 & 10 & 1 & 0 \\ 8 & 64 & 14 & 1 & 0 \end{bmatrix}$$

The first two columns are the data. The polynomial using Newton's interpolatory divided differences would be

$$P(x) = 0 + 2(x - 0) + 1(x - 0)(x - 2).$$

We can interpolate at 3 by substitution:

$$P(3) = 2 * 3 + 1 * 3 * 1 = 9$$

Example 4. Vehicular Stopping Distance

The divided difference table assists in determining what order polynomial should be used to approximate a set of data. This can be demonstrated using the vehicular stopping distance example.

```
>xdata := [ 20,25,30,35,40,45,50,55,60,65,70,75,80]:
>ydata := [42,56,73.5,91.5,116, 142.5, 173, 209.5, 248,
292.5, 343, 401, 464]:
>ddproc(xdata,ydata);
```

20.	42.	0.	0.	0.	0.	0.	0.	0.	0.	0.	0.	0.
25.	56.	2.800	0.	0.	0.	0.	0.	0.	0.	0.	0.	0.
30.	73.5	3.500	0.07000	0.	0.	0.	0.	0.	0.	0.	0.	0.
35.	91.5	3.600	0.01000	−0.004000	0.	0.	0.	0.	0.	0.	0.	0.
40.	116.	4.900	0.1300	0.008000	0.0006000	0.	0.	0.	0.	0.	0.	0.
45.	142.5	5.300	0.04000	−0.006000	−0.0007000	−0.00005200	0.	0.	0.	0.	0.	0.
50.	173.	6.100	0.08000	0.002667	0.0004333	0.00004533	0.	0.	0.	0.	0.	0.
55.	209.5	7.300	0.1200	0.002667	0.	−0.00001733	0.	0.	0.	0.	0.	0.
60.	248.	7.700	0.04000	−0.005333	−0.0004000	−0.00001600	0.	0.	0.	0.	0.	0.
65.	292.5	8.900	0.1200	0.005333	0.0005333	0.00003733	0.	0.	0.	0.	0.	0.
70.	343.	10.10	0.1200	0.	−0.0002667	−0.00003200	0.	0.	0.	0.	0.	0.
75.	401.	11.60	0.1500	0.002000	0.0001000	0.00001467	0.	0.	0.	0.	0.	0.
80.	464.	12.60	0.1000	−0.003333	−0.0002667	−0.00001467	0.	0.	0.	0.	0.	0.

We notice this table differs from previous tables. The third column is small but has both positive and negative values. Negative values within a divided difference table indicate that we might not be able to use the qualitative assessment of the table. However, in this case, the values are so small we can assume they are all very close to zero (ignoring the signs). We might conclude a quadratic polynomial based upon this assessment. However, if we accepted the changing signs then all future columns cannot be used. We might conclude that the tables did not provide us qualitative information for a low-order polynomial. We again demonstrate the Maple solution of fitting the quadratic polynomial using the least-squares criterion.

```
> xdata:=[20,25,30,35,40,45,50,55,60,65,70,75,80]:
> ydata:=[42,56,73.5,91.5,116,142.5,173,209.5,248,292.5,
343,401,464]:
> xyfit:=fit[leastsquare[[x,y],y=a*x^2+b*x+c,{a,b,c}]]
([xdata,ydata]);
```

5.5 Cubic Splines

The Cubic Spline Model

In this section, we introduce cubic spline interpolation as an alternative method for constructing empirical models. By using difference cubic polynomials between successive pairs of data points and connecting the cubic polynomials together in a smooth fashion, we can capture the trend of the data, regardless of the underlying relationships. Simultaneously, we will reduce the tendency toward oscillation and the sensitivity of the coefficients to changes in the data.

For a full development of the cubic spline model with a discussion of obtaining the equations to solve for all the cubic coefficients, see Chapter 11 (Section 11.2).

Natural Cubic Splines as a System of Equations

In modelling natural cubic splines, we want to fit a third-order polynomial between every pair of data points. Let's assume that we have six data pairs as shown in Table 5.5.

We will need 5 third-order equations.

$$[7-14], S_1 = a_3 x^3 + a_2 x^2 + a_1 x + a_0$$
$$[14-21], S_2 = b_3 x^3 + b_2 x^2 + b_1 x + b_0$$
$$[21-28], S_3 = c_3 x^3 + c_2 x^2 + c_1 x + c_0$$
$$[28-35], S_4 = d_3 x^3 + d_2 x^2 + d_1 x + d_0$$
$$[35-42], S_5 = e_3 x^3 + e_2 x^2 + e_1 x + e_0$$

We note that there are 20 unknowns ($a_3, a_2, \ldots, e_1, e_0$), and we need 20 equations to uniquely solve for these unknowns.

By substituting in the (x, y) data pairs, we obtain 10 equations. We still lack 10 equations.

We force both the first derivative (slope) and the second derivative (concavity) at the interior data points to match. For $i > 1$ and less than n,

TABLE 5.5

Data Pairs

t	7	14	21	28	35	42
y	125	275	800	1200	1700	1650

$$\frac{dS_{i+1}}{dx} = \frac{dS_{i+2}}{dx}$$

$$\frac{d^2S_{i+1}}{dx^2} = \frac{d^2S_{i+2}}{dx^2}$$

Since there are four interior points, this gives eight more equations. Matching the derivatives ensures smoothness of the curves at the data points. The last two equations concern the endpoints. Under natural cubic spline, we want the second derivative to equal zero. This yields two more equations, and we have 20 equations. Note that if we had clamped cubic splines, then the first derivatives at the two endpoints would equal specific constants ($f'(x_0)$ and $f'(x_n)$), and again we would have 20 equations.

We present the 20 equations for the given data:

```
 (1)  343 a3 +  49 a2 +  7 a1 + a0 = 125
 (2)  2744 a3 + 196 a2 + 14 a1 + a0 = 275
 (3)  2744 b3 + 196 b2 + 14 b1 + b0 = 275
 (4)  9621 b3 + 441 b2 + 21 b1 + b0 = 800
 (5)  9621 c3 + 441 c2 + 21 c1 + c0 = 800
 (6)  21952 c3 + 784 c2 + 28 c1 + c0 = 1200
 (7)  21952 d3 + 784 d2 + 28 d1 + d0 = 1200
 (8)  42875 d3 + 1225 d2 + 35 d1 + d0 = 1700
 (9)  42875 e3 + 1225 e2 + 35 e1 + e0 = 1700
(10)  74088 e3 + 1764 e2 + 42 e1 + e0 = 1650
(11)  588 a3 + 28 a2 + a1 = 588 b3 + 28 b2+ b1
(12)  1323 b3 + 42 b2 + b1 = 1323 c3 + 42 c2 + c1
(13)  2352 c3 + 56 c2 + c1 = 2352 d3 + 56 d2 + d1
(14)  3675 d3 + 70 d2 + d1 = 3675 e3 + 70 e2 + e1
(15)  84 a3 + 2 a2 = 84 b3 + 2 b2
(16)  126 b3 + 2 b2 = 126 c3 +2 c2
(17)  168 c3 + 2 c2 = 168 d3 + 2 d2
(18)  210 d3 + 2 d2 = 210 e3 + 2 e2
(19)  42 a3 + 2 a2 =0
(20)  252 e3 + 2 e2 =0
```

We will use Maple to solve this system of equations. We increase the number of digits accuracy as well as increase the user interface for the matrix size.

$$B := \begin{bmatrix}
343 & 49 & 7 & 1 & 0 & 0 & 0 & 0 & 0 & 0 & 0 & 0 & 0 & 0 & 0 & 0 & 0 & 0 & 0 & 0 & 125 \\
2744 & 196 & 14 & 1 & 0 & 0 & 0 & 0 & 0 & 0 & 0 & 0 & 0 & 0 & 0 & 0 & 0 & 0 & 0 & 0 & 275 \\
0 & 0 & 0 & 0 & 2744 & 196 & 14 & 1 & 0 & 0 & 0 & 0 & 0 & 0 & 0 & 0 & 0 & 0 & 0 & 0 & 275 \\
0 & 0 & 0 & 0 & 9261 & 441 & 21 & 1 & 0 & 0 & 0 & 0 & 0 & 0 & 0 & 0 & 0 & 0 & 0 & 0 & 800 \\
0 & 0 & 0 & 0 & 0 & 0 & 0 & 0 & 9261 & 441 & 21 & 1 & 0 & 0 & 0 & 0 & 0 & 0 & 0 & 0 & 800 \\
0 & 0 & 0 & 0 & 0 & 0 & 0 & 0 & 21952 & 784 & 28 & 1 & 0 & 0 & 0 & 0 & 0 & 0 & 0 & 0 & 1200 \\
0 & 0 & 0 & 0 & 0 & 0 & 0 & 0 & 0 & 0 & 0 & 0 & 21952 & 784 & 28 & 1 & 0 & 0 & 0 & 0 & 1200 \\
0 & 0 & 0 & 0 & 0 & 0 & 0 & 0 & 0 & 0 & 0 & 0 & 42875 & 1225 & 35 & 1 & 0 & 0 & 0 & 0 & 1700 \\
0 & 0 & 0 & 0 & 0 & 0 & 0 & 0 & 0 & 0 & 0 & 0 & 0 & 0 & 0 & 0 & 42875 & 1225 & 35 & 1 & 1700 \\
0 & 0 & 0 & 0 & 0 & 0 & 0 & 0 & 0 & 0 & 0 & 0 & 0 & 0 & 0 & 0 & 74088 & 1764 & 42 & 1 & 1650 \\
588 & 28 & 1 & 0 & -588 & -28 & -1 & 0 & 0 & 0 & 0 & 0 & 0 & 0 & 0 & 0 & 0 & 0 & 0 & 0 & 0 \\
0 & 0 & 0 & 0 & 1323 & 42 & 1 & -1323 & -42 & -1 & 0 & 0 & 0 & 0 & 0 & 0 & 0 & 0 & 0 & 0 & 0 \\
0 & 0 & 0 & 0 & 0 & 0 & 0 & 0 & 2352 & 56 & 1 & 0 & -2352 & -56 & -1 & 0 & 0 & 0 & 0 & 0 & 0 \\
0 & 0 & 0 & 0 & 0 & 0 & 0 & 0 & 0 & 0 & 0 & 0 & 3765 & 70 & 1 & 0 & -3765 & -70 & -1 & 0 & 0 \\
84 & 2 & 0 & 0 & -84 & -2 & 0 & 0 & 0 & 0 & 0 & 0 & 0 & 0 & 0 & 0 & 0 & 0 & 0 & 0 & 0 \\
0 & 0 & 0 & 0 & 126 & 2 & 0 & 0 & -126 & -2 & 0 & 0 & 0 & 0 & 0 & 0 & 0 & 0 & 0 & 0 & 0 \\
0 & 0 & 0 & 0 & 0 & 0 & 0 & 0 & 168 & 2 & 0 & 0 & -168 & -2 & 0 & 0 & 0 & 0 & 0 & 0 & 0 \\
0 & 0 & 0 & 0 & 0 & 0 & 0 & 0 & 0 & 0 & 0 & 0 & 210 & 2 & 0 & 0 & -210 & -2 & 0 & 0 & 0 \\
42 & 2 & 0 & 0 & 0 & 0 & 0 & 0 & 0 & 0 & 0 & 0 & 0 & 0 & 0 & 0 & 0 & 0 & 0 & 0 & 0 \\
0 & 0 & 0 & 0 & 0 & 0 & 0 & 0 & 0 & 0 & 0 & 0 & 0 & 0 & 0 & 0 & 252 & 2 & 0 & 0 & 0
\end{bmatrix}$$

RREF

$$B := \begin{bmatrix}
1 & 0 & 0 & 0 & 0 & 0 & 0 & 0 & 0 & 0 & 0 & 0 & 0 & 0 & 0 & 0 & 0 & 0 & 0 & 0 & 0.3285032886 \\
0 & 1 & 0 & 0 & 0 & 0 & 0 & 0 & 0 & 0 & 0 & 0 & 0 & 0 & 0 & 0 & 0 & 0 & 0 & 0 & -6.898569060 \\
0 & 0 & 1 & 0 & 0 & 0 & 0 & 0 & 0 & 0 & 0 & 0 & 0 & 0 & 0 & 0 & 0 & 0 & 0 & 0 & 53.62189371 \\
0 & 0 & 0 & 1 & 0 & 0 & 0 & 0 & 0 & 0 & 0 & 0 & 0 & 0 & 0 & 0 & 0 & 0 & 0 & 0 & -25. \\
0 & 0 & 0 & 0 & 1 & 0 & 0 & 0 & 0 & 0 & 0 & 0 & 0 & 0 & 0 & 0 & 0 & 0 & 0 & 0 & -0.549221982 \\
0 & 0 & 0 & 0 & 0 & 1 & 0 & 0 & 0 & 0 & 0 & 0 & 0 & 0 & 0 & 0 & 0 & 0 & 0 & 0 & 29.96589232 \\
0 & 0 & 0 & 0 & 0 & 0 & 1 & 0 & 0 & 0 & 0 & 0 & 0 & 0 & 0 & 0 & 0 & 0 & 0 & 0 & -462.4805656 \\
0 & 0 & 0 & 0 & 0 & 0 & 0 & 1 & 0 & 0 & 0 & 0 & 0 & 0 & 0 & 0 & 0 & 0 & 0 & 0 & 2383.478143 \\
0 & 0 & 0 & 0 & 0 & 0 & 0 & 0 & 1 & 0 & 0 & 0 & 0 & 0 & 0 & 0 & 0 & 0 & 0 & 0 & 0.410658693 \\
0 & 0 & 0 & 0 & 0 & 0 & 0 & 0 & 0 & 1 & 0 & 0 & 0 & 0 & 0 & 0 & 0 & 0 & 0 & 0 & -30.50659023 \\
0 & 0 & 0 & 0 & 0 & 0 & 0 & 0 & 0 & 0 & 1 & 0 & 0 & 0 & 0 & 0 & 0 & 0 & 0 & 0 & 807.4415678 \\
0 & 0 & 0 & 0 & 0 & 0 & 0 & 0 & 0 & 0 & 0 & 1 & 0 & 0 & 0 & 0 & 0 & 0 & 0 & 0 & -6505.97679 \\
0 & 0 & 0 & 0 & 0 & 0 & 0 & 0 & 0 & 0 & 0 & 0 & 1 & 0 & 0 & 0 & 0 & 0 & 0 & 0 & -0.4373436113 \\
0 & 0 & 0 & 0 & 0 & 0 & 0 & 0 & 0 & 0 & 0 & 0 & 0 & 1 & 0 & 0 & 0 & 0 & 0 & 0 & 40.7333735 \\
0 & 0 & 0 & 0 & 0 & 0 & 0 & 0 & 0 & 0 & 0 & 0 & 0 & 0 & 1 & 0 & 0 & 0 & 0 & 0 & -1187.277417 \\
0 & 0 & 0 & 0 & 0 & 0 & 0 & 0 & 0 & 0 & 0 & 0 & 0 & 0 & 0 & 1 & 0 & 0 & 0 & 0 & 12111.4004 \\
0 & 0 & 0 & 0 & 0 & 0 & 0 & 0 & 0 & 0 & 0 & 0 & 0 & 0 & 0 & 0 & 1 & 0 & 0 & 0 & 0.247496114 \\
0 & 0 & 0 & 0 & 0 & 0 & 0 & 0 & 0 & 0 & 0 & 0 & 0 & 0 & 0 & 0 & 0 & 1 & 0 & 0 & -31.18451036 \\
0 & 0 & 0 & 0 & 0 & 0 & 0 & 0 & 0 & 0 & 0 & 0 & 0 & 0 & 0 & 0 & 0 & 0 & 1 & 0 & 1290.479268 \\
0 & 0 & 0 & 0 & 0 & 0 & 0 & 0 & 0 & 0 & 0 & 0 & 0 & 0 & 0 & 0 & 0 & 0 & 0 & 1 & -15877.14509
\end{bmatrix}$$

```
> xdata := [7, 14, 21, 28, 35, 42];
  xdata := [7, 14, 21, 28, 35, 42]
```

```
> ydata := [125,  275,  800,  1200,  1700,1650];
  ydata := [125,  275,  800,  1200,  1700,1650]
```

$> p1 := \text{point } plot\left(\left\{seq\left(\left[xdata[i], yadat[i]\right], i = 1..6\right)\right\}\right):$

$> p2 := plot\left(C[1,21]\cdot x^3 + C[2,21]\cdot x^2 + C[3,21]\cdot x\right.$
$\left.+C[4,21], x = 7..14\right):$

$> p3 := plot\left(C[5,21]\cdot x^3 + C[6,21]\cdot x^2 + C[7,21]\cdot x\right.$
$\left.+C[8,21], x = 14..21\right):$

$> p4 := plot\left(C[9,21]\cdot x^3 + C[10,21]\cdot x^2 + C[11,21]\cdot x\right.$
$\left.+C[12,21], x = 21..28\right):$

$> p5 := plot\left(C[13,21]\cdot x^3 + C[14,21]\cdot x^2 + C[15,21]\cdot x\right.$
$\left.+C[16,21], x = 28..35\right):$

$> p6 := plot\left(C[17,21]\cdot x^3 + C[18,21]\cdot x^2 + C[19,21]\cdot x\right.$
$\left.+C[20,21], x = 35..42\right):$

```
> display( {p1, p2, p3, p4, p5, p6} );
```

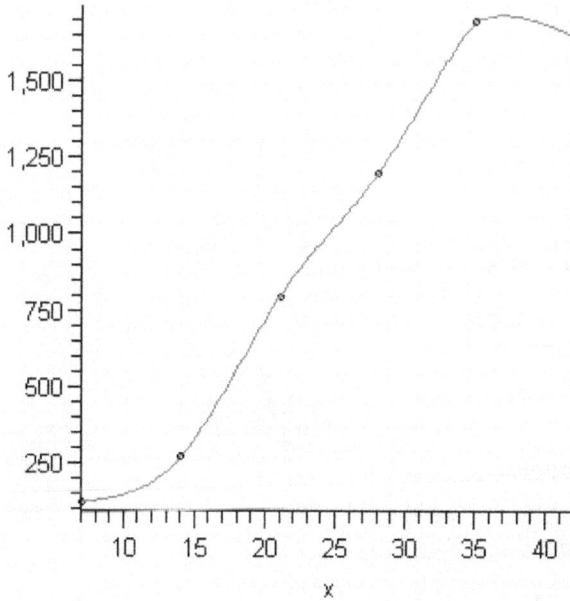

FIGURE 5.3b
Cubic spline model.

We note that in general with N pairs of data we will need $N - 1$ cubic equations. To set up the system of equations, we will need the following:

1. $2N - 2$ cubic equations (2 for each successive pairs of points)
2. For each interior point $(N - 2)$ points, we obtain two equations: One equation is the first derivative being set equal at each interior point, and the second is the second derivative being set equal at each interior point. This yields another $2(N - 2)$ equations.
3. Finally, if we need clamped cubic splines, then we use the fact that the first derivatives at the two end points equal specific constants $(f'(x_0)$ and $f'(x_n))$ to obtain the last two equations. If we need natural cubic splines, we assume the slopes are unknown but constants, and then we make the second derivatives equal to zero at the two endpoints to obtain the last two equations.

Before we move on to applications, we provide some simple examples of these methods with Python.

Python

Cubic splines using linear algebra methods

```
b = np.array([1, 3, 3, 2, 0, 0, 0, 0])
b = b[:, np.newaxis]
A = np.array([[0, 0, 0, 1, 0, 0, 0, 0], [0, 0, 0, 0, 1, 1, 1,
1], [1, 1, 1, 1, 0, 0, 0, 0], \
              [0, 0, 0, 0, 8, 4, 2, 1], [3, 2, 1, 0, -3, -2, -1,
0], [6, 2, 0, 0, -6, -2, 0, 0],\
              [0, 2, 0, 0, 0, 0, 0, 0], [0, 0, 0, 0, 12, 2, 0, 0]])
np.dot(np.linalg.inv(A), b)
array([[-0.75],
       [ 0. ],
       [ 2.75],
       [ 1. ],
       [ 0.75],
       [-4.5 ],
       [ 7.25],
       [-0.5 ]])
```

Our two cubic spline equations are

$S_1(x) = -.75x^3 + 2.75x + 1$ for $0 \leq x \leq 1$ and
$S_2(x) = .75x^3 - 4.5x^2 + 7.25x - 0.5$ for $1 \leq x \leq 2$.

To find $x = 15$, we use $S_2(1.5) = 2.7813$.

Lagrange Polynomials in Python

Code

```
#Lagrange Polynomials
import numpy as np
import numpy.polynomial.polynomial as poly
import matplotlib.pyplot as plt
plt.style.use('seaborn-poster')
x = [0, 1, 2]
y = [1, 3, 2]
P1_coeff = [1,-1.5,.5]
P2_coeff = [0, 2,-1]
P3_coeff = [0,-.5,.5]
# get the polynomial function
P1 = poly.Polynomial(P1_coeff)
P2 = poly.Polynomial(P2_coeff)
P3 = poly.Polynomial(P3_coeff)
x_new = np.arange(-1.0, 3.1, 0.1)
fig = plt.figure(figsize = (10,8))
plt.plot(x_new, P1(x_new), 'b', label = 'P1')
plt.plot(x_new, P2(x_new), 'r', label = 'P2')
plt.plot(x_new, P3(x_new), 'g', label = 'P3')
plt.plot(x, np.ones(len(x)), 'ko', x, np.zeros(len(x)), 'ko')
plt.title('Lagrange Basis Polynomials')
plt.xlabel('x')
plt.ylabel('y')
plt.grid()
plt.legend()
plt.show()
L = P1 + 3*P2 + 2*P3
fig = plt.figure(figsize = (10,8))
plt.plot(x_new, L(x_new), 'b', x, y, 'ro')
plt.title('Lagrange Polynomial')
plt.grid()
plt.xlabel('x')
plt.ylabel('y')
plt.show()
import numpy as np
>>> from scipy.interpolate import lagrange
>>> x = np. array([0,1,2])
>>> y = x ** 3
>>> poly = lagrange(x,y)
```

Since there are only three points given, our Lagrange polynomial has 2 degrees.

```
>>> from numpy.polynomial.polynomial import Polynomial
>>> Polynomial(poly.coef[::-1]).coef
array([ 0., -2., 3.])
```

```
>>> import matplotlib.pyplot as plt
>>> x_new = np.arange(0, 2.1, 0.1)
>>> plt.scatter(x, y, label='data')
>>> plt.plot(x_new, Polynomial(poly.coef[::-1])(x_new),
label='Polynomial')
>>> plt.plot(x_new, 3*x_new**2 - 2*x_new + 0*x_new,
...         label=r"$3 x^2 - 2 x$", linestyle='-.')
>>> plt.legend()
>>> plt.show()
```

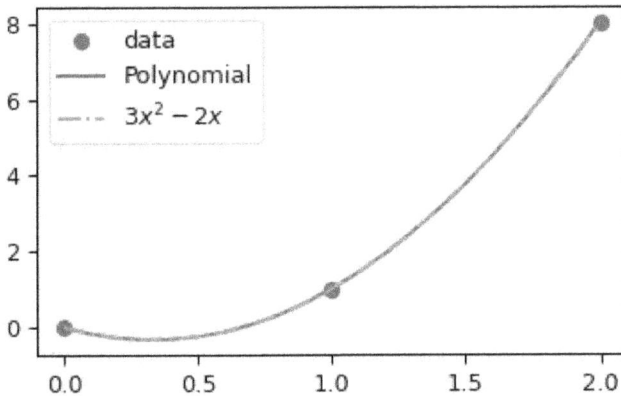

5.6 Telemetry Modelling and Lagrange Polynomials

Your team is responsible for collecting, analyzing, and coordinating the use of certain telemetry data from an experimental satellite (being piggy-backed on a rocket) being used by the NASA and the Jet Propulsion Laboratory to collect special space data. The data represents the velocity of the rocket. From rocket launching through the first 6.5 seconds, these data are transmitted by the satellite back to the ground at varying intervals. Occasionally, radio interferences cause some of the transmitted data to be garbled and consequently lost. The data from one experiment are listed in Table 5.6. Based on the telemetry data in the Table 5.6, we need to perform the following tasks:

a. Estimate what the telemetry data should be for the two garbled transmissions.

b. Estimate the telemetry data at 7 seconds.

c. Estimate the flight time for telemetry data values of 50 feet per second and 800 feet per second.

We will begin by constructing a Lagrange interpolating polynomial for the given data using only four data points centered on each garbled transmission.

Lagrange Polynomial Theorem

TABLE 5.6

Flight Data

Flight Time (s)	Telemetry Data (ft per s)
0.0	0
0.25	1.295
0.50	1.955
0.75	3.288
1.00	5.695
1.25	9.579
1.50	15.340
1.75	23.370
2.00	34.05
2.25	47.83
2.50	garbled
2.75	86.20
3.00	111.70
3.25	141.60
3.57	187.40
3.70	208.60
3.86	240.80
4.16	296.70
4.40	351.40
4.73	436.70
4.93	495.00
5.13	557.10
5.43	662.10
5.65	746.40
6.11	garbled
6.20	987.80
6.43	1102.10

If $x_0, x_1, x_2, \ldots, x_n$ are $n + 1$ distinct numbers and f is a function whose values are given at these numbers, then there exists a unique polynomial $P(x)$ of degree of at most n with the property that

$$f(x_k) = P(x_k) \text{ for each } k, k = 0, 1, 2, 3, \ldots, n.$$

This polynomial is given by

$$P(x) = f(x_0) L_{n,0}(x) + \ldots + f(x_n) L_{n,n}(x),$$

where

$$L_{n,k} = \frac{(x - x_0)(x - x_1)\ldots(x - x_k)}{(x_k - x_0)(x_k - x_1)\ldots(x_k - x_n)} \text{ for each } k = 0, 1, 2, \ldots, n.$$

We will reduce each of the interpolating polynomials developed from the four data points to equations of the form: $P(x) = a_0 + a_1 x + a_2 x^2 + a_3 x^3$, where the coefficients a_i will be found for i = 0, 1, 2, 3 and are all constants containing at least four significant digits as in Table 5.7.

First, we use Lagrange Polynomials. Using data on each side of the garbled transmission, we build a Lagrange interpolating polynomial. Finding the first missing data values with Lagrange polynomials, we begin by using the Neville's method. This is Neville's method, which uses the computer to generate the coefficients for the Lagrange interpolating polynomial and evaluate the polynomial at a given point. The algorithm runs like as follows.

TABLE 5.7

Data

Flight Time (s)	Telemetry Data (ft per s)
2.0	34.05
2.25	47.83
2.5	garbled
2.75	86.20
3.0	111.7
5.43	662.1
5.65	746.4
6.11	garbled
6.20	987.8
6.43	1102.1

Neville's Method Algorithm

Choice of input method:

1. Input entry by entry from keyboard.
2. Input data from a text file.
3. Generate data using a function F. Choose 1, 2, or 3 please.

> 1

Input (the degree of polynomial desired) n

> 3

Input (the first data point as) X(0) and F(X(0)) separated by a space

>2 34.05 Input (the second) X(1) and F(X(1)) separated by a space > 2.25 47.83
Input (the third) X(2) and F(X(2)) separated by a space

> 2.75 86.20

Input (the fourth) X(3) and F(X(3)) separated by a space

> 3 111.7

Input the point (value) at which the polynomial is to be evaluated

> 2.5 The following is the output of running Neville's Method with these four data points.

NEVILLE'S METHOD Table for P evaluated at X = 2.50000000 , follows:

Entries are XX(I), Q(I,0), ..., Q(I,I) for each I = 0, ..., N where N = 3

2.0000 34.0500
2.2500 47.8300 61.6100
2.7500 86.2000 67.015065.2133
3.0000111.7000 60.7000 64.9100 65.0617

The resulting interpolating polynomial predicts that the missing data element for $t = 2.50$ seconds is the velocity of 65.0617 feet per second.

Not using the computer algorithm but building a third-degree Lagrange polynomial in Maple and then evaluating at time equal to 2.5 second also yields 65.0617 feet per second:

$$f := \frac{((x-2.25)\cdot(x-2.75)\cdot(x-3)\cdot 34.05)}{((2-2.25)\cdot(2-2.75)\cdot(2-3))};$$

which results in

$$f := -181.600\,(x-2.25)\,(x-2.75)\,(x-3)$$

$$f1 := \frac{((x-2)\cdot(x-2.75)\cdot(x-3)\cdot 47.83)}{((2.25-2.)\cdot(2.25-2.75)\cdot(2.25-3))};$$

which results in

$$f1 := 510.1867\,(x-2)\,(x-2.75)\,(x-3)$$

$$> f2 := \frac{((x-2.25)\cdot(x-2)\cdot(x-3))\cdot 86.20}{((2.75-2.)\cdot(2.75-2.25)\cdot(2.75-3))};$$

which results in $f2 := -919.4667\,(x-2.25)\,(x-2)\,(x-3)$

$$> f3 := \frac{((x-2.25)\cdot(x-2.75)\cdot(x-2))\cdot 111.70}{((3-2.25)\cdot(3-2.75)\cdot(3-2))};$$

which results in $f3 := 595.7333\,(x-2.25)\,(x-2.75)\,(x-2)$.

Putting the whole thing together yields the following:

>Poly: $=f + f1 + f2 + f3;$
Poly $:= -181.6000\,(x-2.25)\,(x-2.75)\,(x-3) + 510.1867\,(x-2)\,(x-2.75)$
$(x-3) - 919.4667\,(x-2.25)\,(x-2)\,(x-3) + 595.7333\,(x-2.25)$
$(x-2.75)\,(x-2)$
> s1:=simplify(Poly);

and this simplifies to
$s1 := 4.8533x^3 - 5.1466x^2 + 11.1699x - 6.5299$.
Now we substitute x = 2.5 and get
> subs(x = 2.5, s1);
Finally, Poly(2.5) = 65.0617

```
poly.neville <- function(x, y, x0) {
+      n <- length(x)
+      q <- matrix(data = 0, n, n)
+      q[,1] <- y
+
+      for (i in 2:n) {
+          for (j in i:n) {
+              q[j,i] <- ((x0 - x[j-i+1]) * q[j,i-1] - (x0
- x[j]) * q[j-1,i-1]) / (x[j] - x[j-i+1])
+          }
+      }
+
+      res <- list('Approximated value'=q[n,n], 'Neville
iterations table'=q)
+      return(res)
+ }
> x <- c(8.1, 8.3, 8.6, 8.7)
x <- c(2.0,2.25,2.75,3.0)
> y <- c( 34.05, 47.83, 86.20,111.7)
```

```
poly.neville(x, y, 2.5)
$`Approximated value`
[1] 65.06167
$`Neville iterations table`
        [,1]    [,2]    [,3]      [,4]
[1,]   34.05  0.000  0.00000  0.00000
[2,]   47.83 61.610  0.00000  0.00000
[3,]   86.20 67.015 65.21333  0.00000
[4,] 111.70 60.700 64.91000 65.06167
```

The resulting interpolating polynomial predicts that the missing data element for $t = 2.50$ seconds is the velocity of 65.0617 feet per second.

Not using the computer algorithm but building a third-degree Lagrange polynomial in Maple and then evaluating at time equal to 2.5 second also yields 65.0617 feet per second:

f:=((x − 2.25)* (x − 2.75) * (x − 3))*34.05/((2−2.25)* (2−2.75) * (2 − 3));
which results in f:=−181.600(x − 2.25)(x − 2.75)(x − 3)

> f 1 := ((x−2) * (x−2.75) * (x − 3)) * 47.83 / ((2.25 − 2) * (2.25−2.75) * (2.25 − 3)) ; which results in f1 := 510.1867 (x − 2) (x − 2.75) (x − 3)

> f 2 := ((x−2.25)*(x−2)* (x − 3)) * 86.20 / ((2.75 − 2) * (2.75−2.25) * (2.75−3)) ; which results in f2 := −919.4667 (x − 2.25) (x − 2) (x − 3)

> f 3 := ((x − 2.25) * (x−2.75) *(x − 2)) * 111.70 / ((3 − 2.25) * (3 − 2.75) * (3−2)) ; which results in f3:= 595.7333 (x − 2.25) (x − 2.75) (x − 2).

Putting the whole thing together yields the following:

>Poly:=f+f1+f2+f3;
Poly := −181.6000 (x − 2.25) (x − 2.75) (x − 3) + 510.1867 (x − 2) (x − 2.75) (x −3) − 919.4667 (x − 2.25) (x − 2) (x − 3) + 595.7333 (x − 2.25) (x − 2.75) (x − 2)
> sl:=s i m p l i f y (P o l y);

and this simplifies to $sl = 4.8533x^3 − 5.1466x^2 + 11.1699x − 6.5299$.
Now we substitute $x = 2.5$ and get
> s u b s (x = 2.5, sl) ;

Finally, Poly(2.5) = 65.0617.

So, this Lagrange polynomial predicts that the missing data element is $v = 65.0617$ at $t = 2.5$ seconds. This value is exactly the value we obtained with Neville's method.

Predicting the time when the speed is 50 feet per second can be found by solving the polynomial found equal to 50 feet per second, $4.8533x^3 − 5.1467x^2 + 11.17x − 6.53 = 50$. The solution is $x = 2.2846$, or the time that the velocity is 50 feet per second is $t = 2.2846$ seconds. This is illustrated in Figure 5.4.

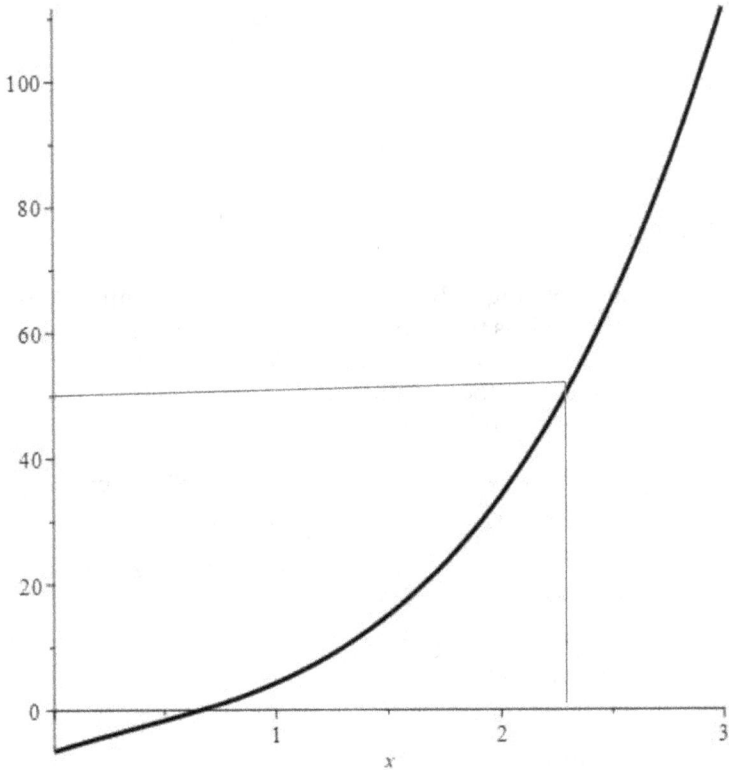

FIGURE 5.4
Plot of Lagrange polynomial.

Finding the second missing value with a Lagrange polynomial. To find the value of the speed for the second garbled time, $t = 6.11$ seconds, we use the following data: [5.43, 662.1], [5.65, 746.4], [6.20, 987.8], and [6.43, 1102.1]. We approximate the missing value by evaluating a Lagrange polynomial at the point $t = 6.11$. As before, we use Neville's method.

NEVILLE'S METHOD Table for P evaluated at $X = 6.11000000$ follows: Entries are XX(I), Q(I,0), . . ., Q(I,I) for each I = 0, . . ., N where $N = 3$ 5.4300 662.1000 5.6500 746.4000 922.6636 6.2000 987.8000 948.2981 945.3019 6.4300 1102.100 943.0739 945.2172 945.2443 This method predicts the value at 6.11 seconds to be 954.2443, so $V(6.11) = 945.2443$. Note that the third-degree polynomial is found the same way as before except that we use different data points. The polynomial is $V_3(x) = 2.0467x^3 + 37.0061x^2 - 215.32x + 412.484$. With this polynomial, we get $V_3(7) = 1420.56$ feet per second, and by setting 800 feet per second $= V_3(x)$, we get $x = 5.7813$ seconds. This is illustrated in Figure 5.5.

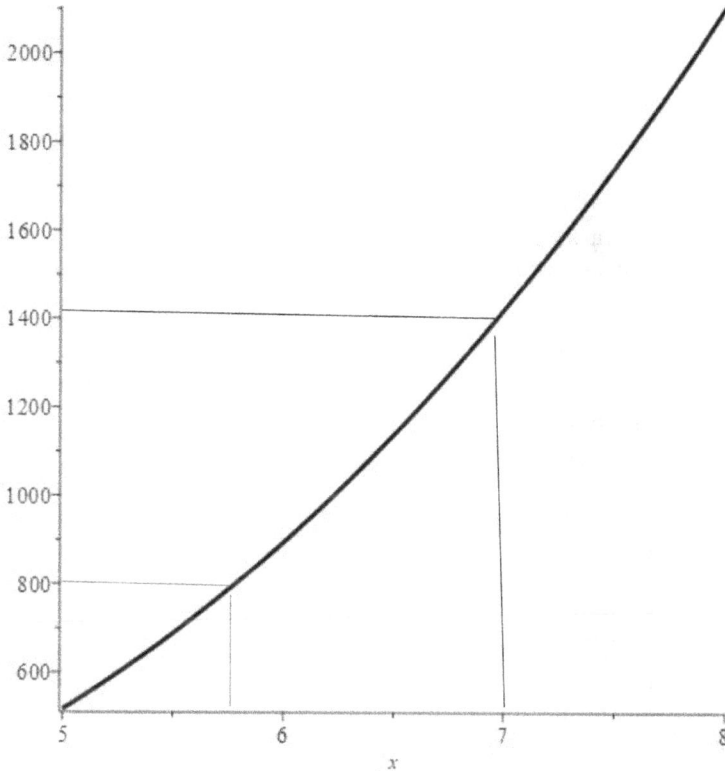

FIGURE 5.5
Plot of Lagrange polynomial for approximating the times when $v = 800$ feet per second and $t = 7$ seconds.

The two missing values have been approximated as follows:

Missing Value	v
$t = 2.5$	65.0617
$t = 6.11$	945.2443

5.7 Method of Divided Differences with Telemetry Data

We can construct a divided difference table for the two garbled transmissions. We will use only seven data points from our original table. We did not center the data about the first garbled transmission. We formed two tables:

one using the point just prior to the garbled transmission and subsequent values after the garbled transmission and the other using the point just after the garbled transmission and the remaining points from prior to the garbled transmission. We centered the data about the second garbled transmission. From the tabulated values, we constructed an interpolating polynomial for each garbled transmission. We then used Maple to reduce the polynomials and we present our results accurate to at least four significant digits. The first missing point is found by building a polynomial using the method of divided differences and then interpolating at $t = 2.5$ seconds.

The input data follow:

```
X(0) = 2.2500 F(X(0)) = 47.8300
X(1) = 2.7500 F(X(1)) = 86.2000
X(2) = 3.0000 F(X(2)) = 111.7000
X(3) = 3.2500 F(X(3)) = 141.6000
X(4) = 3.5700 F(X(4)) = 187.4000
X(5) = 3.7000 F(X(5)) = 208.6000
X(6) = 3.8600 F(X(6)) = 240.8000
```

The resulting coefficients Q(0,0), . . . , Q(N,N) are 47.8300, 76.7400, 33.6800, 1.5200, 4.4582, −5.2708, and 95.3181.

Our function is shown in Equation 5.1:

$$f(x) = 47.8300 + 76.7400\,(x - 2.25) + 33.6800\,(x - 2.25)(x - 2.75), +1.5200$$
$$(x - 2.25)(x - 2.75)(x - 3), 4.4582(x - 2.25)(x - 2.75)(x - 3)(x - 3.25) + -5.2708$$
$$(x - 2.25)(x - 2.75)(x - 3)(x - 3.25)(x - 3.57) + 95.3181\,(x - 2.25)(x - 2.75)$$
$$(x - 3)(x - 3.25)(x - 3.57)(x - 3.7) \tag{5.1}$$

This simplifies to Equation 5.2:

$$f(x) = 77416.0022 - 154286.4576x + 127322.8390x^2 - 55705.519x^3$$
$$+ 13635.2864x^4 - 1770.5628x^5 + 95.3181x^6 \tag{5.2}$$

We evaluate at $x = 2.5$, so $f(2.5) = 61.8573$.

We note that is very different from the 65.0617 that we found from the Lagrange polynomial.

We solve for x, where $f(x) = 50$, with Equation 5.2 and find two solutions: 2.1952 and 2.3449.

Which one is right? Or are there two times when $v = 50$ feet per second? Why didn't we get two answers before with the Lagrange polynomial? And why is our answer so much different from the Lagrange polynomial for $t = 2.5$? I thought they were supposed to be the same. What's wrong? Should we use the Newton's divided difference method or the Lagrange polynomial?

The other missing point is found using the same technique but using the four last points centered on 6.11 and a backward difference method. Input data follows:

```
X(0) = 5.4300
F(X(0)) = 662.1000
X(1) = 5.6500 F(X(1)) = 746.4000
X(2) = 6.2000 F(X(2)) = 987.8000
X(3) = 6.4300 F(X(3)) = 1102.1000
```

```
The coefficients Q(0,0), ..., Q(N,N) are: 662.1000 383.1818
72.3731 2.0467 These result in the following interpolating
polynomial:
```

```
f1:=662.1+383.1818*(x-5.43)+72.3731*(x-5.46)*(x-5.65) +
72.3731 (x - 5.43) (x - 5.65) (x - 6.20)
> simplify(f1); -12952.2338+6771.2281 x-1178.2338x ² +72.3731x ³
> subs(x=6.11,f1);
942.2657
> fsolve(f1-800=0,x); 5.7853
> subs(x=7,f1); 1536.88
```

Now these values for the domain [5.43, 6.43] appear to agree better with those we found using the Lagrange polynomial except the extrapolation value $t = 7$. But extrapolation is always supposed to be risky business especially with polynomials. So, we shouldn't put too much faith in either of our answers for $t = 7$.

5.8 Natural Cubic Spline Interpolation to Telemetry Data

The numbers X(0), ..., X(N) are 2.0000, 2.2500, 2.7500, 3.0000. The coefficients of the spline on the subintervals are for I = 0, ..., N − 1 A(I) B(I) C(I) D(I)
34.0500 52.6450 0.0000 39.6000 47.8300 60.0700 29.7000 7.2800
86.2000 95.2300 40.6200 −54.1600. Using only Maple, we get the
following result: readlib(spline):spline([2,2.25,2.75,3],[34.05,47.
83,86.20,111.7],x,cubic); −388.0400 + 527.8449 x − 237.6000 x
² + 39.6000 x 3, x < 2.25 −19.8951+ 36.9852 x − 19.4401 x 2 +
7.2800 x³, x < 2.75 1257.8654− 1356.9354 x + 487.4401 x 2
− 54.1600 x 3, otherwise Using the second spline (with a domain of [2.25,
2.75]) to approximate V(2.5):

```
.>f:=-19.8951+36.9852*x-19.4401*x^2+7.2800*x^3;
>subs(x=2.5,f);
64.8175
```

> *readlib(spline)* :

> *spline*$\left([2, 2.25, 2.75, 3], [34.05, 47.83, 86.20, 111.7], x, cubic\right)$;

$$\begin{cases} -71.2400000000000 + 52.6450000000000x + 39.6000000000000 \\ (x-2)^3 & x < 2.25 \\[2mm] -87.3275000000000 + 60.0700000000000x + 29.7000000000000 \\ (x-2.25)^2 + 7.28000000000004(x-2.25)^3 & x < 2.75 \\[2mm] -175.682500000000 - 95.2300000000000x + 40.6200000000000 \\ (x-2.75)^2 - 54.1600000000000(x-2.75)^3 & otherwise \end{cases}$$

> *subs*$\left(\begin{array}{l} x = 2.5, -87.3275000000000 + 60.0700000000000x + 29.7000000000000 \\ (x-2.25)^2 + 7.28000000000004(x-2.25)^3 \ 64.81750000 \end{array}\right)$;

And then solving for t at velocity equals 50 feet per second:

```
> fsolve(f=50,x);
2.2855
```

Similarly, for the second garbled point, we use different points:

spline$\left([5.43, 5.65, 6.2, 6.43], [662.1, 746.4, 987.8, 1102.1], x, cubic\right)$;

$$\begin{cases} -1387.28382802908 + 377.418752859867x + 119.071597560974 \\ (x-5.43)^3 & x < 5.65 \\[2mm] -1483.69991086532 + 394.707948825720x + 78.5872543902428 \\ (x-5.65)^2 + 3.23356088838622(x-5.65)^3 & x < 6.2 \\[2mm] -2013.54798799943 + 484.088385161198x + 83.9226298560801 \\ (x-6.2)^2 - 121.626999791420(x-6.2)^3 & otherwise \end{cases}$$

```
> spline([5.43,5.65,6.2,6.43],[662.1,746.4,987.8,1102.1],x,cu
bic);

 -20450.9903 + 10909.8532 x - 1939.6748x 2 + 119.0715 x 3 , x
 < 5.65 441.7724 - 183.6492 x + 23.7769 x 2 + 3.2336 x 3 , x <
 6.2 30199.5877 - 14582.5920 x + 2346.1870 x 2 - 121.6271x 3 ,
 otherwise > f1:=441.7724-183.6492*x+23.7769*x^2+3.2336*x^3
```

simplifies as f 1 := 441.7724 − 183.6492 x + 23.7769 x ^{2} +
3.2336 x ^{3} So for t = 6.11 ,

```
> subs(x=6.11,f1); 944.9095
```

```
> spline([5.43, 5.65, 6.2, 6.43], [662.1, 746.4, 987.8, 1102.1], x, cubic);
```

$$
\begin{cases}
-1387.28382802908 + 377.418752859867x + 119.071597560974 \\
(x-5.43)^3 & x < 5.65 \\[4pt]
-1483.69991086532 + 394.707948825720x + 78.5872543902428 \\
(x-5.65)^2 + 3.23356088838622(x-5.65)^3 & x < 6.2 \\[4pt]
-2013.54798799943 + 484.088385161198x + 83.9226298560801 \\
(x-6.2)^2 - 121.626999791420(x-6.2)^3 & otherwise
\end{cases}
$$

```
> subs
```
$$
\begin{pmatrix}
x = 6.11, -1483.69991086532 + 394.707948825720x + 78.5872543902428 \\
(x-5.65)^2 + 3.23356088838622(x-5.65)^3 \; 944.9094609
\end{pmatrix};
$$

and for velocity equal to 800 feet per second, solving for *t* yields

```
> fsolve(f1=800,x); -11.4641,-1.6712,or 5.7823
```

Again, multiple answers! Which one is the value we seek? And is it right? We wonder what this cubic spline would predict for $v(7)$, $V(7)1430.4209$! This is much closer to the value generated by the Lagrange polynomial, so we might wonder which is better, even though we are not supposed to use extrapolation with these methods.

5.9 Comparisons for Methods

Screenshots from Maple for the three methods for each garbled value are shown in Figure 5.6 and 5.7.

We note they are accurate 10^{-5}.

```
xy := [[2.0, 34.05], [2.25, 47.83], [2.75, 86.2], [3.0, 111.7]];
        xy := [[2.0, 34.05], [2.25, 47.83], [2.75, 86.2], [3.0, 111.7]]

L := PolynomialInterpolation(xy, independentvar = x, method = lagrange) :
Nev := PolynomialInterpolation(xy, independentvar = x, method = neville) :
New := PolynomialInterpolation(xy, independentvar = x, method = newton) :
l1 := expand(Interpolant(L))
```
$$l1 := 4.8533333\,x^3 - 5.146666\,x^2 + 11.17000\,x - 6.530001$$

```
l2 := expand(Interpolant(Nev))
```
$$l2 := 4.85333331\,x^3 - 5.1466666\,x^2 + 11.1700001\,x - 6.5300000$$

```
l3 := expand(Interpolant(New))
```
$$l3 := -6.52999996 + 11.1700000\,x - 5.14666664\,x^2 + 4.853333330\,x^3$$

```
subs(x = 2.5, l1);
```
$$65.06166931$$

```
subs(x = 2.5, l2)
```
$$65.06166697$$

FIGURE 5.6
Maple for the three methods for garbled point at $x = 2.5$.

> *xlyl* := [[5.43, 662.1], [5.65, 746.4], [6.20, 987.8], [6.43, 1102.1]];
> *xlyl* := [[5.43, 662.1], [5.65, 746.4], [6.20, 987.8], [6.43, 1102.1]]

>

> *L* := *PolynomialInterpolation*(*xlyl*, *independentvar* = *x*, *method* = *lagrange*) :
> *Nev* := *PolynomialInterpolation*(*xlyl*, *independentvar* = *x*, *method* = *neville*) :
> *New* := *PolynomialInterpolation*(*xlyl*, *independentvar* = *x*, *method* = *newton*) :
> *l4* := *expand*(*Interpolant*(*L*))

$$l4 := 2.046706\,x^3 + 37.0060\,x^2 - 215.3196\,x + 412.483$$

> *l5* := *expand*(*Interpolant*(*Nev*))

$$l5 := 2.04670140\,x^3 + 37.0060770\,x^2 - 215.319906\,x + 412.483562$$

> *l6* := *expand*(*Interpolant*(*New*))

$$l6 := 412.4836227 - 215.3199242\,x + 37.00607692\,x^2 + 2.046701650\,x^3$$

>

> *subs*(*x* = 6.11, *l4*);

$$945.243797$$

> *subs*(*x* = 6.11, *l5*)

$$945.244314$$

> *subs*(*x* = 6.11, *l6*)

$$945.2443175$$

>

FIGURE 5.7
Maple the three methods for $x = 6.11$.

These estimates are only the same to five significant digits.

5.10 Estimating the Error

We analyze errors in a couple of different ways. In this case, and which is probably true in most cases, we don't have a right answer. So, we are stuck with estimates. Let's look at our approximations for the first missing velocity value when $t = 2.5$. If we assume that our Lagrange value of 945.2443 is the most accurate (a reasonable assumption), then the absolute error estimate for the divided difference generated value is 2.9786 and a relative error estimate of 3.151 × 10⁻³. So, we don't even have our desired four significant digits of agreement let alone with the "right" answer. What about the cubic spline? Here, the absolute error is 0.3343, and the relative error is 3.54 × 10⁻⁴. Indeed, if our assumptions are right, the Lagrange value of 65.6017 is accurate to four significant digits. If any one of our assumptions is wrong, we need to do some more work on our divided difference approximation. But these values were suspect anyway. In general, what would be most valuable is if we could estimate an upper bound for the error. Theoretically, the error at 2.5 seconds can be found using an approximation of the fourth derivative with for for-

mula $\dfrac{(f^{iv}(\varepsilon))(2.5-2)(2.5-2.25)(2.5-2.75)(2.5-3))}{4!} = 6.5104(10^{(-4)})f^{iv}(\varepsilon)$. Now if we can obtain $f^{iv}(\varepsilon)$ in the interval [2, 3], then we can have some guarantee our error is less than the bound. But we don't even know what $f(x)$ is. Maybe our approximations can provide some insight. But the two that we are the most encouraged by are cubic polynomials with a derivative of 0. This provides a favorable but unrealistic result. However, the approximation that we are the most discouraged by is a sixth-order polynomial that may provide some insight into the fourth derivative. This doesn't work either, because $f^{iv}(2.0)$ is somewhere around 40,000. Thus, our error bound would be around 26, a far cry from our desired 5×10^{-2}. Therefore, by all the assumptions we have made and by our error bound estimate, we can only be assured of two significant digits in our divided difference estimate. Before we take this estimate, we had three significant digits of agreement at 2.5. But remember the bound is just an upper estimate, and the actual error will be much less. In the end of this analysis, we still don't have an error estimate, but we have a little more faith in two of our answers and have an insight toward improving our third answer, the divided difference approximation. Let's look at other projects.

5.11 Radiation Dosage Model

This project deals with data collected on radiation given to cancer patients in order to shrink or kill a cancerous tumor. The radiological oncologists do not want to experiment on their patients, and they would like to prescribe the "near correct" dosage (± some acceptable error) based on the patient's size, measured as weight in pounds. Radiation therapy involves the destruction of cancer cells using focused X-rays. When the tumor is localized, radiation may be used alone or in combination with surgery and/or chemotherapy to control the disease at the primary site. Radiation therapy can also be used to relieve symptoms such as pain, bleeding, and other complications from cancer. With radiation, a simulator x-rays the planned treatment area. The resulting films are used to target the delivery of high doses of radiation to the tumor with a minimal dose to the surrounding normal tissues. Computers then display the radiation dose distribution within the tumor and the rest of the body. Radiation therapy begins using the linear accelerator that produces high-energy X-rays and electrons that are delivered to the tumor area. Prior to commencing radiation therapy, customized lead shields, individualized to the patient, are made to protect normal tissue during the therapy. In addition, some radiation therapy centers offer high-dose-rate remote afterloader therapy, where higher doses of radioactive material are inserted through a tube into a catheter that has been placed in or near a cancerous organ. With this technique, doctors can control the dosage of radiation and the amount of time that radiation is exposed to various parts of the tumors. As a result, higher doses of radiation can be administered and delivered more quickly

TABLE 5.8

Oncology Data

Weight	100	110	120	130	140	150	160	170	180	190	200
Dosage	4940	5540	5890	6250	garbled	6930	7544	7900	8550	8700	9500

(www.hhs.org/ccrather.htm). Doctor Glow, our local oncologist, collected the following data on his previous patients shown in Table 5.8.

Two new patients have just been diagnosed: one with a tumor in their right thigh and the other with a tumor in their right arm. The first patient currently weighs 140 pounds, and the second patient weighs 221 pounds. We will assist Dr. Glow by determining how much radiation (in total rads) each patient should be given. We will use cubic splines to obtain our estimates.

Solution with Cubic Splines

We are concerned with precision and accuracy. If we assume that data from Dr. Glow is very precise and accurate, then we can use cubic splines to interpolate and find values for the patient weighing 140 pounds. We will present the solution for the patient who weighs 140 pounds and leave the other for the student. We will use cubic splines to model around this unavailable value. The 140-pound patient was in the fourth interval. The cubic spline for this interval is

```
> r4:=convert(r3,float);  r 4 := - 137622.8682 + 3123.7204 x
- 22.7642 x 2 + 0.5576 x 3
> s u b s ( x = 140, r 4 );
```

We obtain a value of 6524.5026 rads. Therefore, we should recommend that our 140-pound patient receive 6524.5 rads. What kind of assurance do we have that this answer is accurate to four significant digits? Is this important? Why or why not? How much radiation should the 221-pound person receive?

This is a student project.

Exercises

1. Find the interpolating function for the coordinates (0, 26), (1, 7), and (4, 25).

2. Given

$$f(x) = \begin{cases} a(x-2)^2 + b(x-1)^3, & x \in (-\infty, 1) \\ c(x-2)^2, & x \in (1, 3) \\ d(x-2)^2 + e(x-3)^3, & x \in (3, \infty) \end{cases}$$

determine the values of the parameters so that the function interpolates:

X	0	1	4
Y	26	7	25

3. Find the two natural cubic splines for

X	-1	0	1
Y	13	7	9

4. Find a natural cubic spline between $f(8.3) = 17.5649$, $f(8.4) = 18.1$, and $f(8.6) = 18.50515$. Use your cubic spline to estimate $f(8.5)$.

5. Construct a clamped cubic spline for the information in Exercise 4 with $f'(8.3) = 3.116256$ and $f'(8.6) = 3.151762$.

6. Given the following data:

x	0	1	2	3	4
y	13	11	6.5	3.2	1.5

a. Build a Lagrange polynomial and estimate $f(2.5)$.

b. Build a Newton's divided difference table and polynomial to estimate $f(2.5)$.

c. Build a natural cubic spline and estimate $f(2.5)$.

d. Which estimate do you think is best? Why?

7. Given the following data for x and y, build interpolating polynomials using each method to estimate $f(8.5)$.

$$x:=[8.1, 8.3, 8.6, 8.7]$$
$$y:=[16.9446, 17.56492, 18.51505, 18.08921]$$

Projects

1. Population of the United States. Consider the following data. These data give the population of the United States on January 1 of the given year.

Year	Population
2023	339,996,563
2022	338,289,857
2021	336,997,624
2020	335,942,003
2019	334,319,671
2018	332,140,037
2017	329,791,231
2016	327,210,198
2015	324,607,776
2014	322,033,964

Year	Population
2013	319,375,166
2012	316,651,321
2011	313,876,608
2010	311,182,845
2009	308,512,035
2008	305,694,910
2007	302,743,399
2006	299,753,098
2005	296,842,670
2004	293,947,885
2003	291,109,820
2002	288,350,252
2001	285,470,493
2000	282,398,554
1999	279,181,581
1998	275,835,018
1997	272,395,438
1996	268,984,347
1995	265,660,556
1994	262,273,589
1993	258,779,753
1992	255,175,339
1991	251,560,189
1990	248,083,732
1989	244,954,094
1988	242,287,814
1987	239,853,168
1986	237,512,783
1985	235,146,182
1984	232,766,280
1983	230,389,964
1982	228,001,418
1981	225,654,008
1980	223,140,018
1979	220,463,115
1978	217,881,437
1977	215,437,405
1976	213,270,022
1975	211,274,535
1974	209,277,968
1973	207,314,764
1972	205,238,390
1971	202,907,917

Year	Population
1970	200,328,340
1969	197,859,329
1968	195,743,427
1967	193,782,438
1966	191,830,975
1965	189,703,283
1964	187,277,378
1963	184,649,873
1962	181,917,809
1961	179,087,278
1960	176,188,578
1959	173,324,608
1958	170,147,101
1957	166,949,120
1956	not available
1955	161,136,449
1954	158,205,873
1953	155,451,199
1952	152,941,727
1951	150,598,453
1950	148,281,550

Scenario-specific questions

a. Would you eliminate any data points? Why or why not?

b. If forced to predict, what do you expect the population of the United States to be on January 1, 2024?

c. What is the population for January 1, 1956, as expressed from each model developed? What do you think the population was on January 1, 1956? Why?

2. Artillery Firing Tables

The data contained in the following table have been obtained through test firing of M201 cannon tubes under controlled conditions. The test projectile was the high-explosive projectile M106, the propellant was XM188 MOD S-Base Section 8, and the table gives the range, in meters, achieved by the stated projectile/propellant combination at the stated elevation in mils (note: 6400 mils = 360 degrees). Elevation is the angle measured between the horizontal and the gun tube as the projectile leaves the tube; the range given is the map distance between the gun and the point of impact corrected for nonstandard conditions (wind, air density, rotation of the earth, etc.)

Range	Elevation
5000	62.4
10,000	169.2
13,000	271.3
15,000	358.3
17,000	463.5
20,000	691.4

Scenario-specific questions

a. Would you eliminate any data points? Why or why not?

b. What elevation does each model predict is needed to attain a range of 14,700 meters? What elevation would you use?

c. If forced to make a prediction, what would you predict as the elevation needed to attain a range of 21,000 meters? Are you comfortable with this prediction? On which model did you base your prediction? Why did you choose that model?

3. Medicare

The following article was extracted from the *New York Times*, 8 March 1985: Because of federal efforts to reduce soaring Medicare costs, elderly beneficiaries are seeing sharp increases in the out-of-pocket expenses for a stay in the hospital. Both legislators and lobbyists for the elderly were caught unaware by these increases. President Reagan has asked Congress to cut $19 billion from the Medicare budget over a three year period. When the Medicare health and hospital program was established in 1965, the cost of a days' hospitalization was $40 and the average length of stay was 14 days. In 1984 the cuts have caused the cost to rise to $360 a day with the average stay being reduced to 7.5 days. By compressing the period of hospitalization into fewer days, the average daily cost rose sharply. These payments totaled over $4 billion in 1984 alone. The following table tracks the history of the Medicare program and the cost per day paid by the patientpatient.

Year	Cost/Day
1966	40
1967	40
1968	41
1969	49
1970	51
1971	65
1972	75
1973	77
1974	80

Year	Cost/Day
1975	95
1976	110
1977	125
1978	145
1979	165
1981	210
1982	Not found
1983	310
1984	360
1985	400

Scenario-specific questions

a. Would you eliminate any data points? Why or why not?

b. If forced to make a prediction, what would you predict the cost per day to be in 1982? Are you comfortable with your prediction? Which model did you use, and why?

c. What cost per day does each of your models predict occurred in 1982? Which prediction would you use, and why?

4. Postage Stamps

When will the American public be subjected to another increase in postage? The following data are provided:

Date	Cost	
1885	0.02	
1917	0.03	(wartime increase)
1919	0.02	(restored by Congress)
1932	0.03	
1958	0.04	
1963	0.05	
1968	0.06	
1971	0.08	
1974	0.10	
1975	0.13	(temporary)
1976	0.13	
1978	0.15	
1981	0.18	March 22
1981	0.20	November 1
1985	0.22	
1987	0.25	
1991	0.29	

Date	Cost
1993	0.32
1999	0.33
2001	0.34
2002	0.37
2006	0.39
2007	0.41
2008	0.42
2009	0.44
2012	0.45
2013	0.46
2017	Not found
2018	0.50
2019	0.55
2021	0.58
2022	0.60
2023	0.63

Scenario-specific questions

a. Would you eliminate any data points? Why?

b. What does each model predict the cost of postage to be in 2017? Which prediction would you use? Why?

c. Realistically, after 2023, when do you expect the next postage hike? What will be the cost? Provide your rationale.

d. Predict the year in which the cost of a postage stamp will reach $1.00. Which model did you use, and why?

5. Vehicle Acceleration

t	$a(t)$
0	0
0.5	0.125
1	0.25
1.5	0.375
2	0.5
2.5	0.625
3	0.75
3.5	0.875
4	1
4.5	1.125
5	1.25
5.5	1.375
6	1.5

Use different data points and approximate the missing data to find an approximation for $f(2.75)$. Prepare a table reflecting your results using different methods of interpolating polynomials and determine the value for $f(2.75)$ using the Lagrange interpolating polynomial and divided difference method for the first garbled transmission (using forward- or backward-divided differences is a viable option). Verify that the interpolating polynomial generated by Newton's method is identical to the Lagrange interpolating polynomial.

6. Nuclear Blasts

Situation: You are responsible for reanalyzing blast effect data of a tactical nuclear weapon possibly to be used by third world armies. The data were collected in 1961 during an air burst test of a weapon equivalent to 10 kilotons of TNT. At varying distances from ground zero, special remote measurement stations recorded the peak (maximum) side on blast pressure of the nuclear explosion. These peak pressures were reduced by 14.7 psi to obtain peak overpressure, that is, the amount of pressure greater than that produced by the weight of the earth's atmosphere. **Requirements:** You are given the following table relating distance from the air burst of the tactical nuclear weapon to peak overpressure:

Distance from Air Burst (1 unit = 1000 ft)	Peak Overpressure (psi)
0.50	44.704
1.00	36.596
1.50	29.814
2.00	24.189
2.50	19.571
3.00	15.818
3.50	12.803
4.00	10.404
4.50	8.5130
5.00	608373
5.50	5.8900
5.95	5.0772
6.35	4.4912
6.86	3.8844
7.48	3.2992
8.01	2.8912
8.61	2.5028
8.93	2.3246
9.65	2.0058
10.07	1.8896

Based on this table, you are given the following tasks:

a. To establish whether the reported overpressure of 6.8373 psi at 5000 feet is an erroneous measurement. (If you find it to be questionable, you should treat it accordingly throughout the rest of the tasks.)

b. To estimate the overpressure at 11,000 feet. (Is this a meaningful or valid estimate?)

c. To estimate the distance from the air burst where 20-psi peak overpressure occurred. (Virtually complete destruction of all non-earthquake-proof buildings should occur at this distance, based on the Nagasaki and Hiroshima experiences.)

d. To estimate the distance from the air burst where 2.8 psi peak overpressure occurred. (Moderate damage should occur at this distance, based on the Nagasaki and Hiroshima experiences.) Additional Guidance:

   i. You may use any and/or all functional approximation methods learned, but you must justify your use of each method in minimizing error.

   ii. For ease of computation, interpolating polynomials, if used, may be left in an unsimplified form.

   iii. Answers should be consistent with the estimated number of significant digits present.

References and Further Readings

Burden, R. and D. Faires (1997). *Numerical Analysis*. Brooks-Cole, Pacific Grove, CA.

Fox, W. (2018). *Mathematical Modeling for Business Analytics*. Taylor and Francis Publishers, Boca Raton, FL.

Giordano, F., W. Fox and S. Horton (2013). *A First Course in Mathematical Modeling*, 5th ed. Cengage Publishers, Boston, MA.

6

Numerical Differentiation and Integration

6.1 Introduction and Scenario

In this chapter, we want to illustrate the techniques of numerical differentiation and integration. As related to our previous chapter where we built approximating polynomials, now we can take these polynomials and approximate the derivatives or integrals of them. Realistically, functions are not handed to us to solve. In the real world, we most likely have data and then build functions. Perhaps the function built by modelling is not easily differentiable or integrable or even a smooth continuous function.

In this chapter, we will see numerical differentiation and numerical integration methods and uses of these in approximating solutions to modelling problems.

Situation: Your team has collected, analyzed, and coordinated the use of certain telemetry data from an experimental satellite (being piggy-backed on a rocket) being used by NASA and the Jet Propulsion Laboratory (JPL) to collect special space data. The data represents the scaled velocity of the rocket. From rocket launching through the first 24 seconds, these data are transmitted by the satellite back to the ground at varying intervals. The accelerometer, which measures the rocket's instantaneous acceleration every second, was the only data collection device that was properly operating. It is now up to your team to answer questions about the rocket. We will return to this scenario later in the chapter.

We start with numerical differentiation and integration methods.

6.2 Numerical Differentiation

A derivative is defined as $df/dx \lim_{h \to 0} \dfrac{f(x_0 + h) - f(x_0)}{h}$.

A good approximation for $f'(x_0)$ is $\dfrac{f(x_0 + h) - f(x_0)}{h}$.

One possible issue with this approximation is rounding-off error.

There are several algorithms from a two-point to a five-point method. These are shown in Table 6.1.

DOI: 10.1201/9781032703671-6

TABLE 6.1

Numerical Differentiation Methods

Method	Interval	Formula	Rule	Error
Two-point formula	$[x_0, x_0+h]$	$f'(x_0) = \dfrac{f(x_0+h)-f(x_0)}{h} - \dfrac{h}{2}f''(\varepsilon)$	f'' must exist on the interval.	$\lvert h\rvert/2(x_0)^2$
Three-point endpoint	$[x_0, x_0+2h]$	$f'(x_0) = \dfrac{1}{2h}[-3f(x_0)+4f(x_0+h)-f(x_0+2h)] + \dfrac{h^2}{3}f'''(\varepsilon)$	f'' must exist on the interval.	
Three-point midpoint	$[x_0-h, x_0+h]$	$f'(x_0) = \dfrac{1}{2}h[f(x_0+h)-f(x_0-h)] - \dfrac{h^2}{6}f'''(\varepsilon) =$	f'' must exist on the interval.	
Five-point midpoint	$[x_0-2h, x_0+2h]$	$f'(x_0) = \dfrac{1}{12h}[f(x_0-2h)-8(f(x_0-h)+8f(x_0+h) -f(x_0+2h)] + \dfrac{h^4}{30}f^{(5)}(\varepsilon)$	$f^{(5)}$ must exist on the interval.	
Five-point endpoint	$[x_0, x_0+4h]$	$f'(x_0) = \dfrac{1}{12h}[-25f(x_0)+48f(x_0+h)-36f(x_0+2h) +16(f(x_0+3h)-3f(x_0+4h)] + \dfrac{h^4}{5}f^{(5)}(\varepsilon)$	$f^{(5)}$ must exist on the interval.	

Example 1. $f(x) = \ln(x^2)$ **Defined over [1, 4] with Data as Follows**

We want an approximation $f'(1.5)$.

$f(1.5)$ 0.810930216

h	$f(1.5 + h)$	$[f(1.5 + h) - f(1.5)]/h$	Error
1	1.832581464	1.021651248	0.22222
0.1	0.940007258	1.290770423	0.02222
0.01	0.824219302	1.328908544	0.00222
0.001	0.812263105	1.332889086	0.00022

Three-Point Method

Three-point method				
h				
1	0.810930216	1.832581464	2.505526	1.196005
0.1	0.810930216	0.940007258	1.061257	1.329909
0.01	0.810930216	0.824219302	0.837421	1.333294
0.001	0.810930216	0.812263105	0.813595	1.333333

Five-Point Endpoint Method

Five-point endpoint method						
h						
1	−20.27325541	87.96391026	−90.1989	48.13048	−10.2285	1.282809
0.1	−20.27325541	45.12034841	−38.2052	18.80917	−3.85112	1.333257
0.01	−20.27325541	39.56252648	−30.1471	13.60857	−2.59069	1.333333
0.001	−20.27325541	38.98862905	−29.2894	13.03882	−2.44877	1.333333

We actually know the exact solution, $f'(x) = 2/x$, so $f(1.5) = 1.3333333$. The error bounds are close to the true approximate error.

6.3 Numerical Integration

The numerical integration methods that we will discuss are the Riemann sum, the trapezoidal rule, and the Simpson's rule.

Let's consider the function that was developed by a team of analysts for money made for the use of a single piece of machinery, $f(t) = -t * e^{-t} * \cos(2t)$.

First, we plot the function $f(t)$ from $t = [0,4]$.

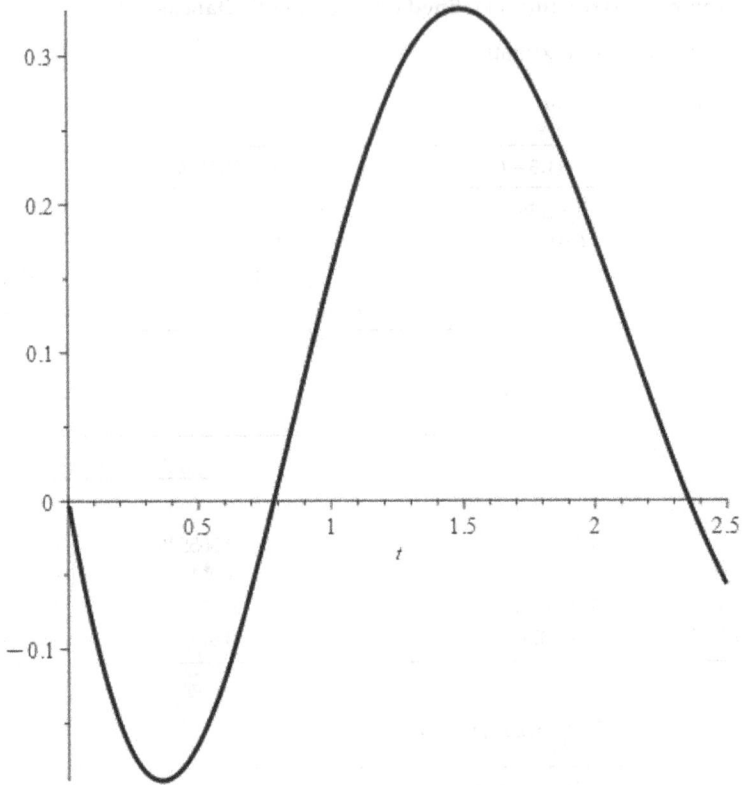

FIGURE 6.1
Plot of *f(t)*.

We see from Figure 6.1 that the plot is negative over two intervals and positive over one interval. There are two roots (other than $t = 0$). We need to find these roots, t_1 and t_2. After finding these roots we can integrate the function from t_1 to t_2 to determine the revenue generated by the machine.

We use Newton's method from Chapter 4. We find the two roots are approximately $t_1 = 0.7853981635$ and $t_2 = 2.356194490$.

Now, we can integrate the function over the nonnegative region using our numerical integration techniques.

Reimann sum rule: $n\sum_{i=1}^{n} f(x_i)\Delta x_i$

Trapezoidal rule: $T_n = \dfrac{\Delta x}{2}(f(x_0) + 2f(x_1) + 2f(x_2) + \ldots + 2f(x_n - 1) + f(x_n))$

Simpson's rule: $S_n = \dfrac{\Delta x}{3}(f(x_0) + 4f(x_1) + 2f(x_2) + \ldots + 4f(x_n - 1) + f(x_n))$

We apply these formulas and find

the midpoint Reimann sum area = .32188166359,

the trapezoidal rule area = .3182828204, and

Simpson's rule = 0.3206820308.

In this example, the function has no closed form for us to compare results. However, we can compute the error bound. We ask you to find the error bounds and to compute the areas in the exercise set.

Now let's proceed with a simpler function, $f(t) = e^{-t}$, $0 \le t \le 1$. We want to find the approximate area under this curve from [0, 1].

We plot this $f(t)$ in Figure 6.2

In this example, we can get an exact answer.

$$\int_0^1 e^{-t}dt = \frac{-1}{e} + 1 = 0.63212.$$

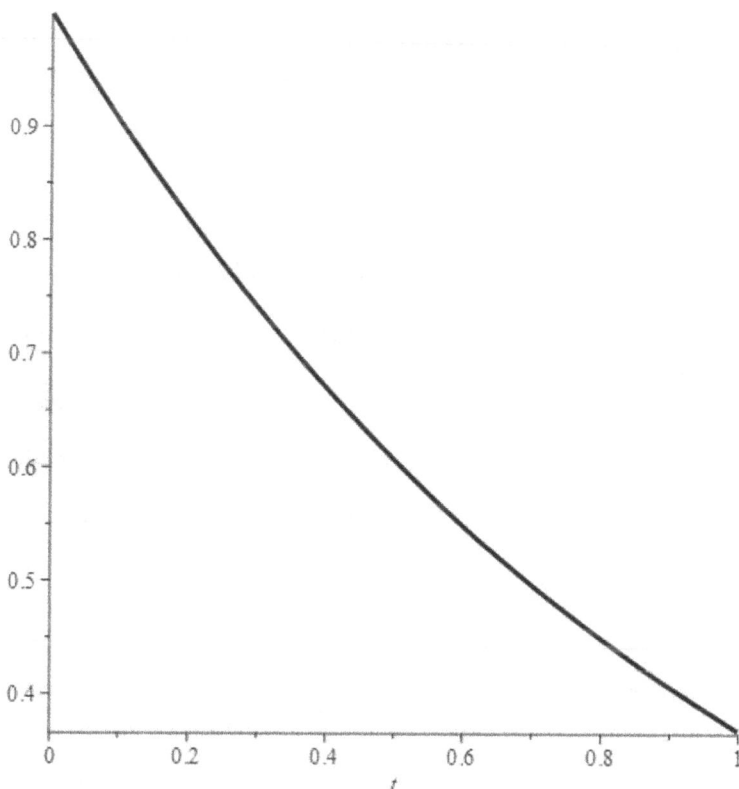

FIGURE 6.2
Plot of e^{-t}.

Our approximations by our numerical integration methods are

the Reimann sum area = 0.63186,

the trapezoidal rule area =0.63265, and

the Simpson's rule area =0.63212.

Now, let's check the error bounds. The maximum for the absolute value of the derivatives between [0, 1] is 1.

Riemann sum rule: $|\dfrac{f''(\varepsilon)}{24}(b-a)^3 \le \dfrac{1}{24}(1)^3 = 4.1667\times10^{-2}$

Trapezoidal rule: $|\dfrac{f''(\varepsilon)}{2412}(b-a)^3 \le \dfrac{1}{2412}(1)^3 = 8.3333\times10^{-2}$

Simpson's rule: $|\dfrac{f^{iv}(\varepsilon)}{2880}(b-a)^5 \le \dfrac{1}{2880}(1)^{35} = 3.4722\,x\,10^{-4}$

Table 6.2 summarizes these results.

One fact that we have observed is that Simpson's rule is more accurate. Thus far, we have only dealt with functions. How do we proceed with data?

6.4 Car Traveling Problem

Here we have collected data on the velocity of a traveling car over 24 time periods.

We use R in this example because it does a good job with the trapezoidal rule with data.

The data:

```
vel=(0, 0.45, 1.79, 4.02, 7.15, 11.18, 16.09, 21.90, 29.05,
29.05,
29.05, 29.05, 29.05, 22.42, 17.9, 17.9, 17.9, 17.9, 14.34, 11.01,
8.9, 6.54, 2.03, 0.55)
time =0, 1, 2,…, 23,24
```

TABLE 6.2

Comparison of Methods

Actual	Midpoint	Trapezoidal	Simpson	Error Bound	Actual Error
.63212	0.63186			0.041667	0.00026
0.63212		0.63265		0.08333	0.00053
0.63212			0.63212	0.00034722	0 (5 decimals places)

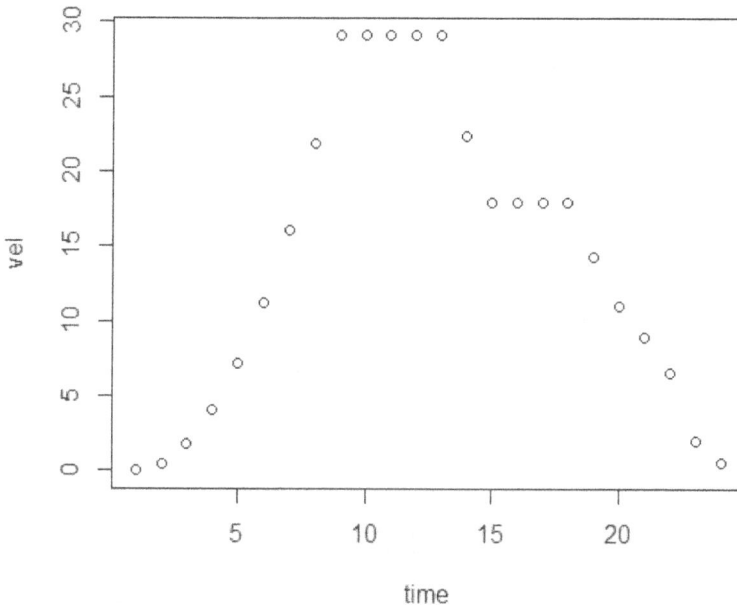

FIGURE 6.3
Plot of car velocity data.

We plot the velocity data shown in Figure 6.3.

We note several pieces of the graph that are horizontal. We assume these are discontinuities, so we do not pursue a continuous curve.

We want to calculate the distance traveled. We will use R.

We open the package pracma and then open the library(pracma). The routine trapz performs discrete integration by using the data points to create trapezoids, so it is well suited for handling data sets with discontinuities. This method assumes linear behavior between the data points, and accuracy may be reduced when the behavior between data points is nonlinear. To illustrate, you can draw trapezoids onto the graph using the data points as vertices.

We plot the distance points shown in Figure 6.4.

The individual values are

```
cdistance
          [,1]
  [1,]   0.000
  [2,]   0.225
  [3,]   1.345
  [4,]   4.250
  [5,]   9.835
```

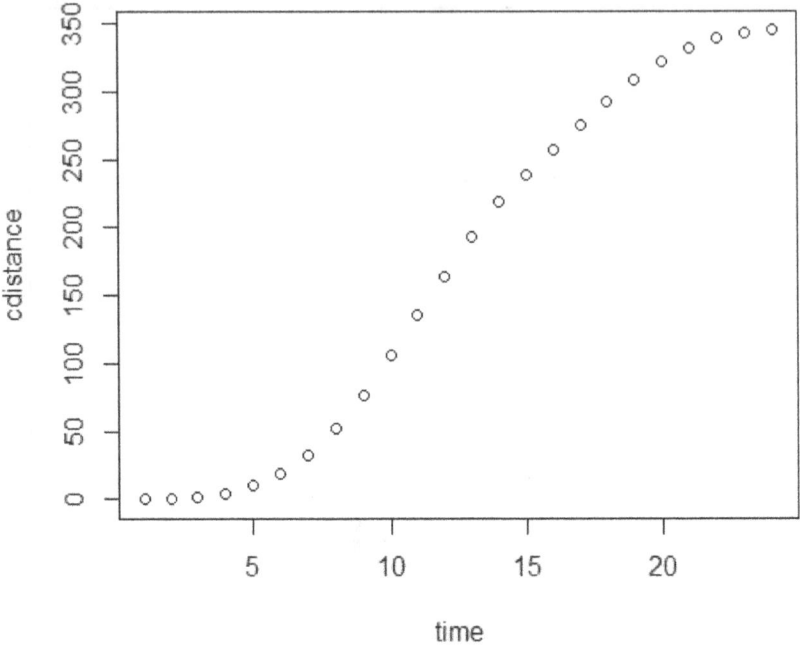

FIGURE 6.4
Plot of cumulative distance over time.

```
 [6,]   19.000
 [7,]   32.635
 [8,]   51.630
 [9,]   77.105
[10,]  106.155
[11,]  135.205
[12,]  164.255
[13,]  193.305
[14,]  219.040
[15,]  239.200
[16,]  257.100
[17,]  275.000
[18,]  292.900
[19,]  309.020
[20,]  321.695
[21,]  331.650
[22,]  339.370
[23,]  343.655
[24,]  344.945
```

6.5 Revisit a Telemetry Model

We return to the scenario in Section 6.1. Situation: Your team has collected, analyzed, and coordinated the use of certain telemetry data from an experimental satellite (being piggy-backed on a rocket) being used by NASA and the JPL to collect special space data. The data represents the scaled velocity of the rocket. From rocket launching through the first 22 seconds, these data are transmitted by the satellite back to the ground at varying intervals. The accelerometer, which measures the rocket's instantaneous acceleration every second, was the only data collection device that was properly operating. It is now up to your team to answer questions about the rocket.

Data for telemetry

```
time1 in secs
 [1]  1  2  3  4  5  6  7  8  9 10 11 12 13 14 15 16 17 18 19
20 21 22
> accel1 in m/sec²
 [1] 29.30307 29.21165 29.03481 28.69494 28.04974 26.85331
24.72752
 [8] 21.22255 16.10165 9.80000 3.49835 -1.62255 -5.12725
-7.25321
[15] -8.44975 -9.09494 -9.43481 -9.61165 -9.00000 9.70307
-9.75018
[22] -9.77400
>
```

Time (s)	Accel (m/s²)	Time (s)	Accel (m/s²)
0	29.30307	38	-9.79997
1	29.21165	39	-9.79998
2	29.03481	40	-9.79999
3	28.69494	41	-9.79999
4	28.04975	42	-9.80000
5	26.85331	43	-9.80000
6	24.72725	44	-9.80000
7	21.22255	45	-9.80000
8	16.10165	46	-9.80000
9	9.8000	47	-9.80000
10	3.49835	48	-9.80000
11	-1.62255	49	-9.80000
12	-5.12725	50	-9.80000
13	-7.25321	51	-9.80000

(*Continued*)

Time (s)	Accel (m/s²)	Time (s)	Accel (m/s²)
14	−8.44975	52	−9.80000
15	−9.09494	53	−9.80000
16	−9.43481	54	−9.80000
17	−9.61165	55	−9.80000
18	−9.70307	56	−9.80000
19	−9.75018	57	−9.80000
20	−9.77440	58	−9.80000
21	−9.78685	59	−9.80000
22	−9.79325	60	−9.80000
23	−9.79653	61	−9.80000
24	−9.79822	62	−9.80000
25	−9.79909	63	−9.80000
26	−9.79953	64	−9.80000
27	−9.79976	65	−9.80000
28	−9.79988	66	−9.80000
29	−9.79994	67	−9.80000
30	−9.79997	68	−9.80000
31	−9.79998	69	−9.80000
32	−9.79999	70	−9.80000
33	−9.80000	71	−9.80000
34	−9.80000	72	−9.80000
35	−9.80000	73	−9.80000
36	−9.8000	74	−9.80000

```
time1<-c(1:20)
accel1<-c(29.30307,29.21165,29.03481,28.69494,28.04974,
26.85331,24.72752,21.22255,16.10165,9.8,3.49835,-1.62255,-
5.12725,-7.25321,-8.44975,-9.09494,-9.43481,-9.61165,-
9,9.70307,-9.75018,-9.774)
```

First, we plot the time versus acceleration (Figure 6.5).
Next, we perform the trapezoidal rule to obtain the velocities:

```
> cvelocity
            [,1]
 [1,]   0.00000
 [2,]  29.25736
 [3,]  58.38059
 [4,]  87.24546
 [5,] 115.61781
 [6,] 143.06933
 [7,] 168.85975
 [8,] 191.83478
 [9,] 210.49688
[10,] 223.44770
```

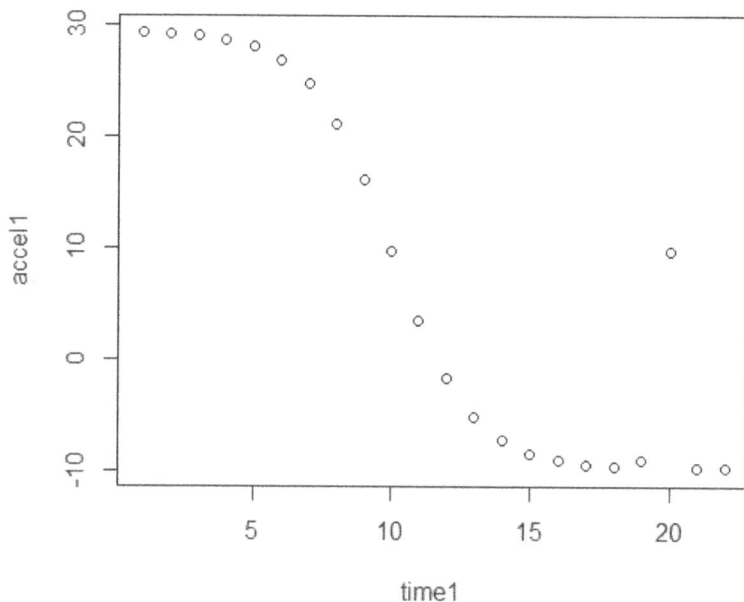

FIGURE 6.5
Plot of time versus acceleration.

```
[11,]  230.09688
[12,]  231.03478
[13,]  227.65988
[14,]  221.46965
[15,]  213.61817
[16,]  204.84582
[17,]  195.58095
[18,]  186.05772
[19,]  176.75189
[20,]  177.10343
[21,]  177.07987
[22,]  167.31778
```

We plot the velocity data shown in Figure 6.6.

Next, we find the cumulative distance traveled by, again, using the trapezoidal rule in R.

```
trapz(cvelocity)
[1] 3553.168 meters
```

We need to determine the following from our telemetry data:

- What were the approximate maximum upward and downward velocities of the rocket, and at what altitude were they attained?

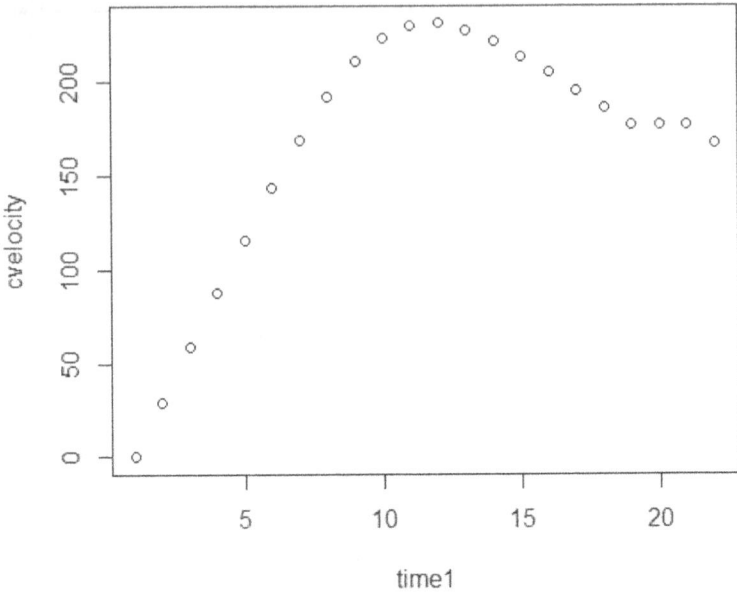

FIGURE 6.6
Plot of velocity.

- What was the approximate maximum altitude attained by the rocket, and at what time did it reach its zenith?
- Calculate the error for your numerical integration methods used.
- Propose a function that represents height of the satellite at any time, t.

Use our approximating function to approximate the acceleration.

From our study of calculus, we recall the relationships that exist between distance, velocity, and acceleration.

$$v(t) = dx/dt$$
$$a(t) = dv/dt = d^2x/dt^2$$

Since we have $a(t)$ data and know $a(t) = dv/dt$, we can integrate to find $v(t)$.

$$v(t) = \int a(t)\,dt$$

Here are our choices, we can build an approximating polynomial for the data and integrate over the domain simply use either Simpson's rule or the trapezoidal rule to approximate the value of the integral. Since we might need all intermediate values of $v(t)$ over the domain, then a numerical technique is the most advantageous.

For example, in Section 6.3, we discussed the trapezoid rule:

$$v_1(t) \approx \frac{(b-a)}{2}\left[f(t_0)+f(t_1)\right]$$

$$v_2(t) \approx \frac{(b-a)}{2}\left[f(x_0)+2f(x_1)+f(x_2)\right]$$

$$\cdots$$

Using this algorithm, we find the following values approximating $v(t)$.

```
0
29.80877084
59.43799565
88.7221445
117.3512257
144.7654528
170.0211765
191.7185571
208.2166606
218.3166606
222.0185571
220.5211765
215.4654528
208.2512257
199.8221445
190.7379957
181.3087708
171.7
161.9984104
152.2489968
142.4749849
132.6883318
122.8951855
113.0987046
103.3005114
93.50143907
83.70191536
73.9021599
64.10228545
54.30234991
44.502383
34.70239999
```

```
24.90240872
15.1024132
 5.302415496
-4.497583323
-14.29758272
-24.09758241
-33.89758225
-43.69758216
-53.49758212
-63.2975821
-73.09758209
-82.89758208
-92.69758208
-102.4975821
-112.2975821
-122.0975821
-131.8975821
-141.6975821
-151.4975821
-161.2975821
-171.0975821
-180.8975821
-190.6975821
-200.4975821
-210.2975821
-220.0975821
-229.8975821
-239.6975821
-249.4975821
-259.2975821
-269.0975821
-278.8975821
-288.6975821
-298.4975821
-308.2975821
-318.0975821
-327.8975821
-337.6975821
-347.4975821
-357.2975821
-367.0975821
-376.8975821
-386.6975821
-396.4975821
```

```
v(t) values are 0 190.738 44.50238-102.498 -249.498 29.80877
181.3088 34.7024 -112.298 -259.298 59.438 171.7 24.90241
-122.098 -269.098 88.72214 161.9984 15.10241 -131.898 -278.898
117.3512 152.249 5.302415 -141.698 -288.698 144.7655 142.475
```

```
-4.49758 -151.498 -298.498 170.0212 132.6883 -14.2976 -161.298
-308.298 191.7186 122.8952 -24.0976 -171.098 -318.098 208.2167
113.0987 -33.8976 -180.898 -327.898 218.3167 103.3005 -43.6976
-190.698 -337.698 222.0186 93.50144 -53.4976 -200.498 -347.498
220.5212 83.70192 -63.2976 -210.298 -357.298 215.4655 73.90216
-73.0976 -220.098 -367.098 208.2512 64.10229 -82.8976 -229.898
-376.898 199.8211 54.30235 -92.6976 -239.698 -386.698
```

From observation of the velocity data, we could the following information, which is left as an exercise:

What was the maximum upward velocity of the satellite, and at what altitude was it attained?

What was the maximum altitude attained by the satellite, and at what time did it reach its zenith? Calculate the error for your numerical integration method used.

Propose a function that represents height of the satellite at any time, t.

Use our approximating function for height to approximate the acceleration.

6.6 Volume of Water in a Tank

The South Carolina state's Water Commission is requiring data from Francis Marion University housing on the rate of water use, in gallons per hour, and the total amount of water used each day. The Department of Engineering and Housing (DEH) in Florence does not have the sophisticated equipment that measures the flow of water in or out of the main water tank. Instead, DEH can measure only the level of water in the tank, within 0.5% accuracy, every hour. More importantly, whenever the level in the tank drops below some minimum level L, a pump fills the tank up to the maximum level, H. However, there is no measurement of the pump flow at these times either. Thus, one cannot readily relate the level in the tank to the amount of water used while the pump is working. This occurs once or twice per day for a couple of hours each time. The following table contains the time, in seconds, since the first measurement and the level of water in the tank in hundredths of a foot. For example, after 3316 seconds, the depth of the water in the tank reached 31.10 feet. The tank is a vertical circular cylinder with a height of 40 feet and a diameter of 57 feet. Usually, the pump starts filling the tank when the level drops to about 27.00 feet (L) and the pump stops when the level rises back to about 35.50 feet (H).

```
Time (s) Level (.01 ft)
0 3175
3316 3110
```

```
6635  3054
10619 2994
13937 2947
17921 2892
21240 2850
25223 2795
28543 2752
32884 2697
35932 pump on
39332 pump on
39435 3550
43318 3445
46636 3350
49953 3260
53936 3167
57254 3087
60574 3012
64554 2927
68535 2842
71854 2757
75021 2697
79254 pump on
82649 pump on
85968 3475
89953 3397
93270 3340
```

Requirements:

1. Build a mathematical equation to estimate flow out of the tank for the DEH. Use some approximation technique of your choice.
2. Use your technique and estimate the flow out of the tank, $f(t)$, at all times, even when the pump is working.
3. Use your technique and estimate the total amount of water used in one day (24 hours). (This problem and solution are a modified version of the MCM-1991 Water Tank Problem submitted by Yves Nievergelt.)

Solution to Water Tank Problem

First, we will use cubic splines from Chapter 5 to estimate the missing four data points. Next, we will use polynomial approximation and obtain a model. We will then integrate the function and assume it is the approximate result. Then using the data, we will use the trapezoidal rule to approximate the area under the curve and compute the error bound.

The cubic splines estimates were

```
35932 3099.9
39332 3593.9
79254 2958.73
82649 3543.97
```

After using cubic splines and estimating the missing data, we then start by plotting the data, shown in Figure 6.7.

We see there are five distinct regions that we could model separately rather than attempt one curve.

We build three polynomial models and use cubic splines to connect then through the interpolated points.

```
trapz(timewater,speedwater)
[1] 11922.16

timewater<-c(0,3316,6635,10619,13937,17921,21240,25223,28543,
32884,35932,39332,39435,43318,46636,49953,53936,57254,60574,
64554,68535,71874,75021,79254,83649,85968,89953,93270)
```

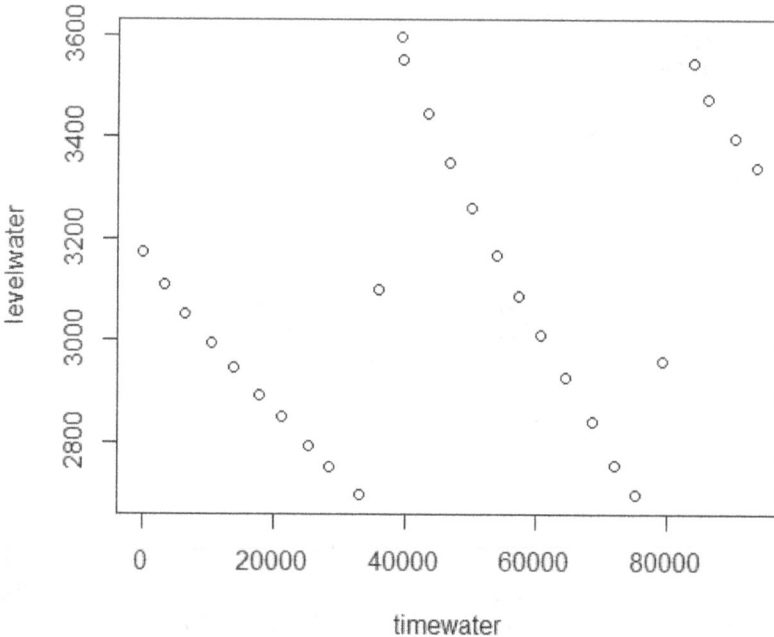

FIGURE 6.7
Plot of water levels as function of time in seconds.

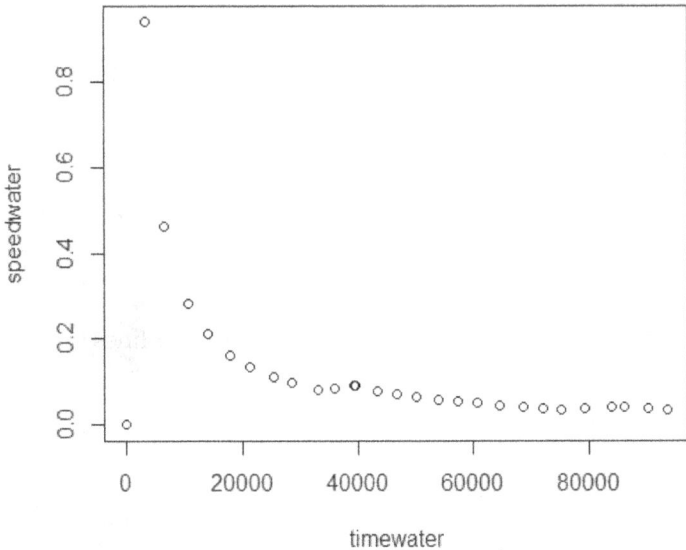

FIGURE 6.8
Plot of water speed and time.

```
> levelwater<-c(3175,3110,3054,2994,2947,2892,2850,2795,2752,
2697,3099.9,3593.9,3550,3445,3350,3260,3167,3087,3012,2927,
2842,2757,2697,2958.73,3543.97,3475,3397,3340)

Error: unexpected symbol in "levelwater<-c(3175,3110,3054,
2994,2947,2892,2850,2795,2752,2697,3099.9,3593.9,3550, 3445,
3350,3260,3167,3087,3012,2927,2842,2757,2697,2958.73,3543.97,
3475,3397"

> levelwater<-c(3175,3110,3054,2994,2947,2892,2850,2795,2752,
2697,3099.9,3593.9,3550,3445,3350,3260,3167,3087,3012,2927,
2842,2757,2697,2958.97,3543.97,3475,3397,3340)

> plot(timewater,levelwater)
> cumlevel=trapz(levelwater)
> cumlevel

[1] 83511

timewater<-c(0,3316,6635,10619,13937,17921,21240,25223,28543,
32884,35932,39332,39435,43318,46636,49953,53936,57254,60574,64
554,68535,71874,75021,79254,83649,85968,89953,93270)

> levelwater<-c(3175,3110,3054,2994,2947,2892,2850,2795,2752,
2697,3099.9,3593.9,3550,3445,3350,3260,3167,3087,3012,2927,
2842,2757,2697,2958.73,3543.97,3475,3397,3340
```

```
+ )

> install.packages("pracma")

WARNING: Rtools is required to build R packages but is not
currently installed. Please download and install the
appropriate version of Rtools before proceeding:

https://cran.rstudio.com/bin/windows/Rtools/
Installing package into 'C:/Users/Fox/AppData/Local/R/
win-library/4.2'
(as 'lib' is unspecified)
trying URL 'https://cran.rstudio.com/bin/windows/contrib/4.2/
pracma_2.4.2.zip'
Content type 'application/zip' length 1729052 bytes (1.6 MB)
downloaded 1.6 MB

package 'pracma' successfully unpacked and MD5 sums checked

The downloaded binary packages are in

C:\Users\Fox\AppData\Local\Temp\RtmpGaVoMr\downloaded_packages
> library(pracma)
Warning message:
package 'pracma' was built under R version 4.2.3
> trapz(timewater)
[1] 1252131
> plot(timewater,levelwater)
> trapz(timewater,levelwater)
[1] 286183487
> cumtrapz(timewater,levelwater)

          [,1]
 [1,]        0
 [2,] 10420530
 [3,] 20649688
 [4,] 32697304
 [5,] 42553423
 [6,] 54184711
 [7,] 63713560
 [8,] 74955578
 [9,] 84163598
[10,] 95990652
[11,] 104825128
[12,] 116204588
[13,] 116572498
[14,] 130153291
[15,] 141426196
[16,] 152388881
```

```
[17,]  165188251
[18,]  175563637
[19,]  185687977
[20,]  197506587
[21,]  208989782
[22,]  218337312
[23,]  226919181
[24,]  238889534
[25,]  253179217
[26,]  261317713
[27,]  275010173
[28,]  286183487
```

We approximate the amount of water as 286,183,487/(3593.9 – 2697) = 319,080.708 gallons. This is an estimate because if we just used smooth curves our estimates were anywhere from 312,000 to 330,000 gallons.

Maple

```
> with(Statistics) :
> X := Vector( [ 1,2, 3,4, 5,6], datatype = float) :
> Y := Vector( [2, 3,4.8,10.2,15.6, 30.9], datatype = float) :
Fit a model that is linear in the parameters.
> Fit( a + bz + cz²,X,Y,z)
6.62999999999999 - 5.37464285714286z + 1.53392857142857z²
> Xt := Vector( [0, 3316, 6635, 10619, 13937, 17921, 21240,
25223, 28543, 32884], datatype
    =float);
```

$$Xt := \begin{bmatrix} 0. \\ 3316. \\ 6635. \\ 10619. \\ 13937. \\ 17921. \\ 21240. \\ 25223. \\ 28543. \\ 32884. \end{bmatrix}$$

```
> Yt := Vector( [3175, 3110, 3054, 2994, 2947, 2892, 2850,
2795, 2752, 2697], datatype
    =float);
```

$$Yt := \begin{bmatrix} 3175. \\ 3110. \\ 3054. \\ 2994. \\ 2947. \\ 2892. \\ 2850. \\ 2795. \\ 2752. \\ 2697. \end{bmatrix}$$

Regression 1 is $f(t) = 3168.65098662311 - 0.01705975132236d\ t + 8.576981979411 \times 10^{-8}\ t^2$. We integrate from $t = 0$ to $t = 32{,}884$ and obtain 9.599069254×10^7.

Then we integrate our cubic spline equation from $t = 32{,}884$ to $t = 39{,}435$.

s2 := int(−3.104815256*10^(−9)*v^3 + 0.0003355311772*v^2 − 11.92303305*v
   + 142350.1177, v = 32884. 39435);

s2 := 2.052176880 10^7

The next polynomial regression equation is

4809.90127123502 − 0.0363310065687629*t + 1.09323252137818*10^(−7)*t^2,
   and the integration result is s3 := 1.103281939*10^8.

Next, we use our cubic spline equation and integrate from $t = 75{,}021$ to 85,968:

s4 := int(−9.169257925*10^(−10)*t^3 + 0.0002216880275*t^2 − 17.76732173*t
   + 475078.7688, t = 75021 .. 85968);

s4 := 3.373531949 10^7

Our last regression equation follows:

part3 := Fit(b*f + a, Xt3, Yt3, t);

part3 := 5066.17439579189 − 0.0185241081142225 t

s5 := int(part3, f = 85968 .. 93270);

s5 := 2.487106728 10^7

We sum up our five integrals to obtain $2.854470420 * 10^8$. We divide by $\Delta l = 869.9$ and obtain 319,080.708 gallons.

```
> part1 := Fit (a + b · r + c · r², Xt, Yt, r);
```

$$part1 := 3168.65098662311 - 0.0170597513224364r + 8.57679819797411 \times 10^{-8}r^2$$

```
> s1 := int( part1, r = 0 ..32884);
```

$$s1 := 9.599069254 \times 10^7$$

```
>s2 := int (-3.104815256 × 10⁻⁹v³ + 0.0003355311772v² - 11.92303305v
           +142350.1177, v = 32884..39435);
```

$$s2 := 2.052176880 \times 10^7$$

```
> Xt1part3 := Vector ([39435, 43318, 46636, 49953, 53936, 57254,
              60554, 64554, 68535, 71854, 75021], datatype = float);
```

$$Xt1 := \begin{bmatrix} 39435. \\ 43318. \\ 46636. \\ 49953. \\ 53936. \\ 57254. \\ 60554. \\ 64554. \\ 68535. \\ 71854. \\ \vdots \end{bmatrix}$$

11 element Vector[column]

```
> Yt1 := Vector ([3550, 3445, 3350, 3260, 3167, 3084, 3012, 2927, 2842,
              2757, 2697], datatype = float);
```

$$Yt1 := \begin{bmatrix} 3550. \\ 3445. \\ 3350. \\ 3260. \\ 3167. \\ 3087. \\ 3012. \\ 2927. \\ 2842. \\ 2757. \\ \vdots \end{bmatrix}$$

11 element Vector[column]

```
> part2 := Fit (a + b · g + c · g², Xtl, Ytl, g);
```
$$part2 := 4809.90127123502 - 0.03633100656876299g$$
$$+ 1.09323252137818 \times 10^{-7}g^2$$

```
> s3 := int (part2, g = 37435..75021);
```
$$s3 := 1.103281939 \times 10^8$$

```
> s4 := int (-9.169257925 × 10⁻¹⁰v³ + 0.0002216880275v² - 17.76732173v
          +47507837688, v = 75021..85968);
```
$$s4 := 3.373531949 \times 10^7$$

```
> Xt3 := Vector ([85968, 89953, 93270], datatype = float);
```
$$Xt3 := \begin{bmatrix} 85968. \\ 89953. \\ 93270. \end{bmatrix}$$

```
> Yt3 := Vector ([3475, 3397, 3340], datatype = float);
```
$$Yt3 := \begin{bmatrix} 3475. \\ 3397. \\ 3340. \end{bmatrix}$$

```
> part3 := Fit (a + b · f, Xt3, Yt3, f);
```
$$part3 := 5066.17439579189 - 0.0185241081142225f$$

```
> s5 := int (part3, f = 85968..93270);
```
$$s5 := 2.487106728 \times 10^7$$

```
> evalf(s1 + s2 + s3 + s4 + s5);
  2.854470420 × 10⁸
```
$$> \frac{(2.854470420 \times 10^8)}{896.9}$$
```
  318259.6075
> part3 := Fit (a + b · f, Xt3, Yt3, f);
```
$$part3 := 5066.17439579189 - 0.0185241081142225f$$
```
> s5 := int(part3, f= 85968 ..93270 ) ;
  s5 := 2.487106728 × 10⁷
> evalf(s1, s2, s3, s4, s5) ;
  2.854470420 × 10⁷
```
$$> \frac{(2.854470420 \times 10^8)}{896.9}$$
```
  318259.6075
> display(d1, d2, d3, d4, d5, title="Water Levels");
```

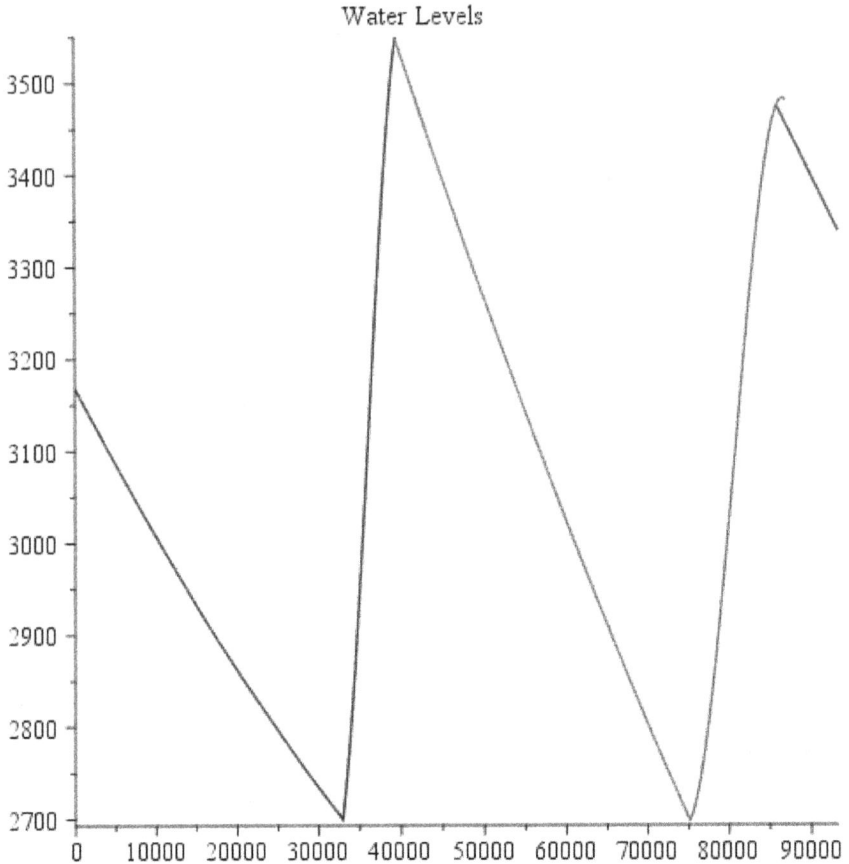

FIGURE 6.9
Plot of water levels using our functions.

Exercises and/or Projects

1. Given the function $f(t) = -4 * t * \exp(-t) * \cos(2 * t)$:
   a. Plot the function and decide on the positive and negative regions.
   b. Use a root finding technique and find the roots.
   c. Use the Riemann sum, trapezoidal, and Simpson's rules to approximate the area in the nonnegative region.
   d. Find the error bounds for each method in part c.
2. Given the following acceleration data, use a numerical integration method to find the velocities at time t.

t	$a(t)$
0	0
0.5	0.125
1	0.25
1.5	0.375
2	0.5
2.5	0.625
3	0.75
3.5	0.875
4	1
4.5	1.125
5	1.25
5.5	1.375
6	1.5

3. Given the function $f(t) = 0.25t$, approximate the derivative at $t = 5.2$ seconds.

4. Numerically evaluate the following five integrals:

$$I_1 = \int_{0.5}^{1} \frac{\pi}{x^2} \sin\left(\frac{\pi}{x}\right) dx$$

$$I_2 = \int_{0.01}^{1} \frac{\pi}{x^2} \sin\left(\frac{\pi}{x}\right) dx$$

$$I_3 = \int_{1}^{10} \sin\left(\ln(x)\right) dx$$

$$I_4 = \int_{0}^{\infty} \frac{dx}{e^x + e^{-x}}$$

$$I_5 = \int_{0}^{1} \frac{x}{e^x - 1} dx$$

The last function, I_5, is known as Debye's function. It has an apparent singularity at $x = 0$. We know that the limit of this function as $x \to 0$ is 1.

Requirements:

a. Explicitly evaluate I_1 and I_2.

b. Using the error formulas, compute how many sub-intervals are required to evaluate I_1 to an accuracy of $5.0 \times 10 - 8$ (i.e., 7 significant digits) using both the trapezoidal and Simpson's rules. Then numerically evaluate using the predicted number of sub-intervals by both methods, and then compare to the exact answer to benchmark the techniques. Did the predicted number of intervals give errors as expected for each method?

c. Find I_2 to four significant figures using Simpson's rule. Plot the integrand function on Maple to assist in explaining any difficulties encountered in calculating I_2 that were not present in calculating I_1. Describe a technique to evaluate I_2 to four significant digits using fewer function evaluations.

d. Find I_3 to six significant figures using some numerical method. Why did you choose your method?

e. Find I_4 using Maple. Find I_4 to six significant figures using some numerical method. Devise a methodology that permits you to minimize the number of subintervals required while still obtaining the necessary accuracy.

f. What is the actual value of Debye's function, I_5? Find I_5 using the following closed Newton–Cotes six- and seven-point formulas shown in Figure 6.10.

5. For Example 1, telemetry, find the following:

a. What were the maximum upward velocity of the satellite, and at what altitude was it attained?

b. What was the maximum altitude attained by the satellite, and at what time did it reach its zenith? Calculate the error for your numerical integration method used.

c. Propose a function that represents height of the function at any time, t.

d. Use our height approximating function to approximate the acceleration.

$$\int_a^b f(x)dx = \frac{5h}{288}(19f_0 + 75f_1 + 50f_2 + 50f_3 + 75f_4 + 19f_5) + E$$

where

$$E = -\frac{275h^7 f^{(6)}(\xi)}{12096}, a < \xi < b$$

and

$$\int_a^b f(x)dx = \frac{h}{140}(41f_0 + 216f_1 + 27f_2 + 272f_3 + 27f_4 + 216f_5 + 41f_6) + E$$

where

$$E = -\frac{9h^9 f^{(8)}(\xi)}{1400}, a < \xi < b$$

FIGURE 6.10
Formulas for Exercise 4f.

Further Reading

Burden, R. and D. Faires (1997). *Numerical Analysis*. Brooks-Cole, Pacific Grove, CA.

Fox, W. (2018). *Mathematical Modeling for Business Analytics*. Taylor and Francis Publishers, Boca Raton, FL.

Giordano, F., W. Fox and S. Horton (2013). *A First Course in Mathematical Modeling*, 5th ed. Cengage Publishers, Boston, MA.

7

Modelling with Numerical Solutions to Differential Equations—Initial Value Problems for Ordinary Differential Equations

7.1 Introduction and Scenario

The study of differential equations is a key course for applied mathematics, science, and engineering. It is important to motivate the ideas and results of ordinary differential equations through real-world applications that a student may encounter in their discipline. Rapidly changing technologies put more computing power in the hands of the students. Much more difficult problems can now be solved by using numerical techniques. In this chapter, we present several modelling scenarios representative of modelling change. These ordinary differential equations with initial values are the type problems that permeate across disciplines.

In real-world modelling with differential equation, the equation developed may not be solvable in closed form. Numerical methods might be the only choice available. We will discuss Euler's method, the improved Euler's method, and the Runge–Kutta 4 (RK4) method.

Bridge Bungee Jumping

Consider bridge jumping, a participant attaches one end of a bungee cord to himself (or herself), attaches the other end to a bridge railing, and then drops off the bridge. In our initial problem, the jumper will be dropping off the Royal Gorge Bridge, a suspension bridge that is 1053 feet above the floor of the Royal Gorge in Colorado. The jumper will use a 200-foot-long bungee cord. It would be nice if the jumper has a safe jump, meaning that the jumper does not crash into the floor of the gorge or run into the bridge on the rebound. In this project, you will do some analysis of the fall.

Assume the jumper weighs 160 pounds. The jumper will free fall until the bungee cord begins to exert a force that acts to restore the cord to its natural (equilibrium) position. In order to determine the spring constant of the bungee cord, we found that a mass weighing 4 pounds stretches the cord 8 feet.

DOI: 10.1201/9781032703671-7

Hopefully, this spring force will help slow down the descent sufficiently so that the jumper does not hit the bottom of the gorge.

Throughout this problem we will assume that DOWN is the POSITIVE direction. Before the bungee cord begins to retard the fall of the jumper, the only forces that act on the jumper are his weight and the force due to wind (air) resistance.

If the force due to the wind resistance is 0.9 times the velocity of the jumper, then use Newton's second law ($\Sigma F = MA$) to write a differential equation that models the fall of the jumper. Be sure to include the initial conditions for the jumper for your differential equation. (Hint: This problem can be formulated as a second-order ODE in position, or as a first-order differential equation (DE) in velocity—use the first order ODE model in this problem).

We will return to this problem later in the chapter.

Spread of a Contagious Disease

Consider modelling the spread of a contagious disease.

Given the following ODE model for the spread of a communicable disease as

$$\frac{dN}{dt} = .2.5N(10 - N), N(0) = 2,$$

where N is the number of infected people in 100s and t is the time variable.

We want to completely analyze the behavior of this ODE. We return to this problem later as well.

First, we are going to discuss numerical methods to solve differential equations.

7.2 Numerical Methods to Solve ODEs

Euler's Method

From the point of view of a mathematician, the *ideal* form of the solution to an initial value problem would be a **formula** for the solution function. After all, if this formula is known, it is usually relatively easy to produce any other form of the solution you may desire, such as a **graphical solution** or a **numerical solution** in the form of a table of values. You might say that a formulaic solution contains the recipes for these other types of solution within it. Unfortunately, as we have seen in our studies already, obtaining a formulaic solution is not always easy and, in many cases, is absolutely impossible.

So we often have to "make do" with a numerical solution, that is, a table of values consisting of points that lie along the solution's curve. This can be a perfectly usable form of the answer in many applied problems, but before we go too much further, let's make sure that we are aware of the shortcomings of this form of solution.

By its very nature, a numerical solution to an initial value problem consists of a **table of values** that is **finite** in length. However, the true solution of the initial value problem is most likely a whole continuum of values; that is, it consists of an *infinite* number of points. Obviously, the numerical solution is actually leaving out an infinite number of points. A question might arise: "With this many holes in it, is the solution good for anything at all?" To make this comparison a little clearer, let's look at a very simple specific example starting with Euler's method.

Given the differential equation:

$$\frac{dy}{dt} = g(t, y), y(t_0) = y_0, t_0 \le t \le b$$

Euler's Method

Step 1. Pick a step size, h, so that the interval $[b - t_0]/n = h$ is divided evenly.

Step 2. Start at $y(t_0) = y_0$.

Step 3. Compute $y_{n+1} = y_n + h * g(t_n, y_n)$.

Step 4. Let $t_{n+1} = t_n + h, n = 0, 1, 2, \ldots$

Step 5. Continue until $t_n = b$.

STOP

Example 1. *$dy/dt = 0.25 * t * y(t)$, $y(0) = 2$*

We want to estimate the solution to $y(3)$ using a numerical approach. The interval is $[0, 3]$.

A few steps by hand:

Let's start with a step size of 1:   point (t, y)
When $t = 0$, $y = 2$, this is the initial condition.   (0,2)
When $t = 0 + h = 1$, $y = 2 + h * 0.25ty = 2 + (1) * (0.25)(0)(2) = 2$.   (1,2)
When $t = 1 + h = 2$, $y = 2 + h * 0.25 * t * y = 2 + (1) * (0.25)(1)(2.5) = 2.5$.   (2,2.5)
When $t = 3$, $y = 2.5 + h * 0.25 * t * y = 2.5+(1) * (0.25)(2)(2.5) - 3.75$.   (3,3.75)

Geometrically, the blue, black, and green lines in Figure 7.1 represent our numerical solution while the red curve represents the actual closed-form solution.

Note that we have underestimated every value between $t = 0$ and $t = 3$ shown in Figure 7.1.

The exact solution is $y(t) = 2 \cdot e^{\frac{t^2}{8}}$.

Let's make a table (see Table 7.1) and see how we did.

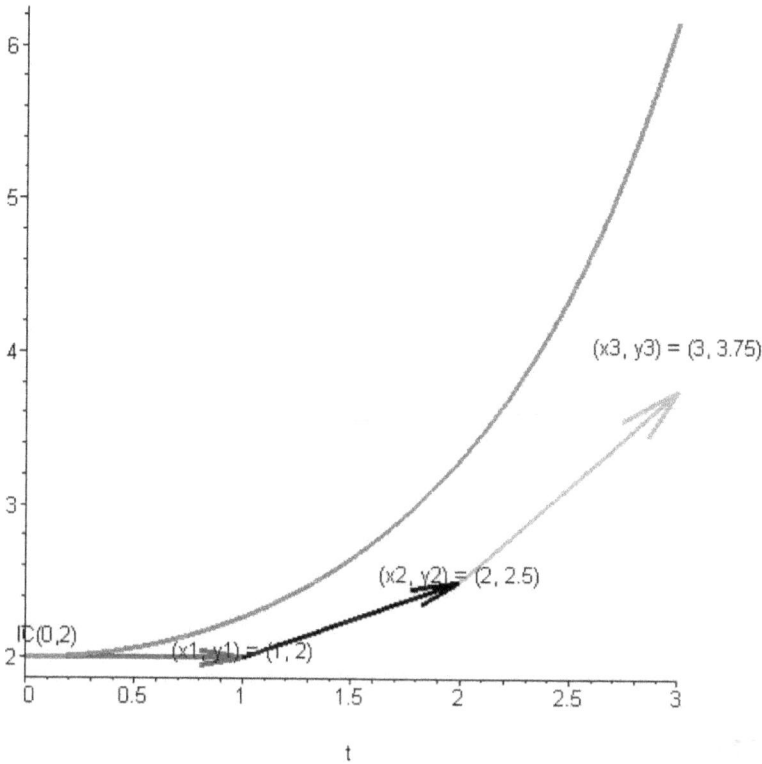

FIGURE 7.1
Euler's method geometrically.

TABLE 7.1

Comparison of Euler's Method to Exact Solution for Example 1

T	Y (approximate, Ya)	Y (exact, Ye)	Error = $|Ye - Ya|$	Percent Error = $\dfrac{100 \cdot |Ye - Ya|}{Ye}$
0	2	2	0	0
1	2.	2.266	.266	11.73
2	2.5	3.297	.797	24.17
3	3.75	6.16	2.41	39.12

Notice that as we move farther away from the initial condition $y(0) = 2$, the worse our estimate becomes.

Can we do better?

Maybe, we might improve by changing the step size. In this example, let us make the step size 0.5 instead of 1.

We may use technology, in this case, Maple, Excel, or R (or any technology you choose) to obtain our output.

```
> ans2 := dsolve({eqn, y(0)=2}, numeric,
      method=classical[foreuler],
      output=array([0,0.5,1,1.5,2,2.5,3]),
      stepsize=0.5);
```

$$ans2 := \begin{bmatrix} \left[t, y(t) \right] \\ \begin{bmatrix} 0 & 2. \\ 0.5 & 2. \\ 1 & 2.1250 \\ 1.5 & 0.3906250 \\ 2 & 2.83886718750 \\ 2.5 & 3.54858398437500 \\ 3 & 4.657516479492187500 \end{bmatrix} \end{bmatrix}$$

Let's make a table and see how we did (Table 7.2).

TABLE 7.2

Comparisons

t	Ya	Ye	Error = $|Ye - Ya|$	Percent Error
0	2	2	0	0
.5	2.	2.063	0.063	3.05
1	2.125	2.2633	0.1383	6.11
1.5	2.390625	2.6496	0.258975	9.77
2	2.838867	3.2974	0.458533	13.91
2.5	3.54858	4.3684	0.81982	18.76
3	4.65752	6.1604	1.503	24.84

Our plot:

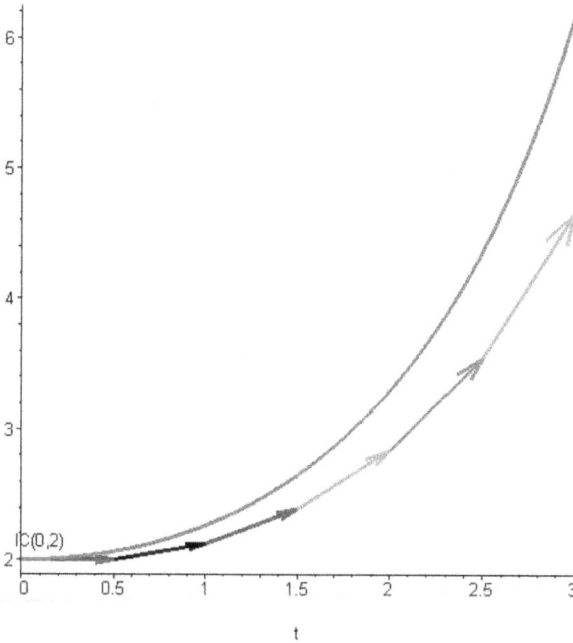

We are still underestimating at all points. We might be able to do better by making this step size small. (Note: It is not always true that the making the step size smaller improves the estimates.)

Example 2. Repeat Example 1 But with Step Size = 0.5

We use Excel this time as our technology.

Euler's Method						
$dy/dt = 0.25 * y(t) * t, y(0) = 2$						
	h	0.5				
				exact	error	% Rel error
t	y	dy/dt	new y			
0	2	0	2	2	0	0
0.5	2	0.25	2.125	2.063487	0.063487	3.076676552
1	2.125	0.53125	2.390625	2.266297	0.141297	6.2347041
1.5	2.390625	0.896484	2.838867	2.64957	0.258945	9.773078825
2	2.838867	1.419434	3.548584	3.297143	0.458575	13.9070006
2.5	3.548584	2.217865	4.657516	4.368402	0.819818	18.76699325
3	4.657516	3.493137	6.404085	6.160434	1.502917	24.39628916

We still did not do well with our estimate.

```
> Digits := 20:
ans2 := dsolve({eqn, y(0)=2}, numeric,
            method=classical[foreuler],
            output=array([0,0.5,1,1.5,2,2.5,3]),
            stepsize=0.5);
```

$$ans2 := \begin{bmatrix} & [t, y(t)] \\ 0 & 2. \\ 0.5 & 2. \\ 1 & 2.1250 \\ 1.5 & 0.3906250 \\ 2 & 2.83886718750 \\ 2.5 & 3.54858398437500 \\ 3 & 4.657516479492187500 \end{bmatrix}$$

The second method that we will discuss is the improved Euler's method, also known as Heun's method.

Improved Euler's Method (Heun's Method)

The accuracy of Euler's method is improved by using an average of two slopes in the tangent line approximation. Use the slope obtained at the beginning of the step and the slope obtained at the end of the step in order to improve our accuracy.

$$\text{Given } \frac{dy}{dt} = g(t, y), y(t_0) = y_0, t_0 \le t \le b.$$

Improved Euler's Method Algorithm

Step 1. Pick a step size, h, so that the interval, $[b - t_0]/n = h$ is divided evenly.

Step 2. Start at $y(t_0) = y_0$.

Step 3. Let $t_{n+1} = t_n + h$.

Step 4. Compute $y_{n+1} = y_n + (h/2) * [g(t_n, y_n) + g(t_{n+1}, y_{n+1})]$.

Step 5. Continue until $t_{n+1} = b$.

STOP

Step 4 uses the formula:

$$y_{n+1} = y_n + \frac{h}{2}(k_1 + k_2)$$
$$k_1 = f(t_n, y_n)$$
$$k_2 = f(t_{n+1}, y_n + h * k_1)$$

Example 3. Improved Euler's Method for $y' = 0.25 \, t \, y(t)$, $y(0) = 2$

We want to estimate the solution to $y(3)$ using the improved Euler's method.

$$k_1 = .25 \, t_n y_n$$
$$k_2 = .25 \, t_{n+1}(y_n + h * k_1)$$

We start with a few steps of the improved Euler's method by hand:

Step size of $h = 1$.	Point (t, y)
When $t_0 = 0$, $y_0 = 2$. This is the initial condition.	$(0, 2)$
When $t_1 = t_0 + h = 1$, $y_1 = 2 + (1/2)* [(.25* (0)* (2) + .25* (1)*$ $(2+.25* (0)* 2)] = 2.25$.	$(1, 2.25)$
When $t_2 = t_1 + h = 2$, $y_2 = 2.25 + (1/2)* (.25(1)(2.25) + .25*(2)$ $(2.25 + (1).25* (1)(2.25)) = 3.23437$.	$(2, 3.234375)$
When $t_3 = 2 + h = 3$, $y_3 = 3.234375 + (1/2)* [(.25)*(2)*(3.234375)$ $+ (.25)* (3)* (3.234375 + .25* (2)* (3.234375)] = 5.862305$	$(3, 5.862305)$

The exact solution is $y(t) = 2 \cdot e^{\frac{t^2}{8}}$.

Table 7.3 shows how we did in our estimates.

Notice that as we move farther away from the initial condition $y(0) = 2$, the worse our estimate becomes. Also note that these estimates are "better" estimates than Euler's method. Think about why.

Figure 7.2 shows the improvement.

TABLE 7.3

Improved Euler's Estimates for Example 1

t	Y (approximate, Ya)	Y (exact, Ye)	Error = $\lvert Ye - Ya \rvert$	Percent Error= $\frac{100 \cdot \lvert Ye - Ya \rvert}{Ye}$
0	2	2	0	0
1	2.25	2.266	0.016	0.706
2	3.234375	3.297	0.062625	1.899
3	5.862305	6.16	0.2977	4.83

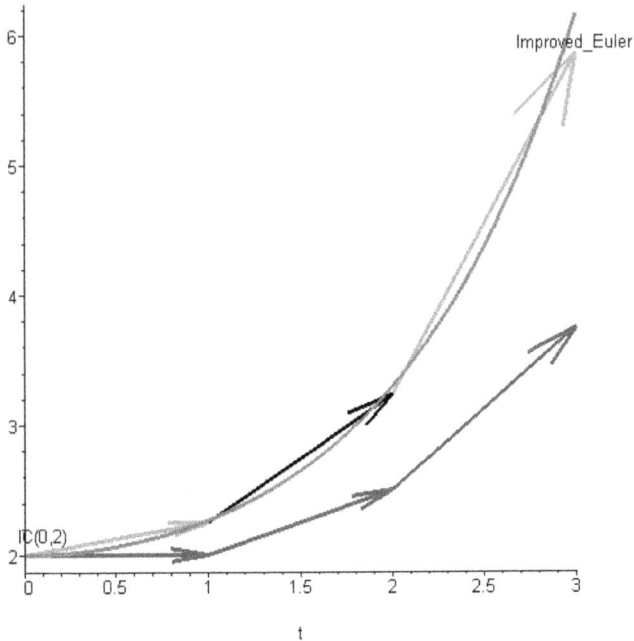

FIGURE 7.2
Improved Euler's estimates of Example 1.

Example in Maple: We note that the only command change is our call within **dsolve({eqn, y(0)=2}, numeric,method=classical[heu nform], output=array([0,0.5,1,1.5,2,2.5,3]), stepsize=0.5);** to the above commands call the heunform improved Euler's algorithm from Maple

```
> restart;
> with(DEtools):with(plots):with(linalg):
```

Enter your ODE

```
> eqn:=diff(y(t),t)=.25*t*y(t);
```

$$eqa := \frac{d}{dx} y(t) = 0.25 t \, y(t)$$

Solve the ODE analytically and plot.

```
> dsolve({eqn,y(0)=2},y(t));
```

$$y(t) = 2e^{\left(\frac{t^2}{8}\right)}$$

```
> plot(2*exp(1/8*t^2),t=0..3, thickness=3);
```

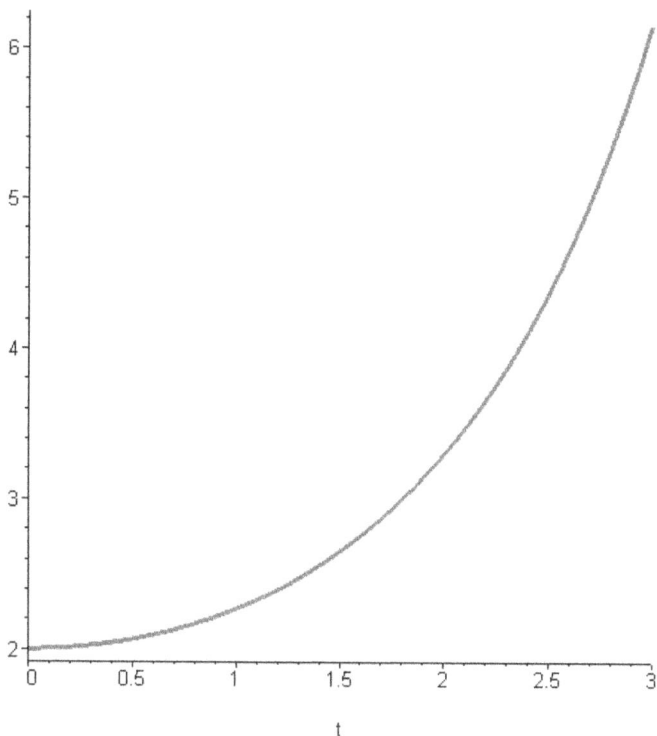

Note you have to copy and paste the output equation from dsolve into the plot command.

Numerical Method—Internal and Classic—OUTPUT Only

You will need to specify both the step size you want and the output array you want to see.

```
> Digits := 20:
ans2 := dsolve({eqn, y(0)=2}, numeric,
            method=classical[heunform],
            output=array([0,0.5,1,1.5,2,2.5,3]),
            stepsize=0.5);
```

$$ans2 := \begin{bmatrix} \begin{bmatrix} t,y(t) \end{bmatrix} \\ \begin{bmatrix} 0 & 2. \\ 0.5 & 2.0625000000000000000 \\ 1 & 2.2639160156250000000 \\ 1.5 & 2.6441831588745117188 \\ 2 & 3.2845712676644325257 \\ 2.5 & 4.3366605018381960691 \\ 3 & 6.0814887506246577688 \end{bmatrix} \end{bmatrix}$$

First, we plot the actual solution (if we have one) then the numerical solution. Second, we will overlay them on one plot.

```
> plot(2*exp(1/8*t^2),t=0..3);
```

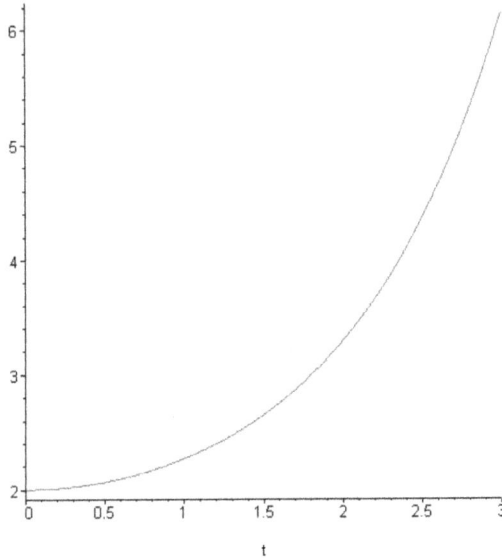

Numerical Plot:

```
> data:=({seq([i,fy2(i)],i=0..3)}):
>
> points:=pointplot(data, symbol=diamond, color=black):
> display( points);
```

Numerical Output for Analysis: Graphical and Percent Error

First, get the numerical output as a procedure as follows:

```
> dsol2 := dsolve({eqn, y(0) = 2},
   numeric,method=classical[heunform],stepsize=1, output=
listprocedure);
dsol2 := [ t = (proc(t) ... end proc ), y(t) = ( proc(t) ... end
proc ) ]
```

Prepare for numerical output:

```
> fy2 := eval(y(t),dsol2);
fy2 := proc(t) ... end proc
```

View output to compare to preceding output (numerical):

```
> seq(fy2(i),i=0..3);
```

```
2., 2.2500000000000000000, 3.2343750000000000000,
5.8623046875000000000
```

Prepare the exact solution for output format:

```
> actual_y2:=evalf(subs(t=i,2*exp(1/8*t^2)));
```

$$actual\_y2 := 2.e^{\left(0.125000000000000000i^2\right)}$$

View actual output
```
> seq(actual_y2(i),i=0..3);
2., 2.2662969061336526336, 3.2974425414002562936,
6.1604336978360624900
```

Put into an array that lists [*t, numerical y(t), actual y(t), Percent Error*]
```
> array([seq([i,fy2(i),evalf(subs(t=i, 2*exp(1/8*t^2))),evalf(10
0*abs(fy2(i)-actual_y2(i))/actual_y2(i))],i=0..3)]);
```

0	2.	2.	0.
1	2.2500000000000000000	2.2662969061336526336	0.71909845923301717516
2	3.2343750000000000000	3.2974425414002562936	1.9126198745975635237
3	5.8623046875000000000	6.1604336978360624900	4.8394159398352818372

Graphical Comparisons

First, we plot the actual solution (if we have one) then the numerical solution. Second, we will overlay them on one plot as shown in Figure 7.3. Otherwise, compare numerical solutions to either qualitative plots or slope field plots.

```
> plot(2*exp(1/8*t^2),t=0..3);
```

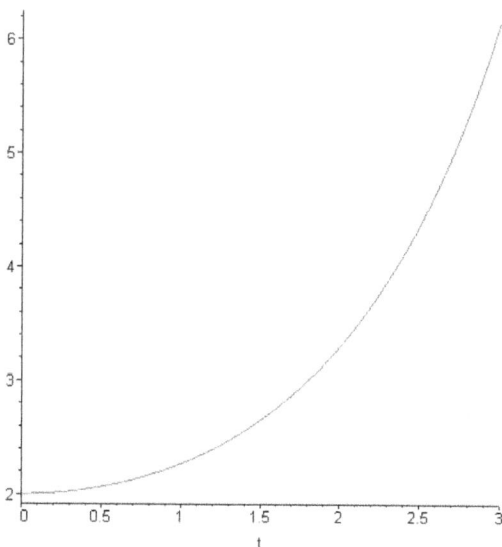

Numerical Plot:

```
>data:=({seq([i,fy2(i)],i=0..3)}):
>points:=pointplot(data, symbol=diamond, color=black):
>display( points);
```

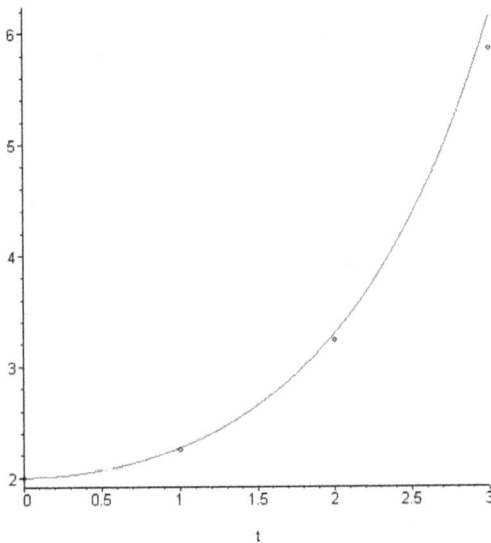

FIGURE 7.3
Plot of the improved Euler's method.

```
Overlay
> with(plots):
> data:=({seq([i,fy2(i)],i=0..3)}):
> curve:=plot(2*exp(1/8*t^2),t=0..3, color=red):
> points:=pointplot(data, symbol=diamond, color=black):
> display(curve, points);
```

Runge–Kutta Methods

When we use the average of the estimates of the derivatives at the endpoints, we can improve the approximation to the solution. A class of approximation techniques that estimate derivatives at various points within an interval and then computes a *weighted average* is the Runge–Kutta method, named for two German mathematicians.

The Runge–Kutta methods are classified by order, where the order depends on the number of slope estimates used at each step. A very popular method is the fourth-order Runge–Kutta (RK4) method.

RK4 Method

For solving $dy/dt = g(t, y)$, $y(t_0) = y_0$ over an interval.

Step 1. First divide the interval $x_0 \leq x \leq b$ into p sub-intervals using equally spaced points. This yields the step size, $h = (b - x_0)/p$.

Step 2. For $n = 1, 2, 3, \ldots, p$, obtain the following sequence of approximations:

$$y_{n+1} = y_n + \frac{K_1 + 2K_2 + 2K_3 + K_4}{6}, where$$

$$K_1 = g(t_n, y_n)h$$
$$K_2 = g(t_n + h/2, y_n + K_1/2)h$$
$$K_3 = g(t_n + h/2, y_n + K_2/2)h$$
$$K_4 = g(t_n + h, y_n + K_3)h$$

We begin with an example by hand.

Example 4. RK4 for $y' = 0.25t$, $y(t)$, $y(0) = 2$

We want to estimate the solution to $y(3)$ using a numerical approach.
   A few steps of RK4 method by hand:

Step size of $h = 1$.

When $t = 0$, $y = 2$. This is the initial condition: $(0, 2)$.
When $t = 0 + h = 1$,

K1 = $h * g(t(n),y(n))$ = (1 * 0.25 * 0 * 2) = 0.
K2 = $h * g(t(n + h/2),y(n) + K1/2)$ = (1 * 0.25 * (0+1/2) * (2 + 0)) =.25.
K3 = $h * g(t(n + h/2),y(n) + K2/2)$ = (1 * 0.25 * (0 + 1/2) * (2 + 0.125)) = 0.265625.
K4 = $h * g(t(n + h),y(n) + K3)$ = (1 * .25 * (0 + 1) * (2 + 0.265625)) = .56640625.
Y(1) = (2 + (1/6) * (0 + 2 * 0.25 + 2 * 0.265625 + 0.56640625)) = 2.266276

When $t = 1 + h = 2$.

K1 = 0.566569   K2 = 0.956085   K3 = 1.029119   K4 = 1.647698
Y(2) = 2.266276 + (1/6) * (0.566569 + 2 * 0.956085 + 2 * 1.029119 + 1.647698) = 3.297055

When $t = 2 + h = 3$,

K1= 1.648528, K2=2.575825, K3=2.865605, K4=4.621995
Y(3) = 3.297055 + (1/6) * (1.648528 + 2 * 2.575825 + 2 * 2.86560 + 4.621995) = 6.155952

The exact solution is $y(t) = 2 \cdot e^{\frac{t^2}{8}}$.

Table 7.4 shows us how we did using the RK4 method.

Notice that as we move farther away from the initial condition $y(0) = 2$, the worse our estimate becomes. Also note that these estimates are much "better estimates" than either Euler's method or the improved Euler's method. Think about why this is true.

The RK4 plot is shown in Figure 7.4

In Maple, we again change the call from within dsolve to RK4 as seen in this sequence of commands:

```
Set UP MAPLE
> restart;
> with(DEtools):with(plots):with(linalg):

Enter your ODE
> eqn:=diff(y(t),t)=.25*t*y(t);
```

$$eqn := \frac{d}{dt} y(t) = 0.25\ t\ y(t)$$

TABLE 7.4

Runge–Kutta 4 Method for Example 4

t	Y (approximate, Ya)	Y (exact, Ye)	Error = $|Ye - Ya|$	Percent Error = $\frac{100 \cdot |Ye - Ya|}{Ye}$
0	2	2	0	0
1	2.266276	2.266297	0.00002	0.002
2	3.297055	3.297443	0.000388	0.0388
3	6.155952	6.160434	0.004482	0.4482

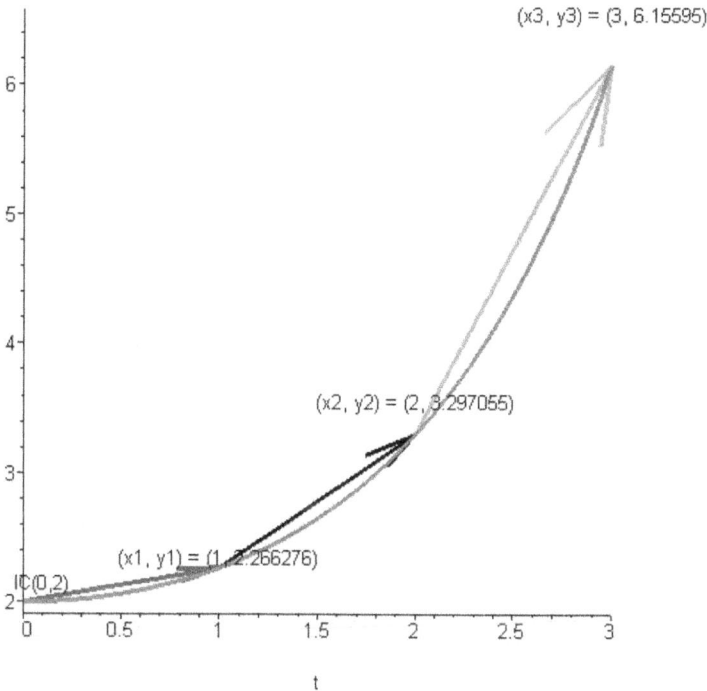

FIGURE 7.4
RK-4 plot for Example 4.

```
>
Solve the ODE Analytically & Plot

> dsolve({eqn,y(0)=2},y(t));
```

$$y(t) = 2e^{\left(\frac{t^2}{8}\right)}$$

```
> plot(2*exp(1/8*t^2),t=0..3, thickness=3);
```

Figure 7.5 shows the solution plot.
Note you have to copy and paste the output equation from dsolve into the plot command.

Numerical Method—Internal and Classic–OUTPUT Only

You will need to specify both the step size you want and the output array you want to see.

```
> Digits := 20:
ans2 := dsolve({eqn, y(0)=2}, numeric,
method=classical[rk4],
```

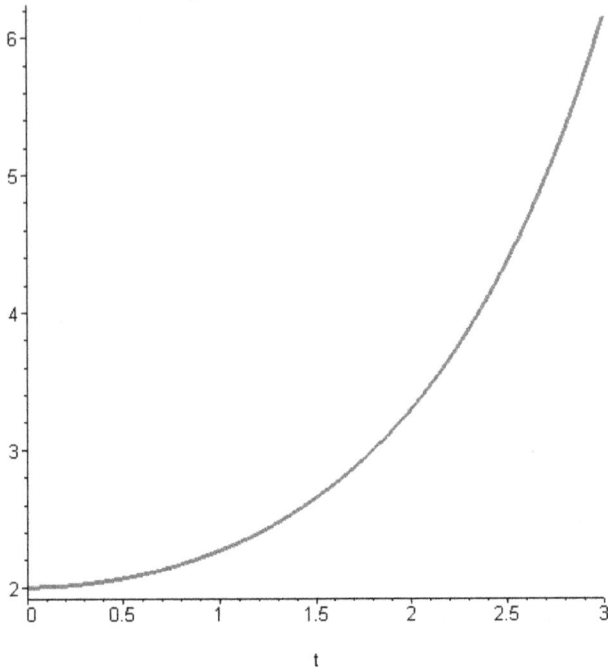

FIGURE 7.5
Plot of solution.

```
output=array([0,0.5,1,1.5,2,2.5,3]),
stepsize=0.5);
```

$$ans2 := \begin{bmatrix} \begin{bmatrix} t, y(t) \end{bmatrix} \\ \begin{bmatrix} 0. & 2. \\ 0.5 & 2.0634867350260416667 \\ 1 & 2.2662959538380770634 \\ 1.5 & 2.6495643470246741905 \\ 2. & 3.2974194721525799553 \\ 2.5 & 4.3683097289253688453 \\ 3. & 6.1600978602067198076 \end{bmatrix} \end{bmatrix}$$

Numerical Output for Analysis: Graphical and Percent Error

First, get the numerical output as a procedure as follows:

```
> dsol2 := dsolve({eqn, y(0) = 2},
     numeric,method=classical[rk4],stepsize=1, output=
listprocedure);
```

```
  dsol2 := [ t = ( proc(t) ... end proc ), y(t) = ( proc(t) ...
end proc ) ]
```

Prepare for numerical output:

```
> fy2 := eval(y(t),dsol2);
```

$$fy2 := \mathrm{proc}(t)...\mathrm{end\,proc}$$

```
View output to compare to preceding output (numerical):
> seq(fy2(i),i=0..3);
2.,2.2662760416666666667,3.2970554033915201823,6.1559523209644
895461
```

Prepare exact solution for output format:

```
> actual_y2:=evalf(subs(t=i,2*exp(1/8*t^2)));
```

$$actual\_y2 := 2.e^{\left(0.1250000000000000000 i^2\right)}$$

View actual output:

```
> seq(actual_y2(i),i=0..3);
2.,2.2662969061336526336,3.2974425414002562936,6.1604336978360
624900
```

Put into an array that lists *[t, numerical y(t), actual y(t), Percent Error]*

```
> array([seq([i,fy2(i),evalf(subs(t=i, 2*exp(1/8*t^2))),evalf(10
0*abs(fy2(i)-actual_y2(i))/actual_y2(i))],i=0..3)]);
```

```
[0, 2. ,2. ,0.]
[1 , 2.2662760416666666667 , 2.2662969061336526336
0.0009206413744597169l041]
[2 , 3.2970554033915201823 , 3.2974425414002562936
0.011740553591927441423]
[3 , 6.1559523209644895461 , 6.1604336978360624900 ,
0.072744502925939929932 ]
```

Graphical Comparisons

First, we plot the actual solution (if we have one) and then the numerical solution. Second, we will overlay them on one plot. Otherwise, compare the numerical solution to either qualitative plots or slope field plots. This is shown in Figures 7.6 and 7.7.

```
> plot(2*exp(1/8*t^2),t=0..3);
```

Numerical Plot:

```
> data:=({seq([i,fy2(i)],i=0..3)}):
> points:=pointplot(data, symbol=diamond, color=black):
> display( points);
```

Overlay

```
> with(plots):
> data:=({seq([i,fy2(i)],i=0..3)}):
> curve:=plot(2*exp(1/8*t^2),t=0..3, color=red):
> points:=pointplot(data, symbol=diamond, color=black):
> display(curve, points);
```

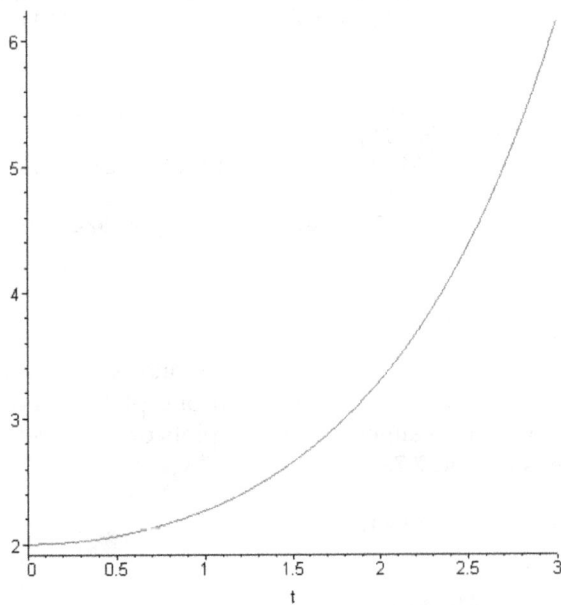

FIGURE 7.6

Plot of $2e^{\frac{t^2}{8}}$ from 0 to 3.

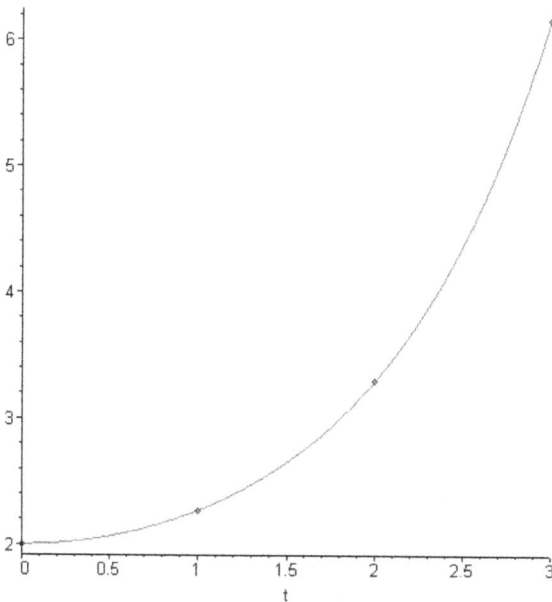

FIGURE 7.7
Plot of the RK4 method solution.

Python and Numerical Methods for ODEs (IVP)

```
Euler's Method CODE
import numpy as np
import matplotlib.pyplot as plt

plt.style.use('seaborn-poster')
%matplotlib inline

# Define parameters
f = lambda t, s: np.exp(-t) # ODE
h = 0.1 # Step size
t = np.arange(0, 1 + h, h) # Numerical grid
s0 = -1 # Initial Condition

# Explicit Euler Method
s = np.zeros(len(t))
s[0] = s0

for i in range(0, len(t) − 1):
    s[i + 1] = s[i] + h*f(t[i], s[i])

plt.figure(figsize = (12, 8))
plt.plot(t, s, 'bo--', label='Approximate')
plt.plot(t, -np.exp(-t), 'g', label='Exact')
plt.title('Approximate and Exact Solution \
```

```
for Simple ODE')
plt.xlabel('t')
plt.ylabel('f(t)')
plt.grid()
plt.legend(loc='lower right')
plt.show()
h = 0.01 # Step size
t = np.arange(0, 1 + h, h) # Numerical grid
s0 = - 1 # Initial Condition

# Explicit Euler Method
s = np.zeros(len(t))
s[0] = s0

for  i in range(0, len(t) − 1):
    s[i + 1] = s[i] + h*f(t[i], s[i])
plt.figure(figsize = (12, 8))
plt.plot(t, s, 'b--', label='Approximate')
plt.plot(t, -np.exp(-t), 'g', label='Exact')
plt.title('Approximate and Exact Solution \
for Simple ODE')
plt.xlabel('t')
plt.ylabel('f(t)')
plt.grid()
plt.legend(loc='lower right')
plt.show()
```

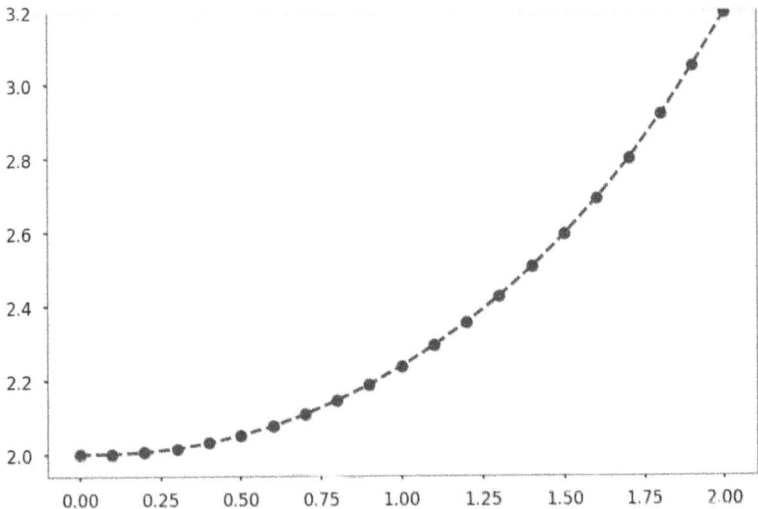

```
Output
Euler:
Enter initial conditions:
```

```
x0 = 0
y0 = 2
Enter calculation point:
xn = 3
Enter number of steps:
Number of steps = 25
```

```
----------------SOLUTION----------------
-----------------------------------------
x0                  y0                  yn
-----------------------------------------
0.0000              2.0000              2.0000
-----------------------------------------
0.1200              2.0000              2.0072
-----------------------------------------
0.2400              2.0072              2.0217
-----------------------------------------
0.3600              2.0217              2.0435
-----------------------------------------
0.4800              2.0435              2.0729
-----------------------------------------
0.6000              2.0729              2.1102
-----------------------------------------
0.7200              2.1102              2.1558
-----------------------------------------
0.8400              2.1558              2.2101
-----------------------------------------
0.9600              2.2101              2.2738
-----------------------------------------
1.0800              2.2738              2.3475
-----------------------------------------
1.2000              2.3475              2.4320
-----------------------------------------
1.3200              2.4320              2.5283
-----------------------------------------
1.4400              2.5283              2.6375
-----------------------------------------
1.5600              2.6375              2.7609
-----------------------------------------
1.6800              2.7609              2.9001
-----------------------------------------
1.8000              2.9001              3.0567
-----------------------------------------
1.9200              3.0567              3.2327
-----------------------------------------
2.0400              3.2327              3.4306
-----------------------------------------
2.1600              3.4306              3.6529
-----------------------------------------
2.2800              3.6529              3.9027
-----------------------------------------
```

2.4000	3.9027	4.1837
2.5200	4.1837	4.5000
2.6400	4.5000	4.8564
2.7600	4.8564	5.2586
2.8800	5.2586	5.7129

```
At x=3.0000, y=5.7129
```

```
RK 4 in Python
```

```python
# RK-4 method python program

# function to be solved
def f(x,y):
    return x+y

# or
# f = lambda x: x+y

# RK-4 method
def rk4(x0,y0,xn,n):

    # Calculating step size
    h =(xn-x0)/n

    print('\n--------SOLUTION--------')
    print('------------------------')
    print('x0\ty0\tyn')
    print('------------------------')
    for i in range(n):
        k1 = h *(f(x0, y0))
        k2 = h *(f((x0+h/2),(y0+k1/2)))
        k3 = h *(f((x0+h/2),(y0+k2/2)))
        k4 = h *(f((x0+h),(y0+k3)))
        k =(k1+2*k2+2*k3+k4)/6
        yn = y0 + k
        print('%.4f\t%.4f\t%.4f'%(x0,y0,yn))
        print('------------------------')
        y0 = yn
        x0 = x0+h

    print('\nAt x=%.4f, y=%.4f'%(xn,yn))

# Inputs
print('Enter initial conditions:')
x0 =float(input('x0 = '))
y0 =float(input('y0 = '))
```

```
print('Enter calculation point: ')
xn =float(input('xn = '))

print('Enter number of steps:')
step =int(input('Number of steps = '))

# RK4 method call
rk4(x0,y0,xn,step)
```

```
Enter initial conditions:
x0 = 0
y0 = 2

Enter calculation point:
xn = 3

Enter number of steps:
Number of steps = 25
```

```
----------------SOLUTION------------------
------------------------------------------
x0                y0                yn
------------------------------------------
0.0000            2.0000            2.0036
------------------------------------------
0.1200            2.0036            2.0145
------------------------------------------
0.2400            2.0145            2.0327
------------------------------------------
0.3600            2.0327            2.0584
------------------------------------------
0.4800            2.0584            2.0921
------------------------------------------
0.6000            2.0921            2.1339
------------------------------------------
0.7200            2.1339            2.1844
------------------------------------------
0.8400            2.1844            2.2442
------------------------------------------
0.9600            2.2442            2.3139
------------------------------------------
1.0800            2.3139            2.3944
------------------------------------------
1.2000            2.3944            2.4867
------------------------------------------
1.3200            2.4867            2.5918
------------------------------------------
1.4400            2.5918            2.7111
------------------------------------------
1.5600            2.7111            2.8461
------------------------------------------
```

1.6800	2.8461	2.9986
1.8000	2.9986	3.1707
1.9200	3.1707	3.3647
2.0400	3.3647	3.5835
2.1600	3.5835	3.8303
2.2800	3.8303	4.1089
2.4000	4.1089	4.4236
2.5200	4.4236	4.7796
2.6400	4.7796	5.1828
2.7600	5.1828	5.6404
2.8800	5.6404	6.1604

At x=3.0000, y=6.1604

Summary in Python Results

Method	Approximate y (3)	Exact y (3)	|Error|
Euler	5.7129	6.16043	0.44753
RK4	6.1604	6.16043	0.00003

7.3 Population Modelling

Suppose a society has a population of $x(t)$ individuals at time t, in years, and that all left-handed people who mate with other left-handed people have offspring who are also left-handed, while a fixed proportion r of all other offspring are also left-handed. If the birth and death rates for all individuals are assumed to be the constants B and D, respectively, and if right-handed and left-handed people mate at random, the problem can be expressed by the following differential equation:

$$\frac{dx(t)}{dt} = (B-D)x(t)$$

$$\frac{dx(t)}{dt} = (B-D)x(t) + rB(x(t) - x_n(t),$$

where $x_n(t)$ denotes the number of left-handed people in the population at time t.

To simplify our modelling and equation building process, let the variable $p(t) = x_n(t)/x(t)$ be introduced to represent the proportion of left-handed people in the population and now may be combined into a single differential equation:

$$dp(t)/dt = r(B(1 - p(t)).$$

Historical data were collected and analyzed to obtain the following coefficient values: $p(0) = 0.01$, $B = 0.02$, $D = 0.015$, and $r = 0.1$. We use these values to approximate the solution $p(t)$ from $t = 0$ to $t = 50$ with a step size (arbitrarily chosen) of $h = 1$ year. We will illustrate several numerical methods in obtaining our solution: Euler's method, the improved Euler's method, and RK4. The plot is shown in Figure 7.8.

Methods of Numerical Solutions using the Numerical Analysis package in MAPLE:

Based on our answer for $y(50)$, Maple appears to be using an RK4 method for its numeric algorithm. We can obtain an exact solution to this differential equation for $p(t)$ exactly and compare our results from the numerical methods when $t = 50$ with the exact value at that time. We can make a recommendation of which method is most accurate. We provide an error bound for each method. The exact solution to

$dp(t)/dt = 0.002\ (1 - p(t))$, $p(0) = 0.01$ can be found by separation of variables or variation of parameters as $p(t) = 1. - 0.9900000000*$ $\exp(-0.002000000000 * t)$. So $p(50) = 0.10241$.

```
Digits := 20 :
ansp := dsolve( { eqnp1, icp1}, numeric, method = classical
[foreuler], output = Array([10, 20, 30, 40, 50]), stepsize = 1);
```

$$ansp := \begin{bmatrix} \begin{bmatrix} t & p(t) \end{bmatrix} \\ \begin{bmatrix} 10. & 0.029622747081570069595 \\ 20. & 0.048856552543920677894 \\ 30. & 0.067709125582024996433 \\ 40. & 0.086188022587236842146 \\ 50. & 0.10430065017600459226 \end{bmatrix} \end{bmatrix}$$

```
ans2 := dsolve({eqnp1, icp1}, numeric, method = classical
[heunform], output = Array([10, 20, 30, 40, 50]), stepsize = 1)
```

$$ans2 := \begin{bmatrix} \begin{bmatrix} t & p(t) \end{bmatrix} \\ \begin{bmatrix} 10. & 0.029603300468266216578 \\ 20. & 0.048818429836280790261 \\ 30. & 0.067653074401731933619 \\ 40. & 0.086114768263516878847 \\ 50. & 0.10421089633547786816 \end{bmatrix} \end{bmatrix}$$

```
ans2 := dsolve( {eqnp1, icp1}, numeric, method = classical[rk4],
output = Array([10, 20, 30, 40, 50]), stepsize = 1)
```

$$ans2 := \begin{bmatrix} \begin{bmatrix} t & p(t) \end{bmatrix} \\ \begin{bmatrix} 10. & 0.0296033134263096658760 \\ 20. & 0.0488184552391949441224 \\ 30. & 0.0676531117515863063342 \\ 40. & 0.0861148170772208105550 \\ 50. & 0.1042109561443880588079 \end{bmatrix} \end{bmatrix}$$

Euler(diff(p(t), t) = 0.002*(-p(t) + 1), p(0) = 0.01, t = 50); 0.1051

Euler(diff(p(t), t) = 0.002*(-p(t) + 1), p(0) = 0.01, t = 50, output = Error);
  0.0009075

Euler(diff(p(t), t) = 0.002*(-p(t) + 1), p(0) = 0.01, t = 50, output = plot);

RungeKutta(diff(p(t), t) = 0.002*(-p(t) + 1), p(0) = 0.01, t = 50, submethod =
  rk4);

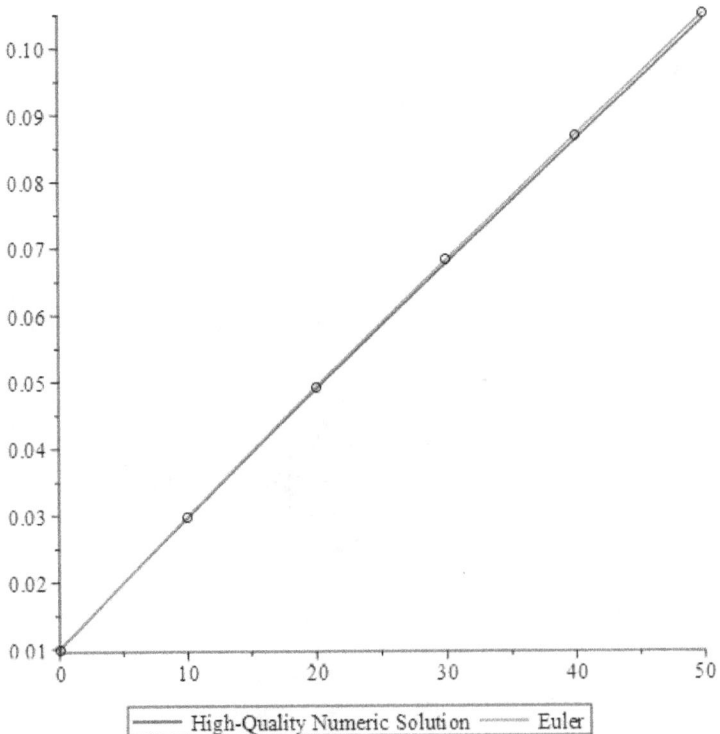

FIGURE 7.8
Plot of solution.

0.1042

RungeKutta(diff(p(t), t) = 0.002*(-p(t) + 1), p(0) = 0.01, t = 50, submethod = rk4, output = plot);

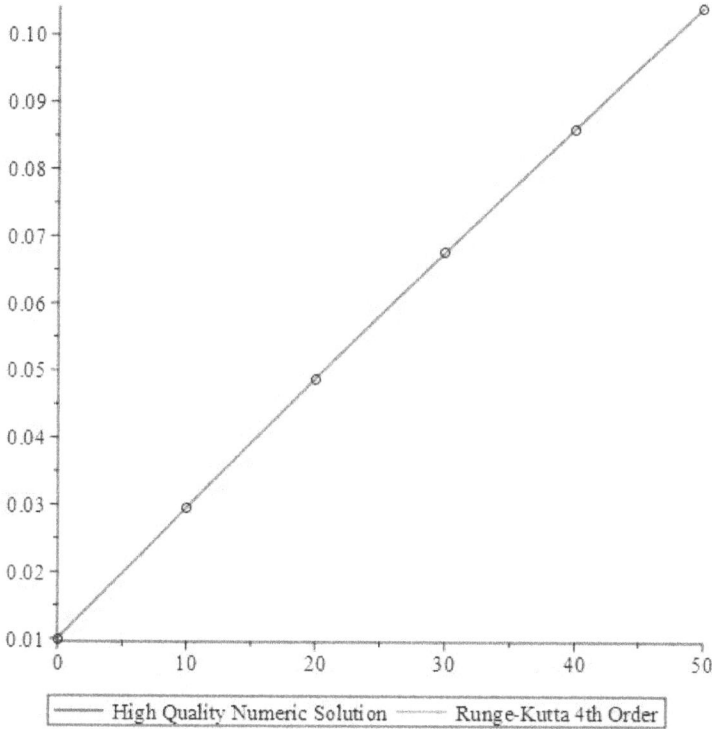

RungeKutta(diff(p(t), t) = 0.002*(-p(t) + 1), p(0) = 0.01, t = 50, submethod = rk4, output = Error);

−10 1.214 10

A comparison of methods is shown in Table 7.5.

It appears that the RK4 method did a better job in approximating the solution.

TABLE 7.5

Comparison of Methods

Method	Approximate for p (50)	Exact Solution	Absolute Error
Euler	0.1043006502	0.10421	8.6988×10^{-4}
Improved Euler	0.1042108964	0.10421	8.6019×10^{-4}
RK4	0.1042109562	0.10421	9.1354×10^{-6}

Euler's error bound is found by

$$|p_i - w_i| \le \frac{hM}{2L}\left(e^{(t_{i-1}-a)L} - 1\right) \le 0.000396$$

,

which we see with our estimates is true.

The improved Euler's method error bound is $O(h^3)$. By using Taylor polynomials (see Chapter 1), we find the local truncation error for $p(50)$ to be

0.104210894 − (0.1024175266 + .5(0.00179564947 + 0.0017958212)) = −6.1 × 10⁻⁹.

For RK4, the error bound is $y^{(5)}(c)\, h^5/5!$ or $O(h^5)$. In our problem,

$$y^{(5)}(c)\, h^5/5! = 2.38877 \times 10^{-16}.$$

We note our errors are less than our bounds.

7.4 Spread of a Contagious Disease

We return to our spread of a contagious disease. Consider modelling the spread of a contagious disease. Given the following ODE model for the spread of a communicable disease as

$$\frac{dN}{dt} = .2.5N(10 - N), N(0) = 2,$$

where N is infected people in 100s and t is some time period in days, months, or years.

We need to analyze the results. We will do this using Python.

```
Enter initial conditions:
x0 = 0
y0 = 2

Enter calculation point:
xn = 15

Enter number of steps:
Number of steps = 25
```

```
---------------SOLUTION----------------
----------------------------------------
x0                y0                yn
----------------------------------------
0.0000            2.0000            5.2639
----------------------------------------
0.6000            5.2639            8.3116
----------------------------------------
1.2000            8.3116            9.5181
----------------------------------------
1.8000            9.5181            9.8668
----------------------------------------
2.4000            9.8668            9.9635
----------------------------------------
3.0000            9.9635            9.9900
----------------------------------------
3.6000            9.9900            9.9973
----------------------------------------
4.2000            9.9973            9.9993
----------------------------------------
4.8000            9.9993            9.9998
----------------------------------------
5.4000            9.9998            9.9999
----------------------------------------
6.0000            9.9999            10.0000
----------------------------------------
6.6000            10.0000           10.0000
----------------------------------------
7.2000            10.0000           10.0000
----------------------------------------
7.8000            10.0000           10.0000
----------------------------------------
8.4000            10.0000           10.0000
----------------------------------------
9.0000            10.0000           10.0000
----------------------------------------
9.6000            10.0000           10.0000
----------------------------------------
10.2000           10.0000           10.0000
----------------------------------------
10.8000           10.0000           10.0000
----------------------------------------
11.4000           10.0000           10.0000
----------------------------------------
12.0000           10.0000           10.0000
----------------------------------------
12.6000           10.0000           10.0000
----------------------------------------
```

13.2000	10.0000	10.0000
13.8000	10.0000	10.0000
14.4000	10.0000	10.0000

At x=15.0000, y=10.0000

We find that 100% of the population have the disease in 6.6 time periods. This means that we need to quickly get a vaccine or some medications to these people.

7.5 Bungee Jumping

In bridge jumping, a participant attaches one end of a bungee cord to himself (or herself), attaches the other end to a bridge railing, and then drops off the bridge. In this project, the jumper will be dropping off the Royal Gorge Bridge, a suspension bridge that is 1053 feet above the floor of the Royal Gorge in Colorado. The jumper will use a 200-foot-long bungee cord. It would be nice if the jumper has a safe jump, meaning that the jumper does not crash into the floor of the gorge or run into the bridge on the rebound. In this project, you will do some analysis of the fall.

Assume the jumper weighs 160 pounds. The jumper will free fall until the bungee cord begins to exert a force that acts to restore the cord to its natural (equilibrium) position. In order to determine the spring constant of the bungee cord, you found that that a mass weighing 4 pounds stretches the cord 8 feet. Hopefully, this spring force will help slow down the descent sufficiently so that the jumper does not hit the bottom of the gorge. Throughout this problem, we assume that down is the positive direction.

Basically, there are two phases and one aspect that simplifies the problem. Of course, the two phases are the free fall and the spring mass problem once the cord engages. Since the two phases are two distinct problems, they can be treated separately as long as we can find the initial conditions. The free-fall phase is easiest in velocity but can be done in position, $x(0) = 0$ and/or $v(0) = 0$ at the bridge. With the solution we need to find the time and velocity at 200 feet below the bridge. In the spring mass phase, it is simplest if you find the equilibrium first. We might reset $x(0) = 0$ and $v(0)$ equals the velocity at 200 feet or reset $x(0) = 0$ and $v(0) = $ velocity at equilibrium (the last option requires an additional phase for the time between 200 feet and equilibrium). Actually, we will try something else, which will simplify our problem immensely.

Finally, there might be discussions about the spring pulling up when moving in the upward direction, and what about slack? This could be answered by assuming that we are only interested in the first oscillation where we determine if we will hit the bottom of the gorge and/or we might bang into the bridge on the upward motion.

Let's start our modelling with the free fall phase. Before the bungee cord begins to retard the fall of the jumper, the only forces that act on the jumper are his weight and the force due to wind (air) resistance.

If the force due to the wind resistance is 0.9 times the velocity of the jumper, then use Newton's second law ($\Sigma F = MA$) to write a differential equation that models the fall of the jumper.

$$\Sigma F = MA$$
$$mx'' = -mg - kx'$$

Once the bungee cord starts to stretch Hooke's law for springs will come into effect and ($F = Ks$) 4 = k * 8. So $k = 4/8 = 0.5$.

Let's also assume that the acceleration of gravity is 32 ft/s². Weight $W = mg$, so $W/g = m$. This implies mass $m = 160/32. = 5$ slugs.

We divide the entire equation by m to get the following equation:

$$x'' = -mg/m - k/m * x' - 0.9/m * x'.$$

We use the term $-k/mx'$ only after bungee cord has stretched out to 200 ft below the bridge and delete the force due to the spring during free fall.

In free fall, we use Newton's second law, and it becomes $5x'' = 160 - 0.9x'$ in terms of position and in terms of velocity is $v' + 0.180v = 32$, with an initial condition of $v(0) = 0$.

Or rewritten as $dv/dt + 0.1800\, v(t) = 32$ and $v(0) = 0$,

$$32 - 0.18\, v(t) = dv/dt.$$

Solving this differential equation numerically with a step size of 1, Table 7.6 provides the results.

We are more likely to use the RK4 results (to the student, why?).

First, we see we eventually reach a terminal velocity of 177.77777778 as the limit as t approaches infinity. If we estimate the velocity of our jumper after

TABLE 7.6

Numerical Results to Free Fall Using Maple's Numerical Analysis Package

	Euler	Error	Improved Euler	Error	RK4	Error	Exact Answer
0	0	0	0	0	0	0	0
2	64	10.25	52.48	1.27	53.74	0.012	53.746
4	105	13.72	89.47	1.772	91.23	0.0077	91.244
6	131.2	13.77	115.5	1.863	117.4	.0071	117.41
8	148	12.25	133.9	1.790	135.6	0.05419	135.66
10	158.7	10.29	146.9	1.540	148.4	0.0187	148.39

RK4 = Runge–Kutta 4 method.

4 seconds, it depends on the method. We find the values for $v(4)$ as 94.177 using Euler's method, 91.154 using the improved Euler's method, and 91.244 using the RK4 method. When the jumper has fallen 200 feet, we find the speed is about 91.244 with RK4.

Be sure to include the initial conditions for the jumper for your differential equation. The initial condition will be $x(0) = 0$, $x'(0) = 0$:

$$160 - 0.9v = 0.5dv/dt.$$

Exact solution as used in the table is $v(t) = (1600/9) - (1600/9) e^{-9t/5}$.

The plot is seen in Figure 7.9.

So, our jumper reaches terminal velocity in about 4 to 5 seconds.

We use the trapezoidal method to integrate $v(t)$ to find $x(t)$ and determine when we have fallen 200 feet.

xdis<-cumtrapz(t,v).

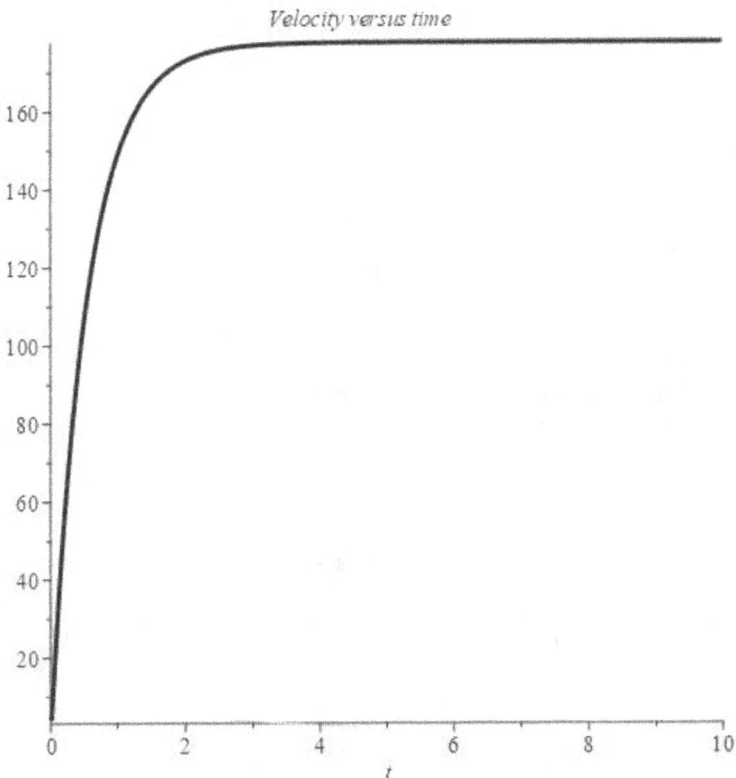

FIGURE 7.9
Plot of velocity versus time with a constant terminal velocity = 177.777.

We find that we reach 200 feet in about 3.96 seconds.

> $ApproximateInt(177.7778(1-\exp(-.18 \cdot t)), t = 0..4, method = trapezoid,$
> $output = plot);$

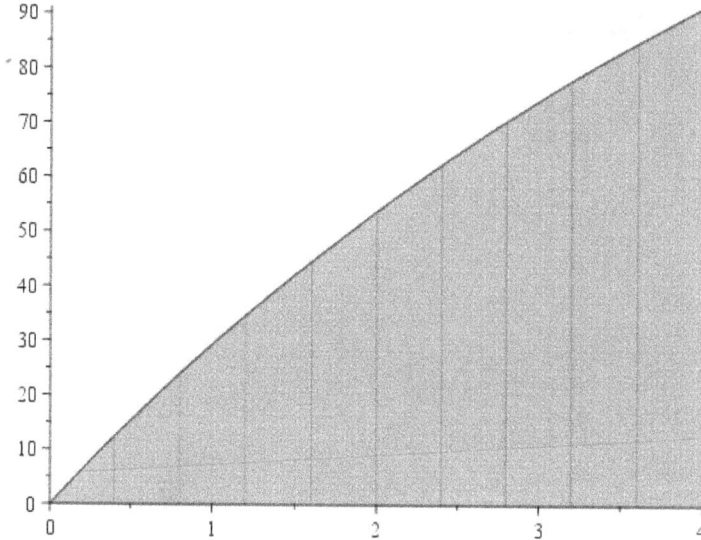

An approximation of $\int_0^4 f(t)\, dt$ using trapezoid rule, where
$f(t) = 177.7778 - 177.7778\ e^{-0.18t}$ and the partition is uniform. The
approximate value of the integral is 203.9808177. Number of subintervals used:
10.

> $ApproximateInt\left(177.7778(1-\exp(-.18 \cdot t)), t = 0..3.96, method = trapezoid,\right.$
> $\left.output = value\right);$

```
200.3493741
```

We check with numerical methods from R from Chapter 6 with
trapz and cumtrapz.

```
n<-51
> x<-seq(0,5,len=n)
> y<-177.777*(1-exp(-.18*x))
> trapz(x,y)
[1] 302.7677
> cumtrapz(x,y)
          [,1]
[1,]   0.0000000
[2,]   0.1585679
```

```
  [3,]     0.6314429
  [4,]     1.4130182
  [5,]     2.4977868
  [6,]     3.8803400
  [7,]     5.5553657
  [8,]     7.5176465
  [9,]     9.7620581
 [10,]    12.2835675
 [11,]    15.0772315
 [12,]    18.1381954
 [13,]    21.4616906
 [14,]    25.0430338
 [15,]    28.8776255
 [16,]    32.9609478
 [17,]    37.2885636
 [18,]    41.8561151
 [19,]    46.6593220
 [20,]    51.6939804
 [21,]    56.9559616
 [22,]    62.4412102
 [23,]    68.1457435
 [24,]    74.0656496
 [25,]    80.1970866
 [26,]    86.5362808
 [27,]    93.0795262
 [28,]    99.8231826
 [29,]   106.7636750
 [30,]   113.8974920
 [31,]   121.2211849
 [32,]   128.7313664
 [33,]   136.4247099
 [34,]   144.2979479
 [35,]   152.3478713
 [36,]   160.5713282
 [37,]   168.9652229
 [38,]   177.5265150
 [39,]   186.2522183
 [40,]   195.1393999
 [41,]   204.1851791
 [42,]   213.3867268
 [43,]   222.7412642
 [44,]   232.2460622
 [45,]   241.8984401
 [46,]   251.6957655
 [47,]   261.6354524
 [48,]   271.7149615
 [49,]   281.9317983
 [50,]   292.2835131
 [51,]   302.7676998
>
```

Again between 3–4 seconds, and we can easily narrow as before to 3.96 seconds to reach 200 feet.

In the second part when the bungee cord acts like a spring, we use Hooke's law, $F = kx$. We continue the solution using the cord as a spring in Section 7.6.

7.6 Revisit Bungee as a Second-Order ODE IVP

Let's do bungee jumping as a second-order ODE with initial conditions We know the length of the cord is 200 feet, and we know the spring coefficient is 0.5 lb/ft (Hooke's law $F = kx$). We can determine that a jumper that weighs 160 pounds will stretch the bungee cord 320 feet (the equilibrium or stationary point. If we position our origin at 520 feet below the bridge (200-foot bungee cord + 320-foot stretch with the jumper's weight) and have down be the positive direction, we can model and obtain the following second-order ODE with new initial conditions.

$5x'' = 160 - .9x' - .5(x + 320)$, with $x(0) = -320$ and $x'(0) = 90.5214$ (the velocity at 200 feet).

We used RK4 method shown in Figure 7.10 (screenshot).

dsolve({ode2, ics1}, numeric, method = classical[rk4], output = array([0, 1, 2, 3, 4, 5, 6, 7, 8, 9, 10, 15, 20, 25, 30, 35, 40, 45, 50, 55, 60, 65, 70, 75, 80, 100, 200]), stepsize = 1);

t	$x(t)$	$\frac{d}{dt}x(t)$
0.	320.	90.5214
1.	386.518518293133	42.8283911015693
2.	407.006155299640	-0.896897798912171
3.	387.184306718624	-37.3398608834392
4.	335.490187372801	-64.3926256508202
5.	261.863904345366	-81.1466087455634
6.	176.598895720402	-87.7774632636120
7.	89.3496212220651	-85.3505800623594
8.	8.36065883408946	-75.5778772914051
9.	-60.0492961883253	-60.5559714248762
10.	-111.743682060565	-42.5125567537891
15.	-114.342664384965	30.0144301390066
20.	37.4123209232077	19.3970343282530
25.	49.1282875414474	-10.8437562486881
30.	-11.7640323408502	-8.64997125766041
35.	-20.8056205257142	3.80234420197770
40.	3.32280629540524	3.78481294476535
45.	8.69455140169078	-1.28048608770667
50.	-0.740617189262695	-1.62913481398827
55.	-3.58789197968448	0.406316217768852
60.	0.0491841826628104	0.691070298425611
65.	1.46256700547694	-0.116699732407725
70.	0.0827262379717796	-0.289239047347827
75.	-0.588981498336091	0.0271433969824741
80.	-0.0750119335547152	0.119533314613274
100.	-0.0255569977123390	0.0196793789452066
200.	-7.75705500287822 × 10⁻⁶	8.14571829093310 × 10⁻⁷

FIGURE 7.10
Screenshot of RK4 solution to the second-order bungee problem.

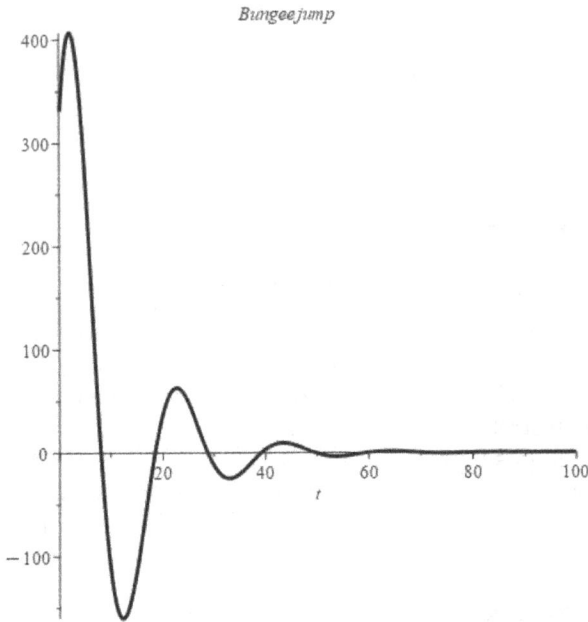

FIGURE 7.11
Plot of bungee jumping.

Our plot of the solution is shown in Figure 7.11 shows the oscillations.

Our analysis shows the jumper oscillates up and down. Since the depth of the gorge is 1053 feet, we never hit the bottom. Do we rebound and hit the bridge? You are asked to solve and interpret this as the Exercises at the end of the chapter.

7.7 Harvesting a Species

Consider harvesting blue crab in Virginia from the Chesapeake Bay. There have been many newspaper reports about the declining populations of blue crabs and the difficulty in harvesting these crabs. Let's model the situation and analyze some "what-ifs."

The basic balance law for harvesting is

$$P'(t) = r(1-\frac{P(t)}{k})P(t) - H(t),$$

where r is the intrinsic rate coefficient (growth rate if $P > 0$) and k represents the carrying capacity (or saturation level).

In the absence of harvesting ($H(t) = 0$), the ODE is autonomous Let $r = 0.3$, $k = 12$, and $P(0) = 5$, and we will solve numerically using Euler's method and obtain a plot of the numerical solution. We want to predict at 400 days.

$$\begin{bmatrix} \begin{bmatrix} t & p(t) \end{bmatrix} \\ \begin{bmatrix} \begin{array}{ll} 0. & 5. \\ 1. & 4.50000000000000 \\ 2. & 4.10625000000000 \\ 3. & 3.78737402343750 \\ 4. & 3.52345332418821 \\ 5. & 3.30117157409959 \\ 6. & 3.11125751941100 \\ 7. & 2.94704087359399 \\ 8. & 2.80359064766800 \\ 9. & 2.67717740086741 \\ 10. & 2.56492486499622 \\ 15. & 2.15106492834448 \\ 20. & 1.88636605706183 \\ 25. & 1.70333349210895 \\ 30. & 1.56998773160613 \end{array} \end{bmatrix} \end{bmatrix}$$

$$\begin{bmatrix} \begin{array}{ll} 35. & 1.46915392065279 \\ 40. & 1.39075888698673 \\ 45. & 1.32849378416310 \\ 50. & 1.27820076204226 \\ 55. & 1.23702553154162 \\ 60. & 1.20294215480541 \\ 65. & 1.17447207211523 \\ 70. & 1.15051052537237 \\ 75. & 1.13021528892347 \\ 80. & 1.11293304897355 \\ 100. & 1.06506226353485 \\ 200. & 1.00487392955600 \\ 300. & 1.00038578638956 \\ 400. & 1.00003066551926 \end{array} \end{bmatrix}$$

```
By Euler's Method, our approximate solution is
P(400)=1.00003066551926.
Compare to the analytical solution which is P(t) = -5./
(4.*exp(-0.02500000000*t) - 5.)
P(400)=1.000036322.
Our absolute error is 5.6654808 x 10 ⁻⁶
```

Now consider light harvesting where `H(t)=0.04` (Assume r and k are the same values as before.)

We solve numerically using the RK4 method with $P(0) = 5$.

We see from our tabulated values that we run out of crabs before time period 10.

$$
\begin{bmatrix}
 & \begin{bmatrix} t & p(t) \end{bmatrix} & \\
0. & 5. & \\
1. & 4.18861631936901 & \\
2. & 3.51465557922920 & \\
3. & 2.93545246693939 & \\
4. & 2.42293896655056 & \\
5. & 1.95758961923879 & \\
6. & 1.52506631852378 & \\
7. & 1.11422679678759 & \\
8. & 0.715854209553363 & \\
9. & 0.321772079138831 & \\
10. & -0.0758492648140658 &
\end{bmatrix}
$$

Exercises

1. Obtain Euler estimates for $y(3)$ using the DE $y' = y + t$ for step sizes of 0.1, 0.05, and 0.01. Compute the percent error for each.

2. Use Euler's method by hand $y' = y + 1$, $y(0) = 1$ with a step size of 0.25 to estimate $y(0.5)$. Then do Euler's method using MAPLE and compare your results to make sure you have the correct solution. Compute the percent error since you can find this exact solution. Then change the step sizes to 0.1 and then 0.05 and use Maple to estimate $y(0.5)$.

3. Consider $v' = 32 - 1.6\,v$, $v(0) = 0$. Using Euler's method in Maple and a step size of 0.05, estimate $v(2)$. What is the terminal velocity? Keep stepping out until you approximate the terminal velocity to three decimal places of accuracy.

4. Try **by hand** $y' = y + 1$, $y(0) = 1$, with step size of 0.25 to estimate $y(0.5)$. Then do the improved Euler's method using Maple and compare your results to make sure you have the correct solution. Compute the percent error since you can find this exact solution.

5. Consider $v' = 32 - 1.6v$, $v(0) = 0$. Using heunform in Maple and a step size of 0.05, estimate $v(2)$. What is the terminal velocity? Keep stepping out until you approximate the terminal velocity to three decimal places of accuracy.

6. Consider $P' = P * (15 - 3 * P)$, $P(0) = 2$.

   Use heunform in Maple and a step size of $h = 1$, $h = 0.5$, and $h = 0.1$.

   Discuss the results as compared to the qualitative solution.

7. Try **by hand** $y' = y + 1$, $y(0) = 1$, with step size of 0.25 to estimate $y(0.5)$. Then do the RK4 method using Maple and compare your results to make sure you have the correct solution. Compute the percent error since you can find this exact solution.

8. Consider $v' = 32 - 1.6 v$, $v(0) = 0$. Using RK4 in Maple and a step size of 0.05, estimate $v(2)$. What is the terminal velocity? Keep stepping out until you approximate the terminal velocity to three decimal places of accuracy.

9. Consider $P' = P * (15 - 3 * P)$, $P(0) = 2$.

   Use RK4 in Maple and a step size of $h = 1$, $h = 0.5$, and $h = 0.1$.

   Discuss the results as compared to the qualitative solution.

10. In the bungee jumper problem, does the jumper rebound and hit the bridge that they jump off?

Projects

1. For the bungee jumper example, do the numerical methods, Euler, improved Euler, and RK4, again using step sizes of 0.5 and then 0.1.

2. Resolve the bungee jumper problems using gravity $= 32.17$ ft/s^2.

7.8 System of ODEs

Consider a friend who recently retired and bought farmland in South Carolina. His desire is to have a fishing pond and his favorite fish to catch are bass and trout. He finds he has a fair size freshwater pond on his land but it contains no fish. He takes a water sample to the local fish and game authority, and it analyzes his water. It concludes that the water can sustain a fish population. He visits the local fish hatcheries, where they provide him the growth rates of bass and trout in isolation; call these values r and s, respectively. The experts tell him that bass and trout have the same food sources in the water and will compete for the oxygen in the water as well as the food for survival. The experts estimate the interactions rates between the bass and trout for survival, and call these rates m and n, respectively. We desire to build a mathematical model to help our friend determine if the pond can sustain both species of fish. This leads to a competitive hunter system of differential equations.

Example 5. Competition between Species

Imagine a small fishpond supporting both trout and bass. Let $T(t)$ denote the population of trout at time t and $B(t)$ denote the population of bass at time t. We want to know if both can coexist in the pond. Although population growth depends on many factors, we will limit ourselves to basic isolated growth and the interaction with the other competing species for the scarce life-support resources.

We assume that the species grow in isolation. The level of the population of the trout or the bass, $B(t)$ and $T(t)$, depend on many variables such as their initial numbers, the amount of competition, the existence of predators, their individual species birth and death rates, and so forth. In isolation, we assume the following proportionality models (following the same arguments as the basic populations models that we have discussed before) to be true where the environment can support an unlimited number of trout and/or bass. Later, we might refine this model to incorporate the limited growth assumptions of the logistics model:

$$\frac{dB}{dt} = mB$$

$$\frac{dT}{dt} = aT.$$

Next, we modify the proceeding differential equations to take into account the competition of the trout and bass for living space, oxygen, and food supply. The effect is that the interaction decreases the growth of the species. The interaction terms for competition led to decay rate that we call n for bass and b for trout. This leads to following simplified model:

$$\frac{dB}{dt} = mB - nBT$$

$$\frac{dT}{dt} = aT - bBT.$$

If we have the initial stockage level, B_0 and T_0, we determine how the species coexist over time.

If the model is not reasonable, we might try logistic growth instead of isolated growth. Logistic growth in isolation was discussed in first-order ODEs models as a refinement.

$B(t)$ = number of bass fish after time t

$T(t)$ = number of trout after time t

Rate of change of growth = rate in isolation + rate in competition for resources

$dB/dt = 0.7B - 0.02B * T$
$dT/dt = 0.5T - 0.01B * T$

Solve $dB/dt = 0$ and $dT/dt = 0$:

$dB/d = 0 = B(.7 - .02T) = 0$, so $B = 0$ or $T = 35$.
$dT/dt = 0 = T(.5 - .01B) = 0$, so $T = 0$ or $B = 50$.

The equilibrium values are $(0, 0)$ and $(50, 35)$.

```
> with(plots):with(DEtools):
> eqn1:=diff(B(t),t)=.7*B(t)-.02*B(t)*T(t);
```

$$eqn1 := \frac{d}{dt}B(t) = 0.7B(t) - 0.02B(t)T(t)$$

```
> eqn2:=diff(T(t),t)=.5*T(t)-.01*B(t)*T(t);
```

$$eqn2 := \frac{d}{dt}T(t) = 0.5T(t) - 0.01B(t)T(t)$$

```
> DEplot([eqn1,eqn2], [B(t),T(t)], t=0..20, B=0..75,
T=0..75);
```

Both equilibrium values are not stable. Depending on the starting values, one species will dominate over time.

Thus, the phase portrait is useful to give us a sense of the possible solutions.

In the previous sections, we discussed the use of numerical solutions (Euler's, improved Euler's, and RK4 methods) to first-order differential equations. In this chapter, we extend the use of numerical solutions to systems of differential equations. We show only Euler's and RK4 methods. Our goal here is to provide a solution method for many models of systems of ODEs that do not have closed form analytical solutions.

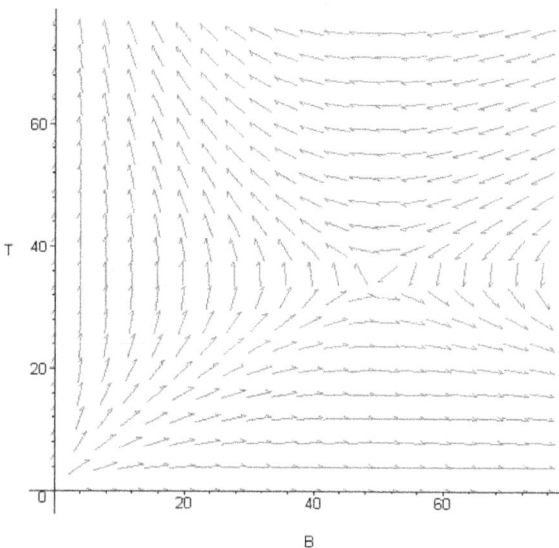

Throughout most of the chapter, we have investigated and modeled autonomous systems of first-order differential equations. A more general form of systems of two ordinary first-order differential equations is given by

$$\frac{dx}{dt} = f(t, x, y)$$

$$\frac{dy}{dt} = g(t, x, y). \tag{7.1}$$

If the variable t appears explicitly in one of the functions f or g, the system is not autonomous. In this section, we present numerical techniques for approximating solutions for $x(t)$ and $y(t)$ subject to initial conditions $x(t_0) = x_0$ and $y(t_0) = y_0$.

We will give the algorithm for each and show the Maple commands to execute a numerical solution. We also show how to obtain both the phase portraits and the plots of approximate numerical solutions.

Euler's method with systems:

Consider the iterative formula for Euler's method for systems as

$$x(n) = x(n-1) + f(t(n-1), x(n-1), y(n-1))\Delta t$$
$$y(n) = y(n-1) + g(t(n-1), x(n-1), y(n-1))\Delta t.$$

We illustrate a few iterations for the following initial value problem with a step size of $\Delta t = 0.1$:

$$x' = 3x - 2y, \ x(0)=3$$
$$y' = 5x - 4y, \ y(0) = 6$$
$$x(0) = 3, \ y(0) = 6 \text{ given}$$

$$x(1) = 3 + (0.1) \cdot (3 \cdot 3 - 2 \cdot 6) = 2.7$$
$$y(1) = 6 + (0.1) \cdot (5 \cdot 3 - 4 \cdot 6) = 5.1$$
and
$$x(2) = 2.7 + (0.1) \cdot (3 \cdot (2.7) - 2 \cdot (5.1)) = 2.49$$
$$y(2) = 5.1 + (0.1) \cdot (5 \cdot (2.7) - 4 \cdot (5.1)) = 4.41$$

and so forth.

In Maple, we enter the system and initial conditions and then use the dsolve with classical numerical methods. Here is the command sequence to obtain the Euler estimates to our example.

> $ode1 := diff(x(t), t) = 3 \cdot x(t) - 2 \cdot y(t);$

$$ode1 := \frac{d}{dt} x(t) - 3x(t) - 2y(t)$$

> $ode2 := diff(y(t), t) = 5 \cdot x(t) - 4 \cdot y(t);$

$$ode2 := \frac{d}{dt} y(t) = 5x(t) - 4y(t)$$

```
>inits := x ( 0 ) = 3, y( 0 ) = 6;
```

$$inits := x\,(0) = 3, y\,(0) = 6$$

```
>eulersol := dsolve( {ode1, ode2, inits}, numeric,
method = classical[ foreuler],
output =array( [0, .1, .2, .3, .4, .5, .6, .7, .8,
.9, 1, 1.1, 1.2,
1.3, 1.4, 1.5, 1.6, 1.7, 1.8, 1.9, 2, 2.1, 2.2] ),
stepsize= 0.1 );
```

$$eulersol := \left[\begin{bmatrix} t & x(t) & y(t) \end{bmatrix}\right], \, [$$

0.	3.	6.
0.1	2.70000000000000016	5.09999999999999964
0.2	2.49000000000000022	4.41000000000000014
0.3	2.35500000000000042	3.89100000000000044
0.4	2.28330000000000056	3.51210000000000022
0.5	2.26587000000000050	3.24891000000000042
0.6	2.29584900000000046	3.08228100000000050
0.7	2.36814750000000052	2.99729310000000070
0.8	2.47913313000000058	2.98244961000000064
0.9	2.62638314700000075	3.02903633100000080
1.	2.80849082490000113	3.13061337210000090
1.1	3.02491539795000097	3.28261343571000142
1.2	3.27586733019300080	3.48202576040100098
1.3	3.56222237717070068	3.72714912133710108
1.4	3.88545926605448954	4.01740066138760987
1.5	4.24761691359331550	4.35317002985981106
1.6	4.65126798169934742	4.73571047471254403
1.7	5.09950628126664274	5.16706027567719950
1.8	5.59594611051119450	5.64998930603964044
1.9	6.14473208245662317	6.18796663887938080
2.	6.75055837941773440	6.78514602455594052
2.1	7.41869668833186857	7.44636680444243292
2.2	8.15503233394294114	8.17716842683139332

The power of Euler's method is twofold. First, it is easy to use, and second, as a numerical method, it can be used to estimate a solution to a system of differential equations that does not have a closed-form solution.

Assume we have the following predator–prey system that does not have a closed-form analytical solution:

$$\frac{dx}{dt} = 3x - xy$$

$$\frac{dy}{dt} = xy - 2y \qquad \cdot$$

$$x(0) = 1, y(0) = 2$$

$$t_0 = 1, \Delta t = .1$$

We will obtain an estimate of the solution using Euler's method:

```
> with(linalg):with(DEtools):
> ode1:=diff(x(t),t)=3*x(t)-x(t)*y(t);
```

$$ode1 := \frac{d}{dt}x(t) = 3x(t) - x(t)y(t)$$

```
> ode2:=diff(y(t),t)=x(t)*y(t)-2*y(t);
```

$$ode2 := \frac{d}{dt}y(t) = x(t)y(t) - 2y(t)$$

```
> inits:=x(0)=1,y(0)=2;
```

$$inits := x(0) = 1, y(0) = 2$$

```
>
> eulersol:=dsolve({ode1,ode2,inits},numeric,
method=classical[foreuler], output=array([0,.1,
.2,.3,.4,.5,.6,.7,.8,.9,1,1.1,1.2,1.3,1.4,1.5,
1.6,1.7,1.8,1.9,2,2.1,2.2]),stepsize=0.1);
```

$$eulersol :=$$

$[t, x(t), y(t)]$		
0.	1.	2.
0.1	1.10000000000000008	1.80000000000000004
0.2	1.23200000000000021	1.63800000000000012
0.3	1.39979840000000010	1.51220160000000026
0.4	1.60806018198425638	1.42143901801574435
0.5	1.86190228798054114	1.36572716301158748
0.6	2.16618792141785876	1.34686678336611476
0.7	2.52428764205455636	1.36925008248155188
0.8	2.93593582846188727	1.44103817219427799
0.9	3.39363701700781206	1.57591009774806334
1.	3.87692143779073328	1.79553476251787370
1.1	4.34388314781755014	2.13254253132470328
1.2	4.72069653578025861	2.63238558144231760
1.3	4.89423614699907182	3.34857781466911942
1.4	4.72363393293951717	4.31773530989456944
1.5	4.10118401049446124	5.49372835024256556
1.6	3.07846012684130698	6.64806176699554552
1.7	1.95541885784630654	7.36502872064382874
1.8	1.10187291030753887	7.33219458140772317
1.9	0.624520125164112150	6.67367032336186838
2.	0.395092020148348210	5.75572040125449292
2.1	0.286215706118782388	4.83198024107766244
2.2	0.233781554289212323	4.00388305652733401

```
> with(plots):
> plot1:=odeplot(eulersol,[t,x(t)],0..10,
color=green,title=`x(t)`):
> plot2:=odeplot(eulersol,[t,y(t)],0..10,color=blue,t
itle=`y(t)`):
> display(plot1,plot2);
```

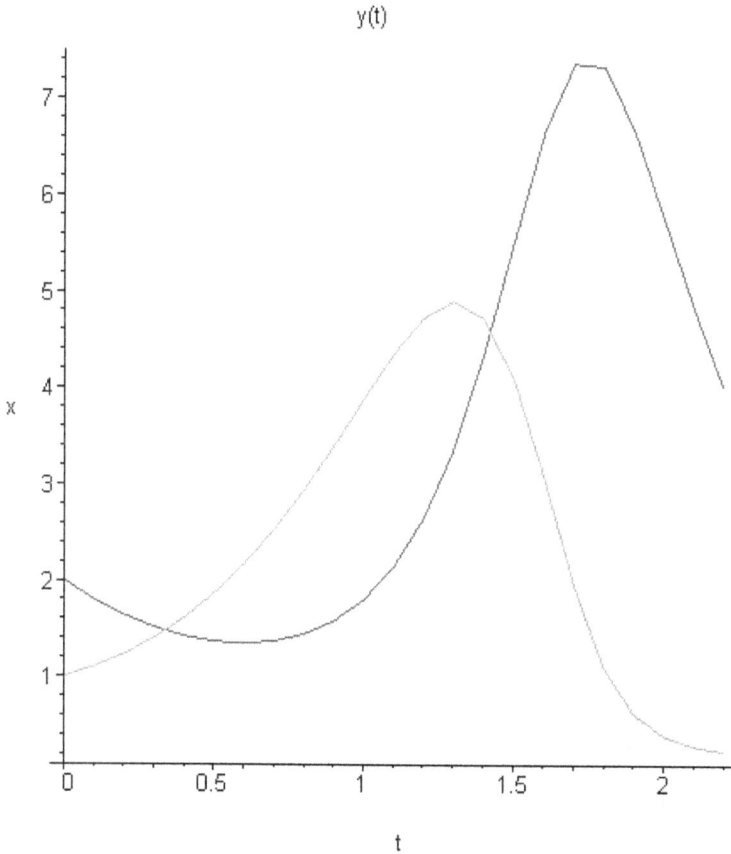

We experiment and find that when we plot $x(t)$ versus $y(t)$, we have approximately a closed loop.

```
> odeplot(eulersol,[x(t),y(t)],0..22, color=green,tit
le=`System`);
```

Another method for numerical estimates is the RK4 applied to systems. We illustrate with the same predator–prey example.

```
> restart;
> with(linalg):with(DEtools):
Warning, the protected names norm and trace have been
redefined and unprotected
```

System

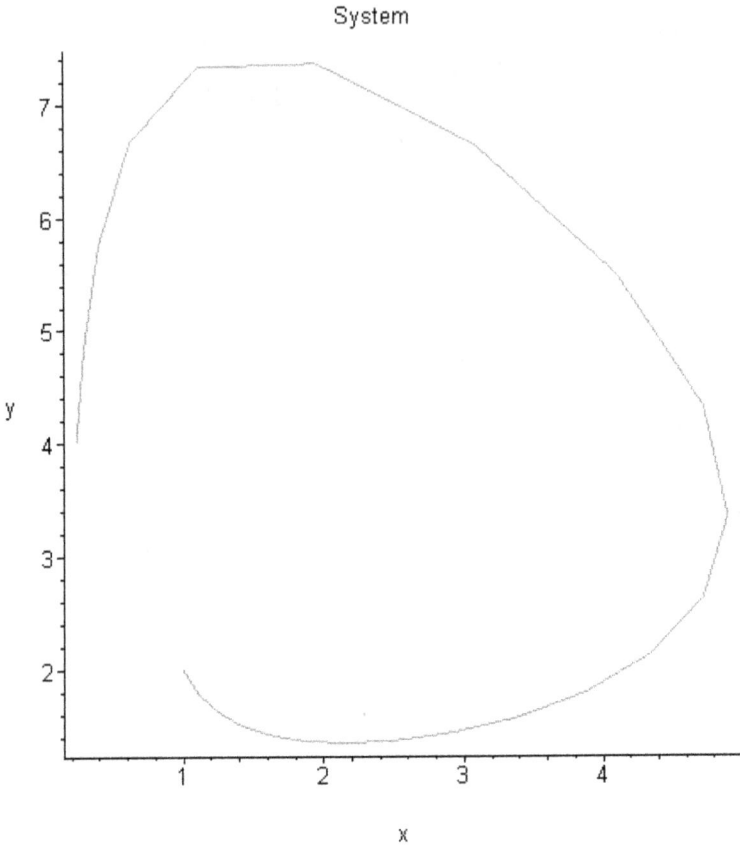

Warning, the previous binding of the name adjoint has been removed and it now has an assigned value
>
> ode1:=diff(x(t),t)=3*x(t)-x(t)*y(t);

$$ode1 := \frac{d}{dt}x(t) = 3x(t) - x(t)y(t)$$

> ode2:=diff(y(t),t)=x(t)*y(t)-2*y(t);

$$ode2 := \frac{d}{dt}y(t) = x(t)y(t) - 2y(t)$$

> inits:=x(0)=1,y(0)=2;

$$inits := x(0) = 1, y(0) = 2$$

```
>
> rk4sol:=dsolve({ode1,ode2,inits},numeric,
method=classical[rk4], output=ar
ray([0,.1,.2,.3,.4,.5,.6,.7,.8,.9,1,1.1,1.2,1.3,1.4,1
.5,1.6,1.7,1.8,1.9,2,2.1,2.2]),stepsize=0.1);
```

$$
rk4sol := \begin{bmatrix} \begin{array}{ccc} & \left[t, x(t), y(t)\right] & \\ 0. & 1. & 2. \\ 0.1 & 1.11554071453956705 & 1.81968188493959970 \\ 0.2 & 1.26463746535620780 & 1.67761981669985926 \\ 0.3 & 1.45146389024689194 & 1.57278931221810914 \\ 0.4 & 1.68031037053259168 & 1.50542579227745321 \\ 0.5 & 1.95456986108324960 & 1.47762563817635396 \\ 0.6 & 2.27500034092209802 & 1.49412980191491540 \\ 0.7 & 2.63687047094120208 & 1.56336399876409616 \\ 0.8 & 3.02561346371586382 & 1.69866644464148696 \\ 0.9 & 3.41107833928229720 & 1.91916110241906601 \\ 1. & 3.74212348522216676 & 2.24852435676575446 \\ 1.1 & 3.94706090545572064 & 2.70772001337807611 \\ 1.2 & 3.95001172191016892 & 3.29658238745704324 \\ 1.3 & 3.70934802583802936 & 3.96638771088000476 \\ 1.4 & 3.25874674267819132 & 4.60727591934640213 \\ 1.5 & 2.70448435236999174 & 5.08419319694337712 \\ 1.6 & 2.16641586882149006 & 5.30768472071020270 \\ 1.7 & 1.71962027608752543 & 5.27274016506954890 \\ 1.8 & 1.38441109626295588 & 5.03717315896791806 \\ 1.9 & 1.14891557150367586 & 4.67751425643222784 \\ 2. & 0.991726731312949417 & 4.25984946993614156 \\ 2.1 & 0.893335577155144112 & 3.83076393278872685 \\ 2.2 & 0.839403883501402824 & 3.41909584046305914 \end{array} \end{bmatrix}
$$

```
> with(plots):
Warning, the name changecoords has been redefined
> plot1:=odeplot(rk4sol,[t,x(t)],0..10,
color=green,title=`x(t)`):
> plot2:=odeplot(rk4sol,[t,y(t)],0..10,color=blue,tit
le=`y(t)`):
> display(plot1,plot2);
```

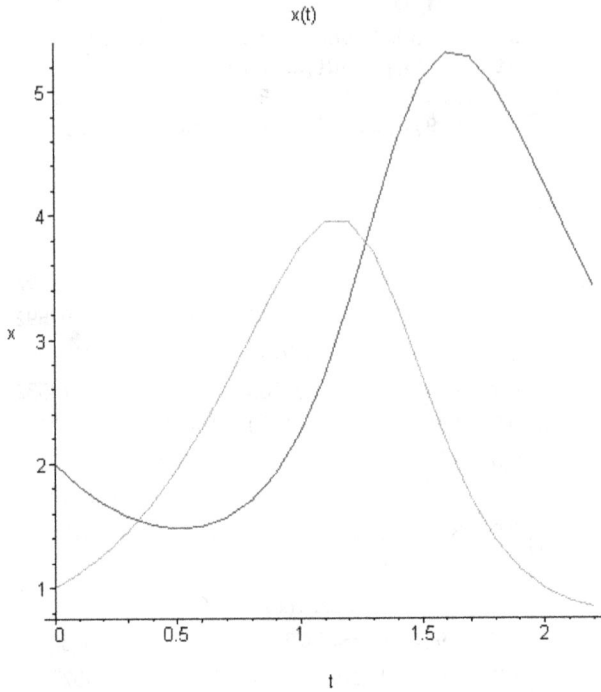

x(t)

```
> odeplot(rk4sol,[x(t),y(t)],0..22, color=green,
title=`System`);
```

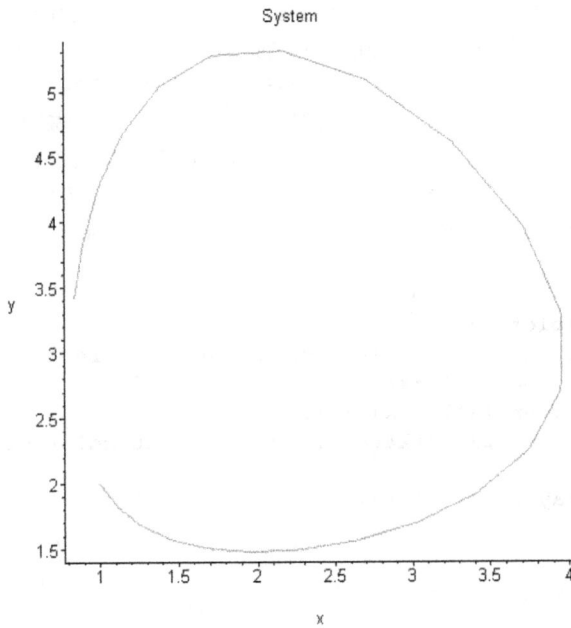

System

Projects

1. Given the following system of linear first-order ODEs of species cooperation (symbiosis):

$$dx_1/dt = -0.5x_1 + x_2,$$
$$dx_2/dt = 0.25x_1 - 0.5x_2,$$

and $x_1(0) = 200$ and $x_2(0) = 500$.

a. Perform Euler's method with step size of $h = 0.1$ to obtain graphs of numerical solutions for $x_1(t)$ and $x_2(t)$ versus t and for x_1 versus x_2. You can put both $x_1(t)$ and $x_2(t)$ versus t on one axis if you want.

b. From the graphs discuss the long-term behavior of the system (discuss stability).

c. Analytically using eigenvalues and eigenvectors solve the system of DEs to determine the population of each species for $t > 0$.

d. Determine if there is a steady state solution for this system.

e. Obtain real plots of $x_1(t)$ and $x_2(t)$ versus t and for $x_1(t)$ versus $x_2(t)$. Compare to the numerical plots. Briefly discuss.

2. A competitive hunter model is defined by the system

$$dx/dt = 15x - x^2 - 2xy = x(15 - x - 2y)$$
$$dy/dt = 12y - y^2 - 1.5xy = y(12 - y - 1.5x).$$

a. Perform a graphical analysis of this competitive hunter model in the x–y plane.

b. Identify all equilibrium points and classify their stability.

c. Find the numerical solutions using Euler's method with step size $h = 0.05$. Try it from two separate initial conditions: First, use $x(0) = 5$ and $y(0) = 4$; then use $x(0) = 3$, $y(0) = 9$. Obtain graphs of $x(t)$, $y(t)$ individually (or on the same axis) and then a plot of x versus y using your numerical approximations. Compare to your phase portrait analysis.

3. Since bass and trout both live in the same lake and eat the same food sources, they are competing for survival. The rates of growth for bass (dB/dt) and for trout (dT/dt) are estimated by the following equations:

$$dB/dt = (10 - B - T)B$$
$$dT/dt = (15 - B - 3T)T$$

Coefficients and values are in thousands.

a. Obtain a "qualitative" graphical solution of this system. Find all equilibrium points of the system and classify each as unstable, stable, or asymptotically stable.

b. If the initial conditions are $B(0) = 5$ and $T(0) = 2$, determine the long term behavior of the system from your graph in part a. Sketch it out.

c. Using Euler's method, $h = 0.1$ and the same initial conditions as earlier, obtain estimates for B and T. Using these estimates, determine a more accurate graph by plotting B versus T for the solution from $t = 0$ to $t = 7$.

Euler's Method:

$$x_{n+1} = x_n + h\,f(x_n, y_n) \text{ and } y_{n+1} = y_n + h\,g(x_n, y_n)$$

d. Compare the graph in part c to the possible solutions found in parts a and b. Briefly comment.

Further Readings

Burden, R. and D. Faires (1997). *Numerical Analysis*. Brooks-Cole, Pacific Grove, CA.

Fox, W. P. (2018). *Mathematical Modeling for Business Analytics*. Taylor and Francis Publishers, Boca Raton, FL.

Fox, W. P. and R. Burks (2021). *Advanced Mathematical Modeling*. Taylor and Francis, CRC Press, Boca Raton, FL.

Fox, W. P. and R. Burks (2022). *Mathematical Modeling Under Change, Uncertainty, and Machine Learning*. Taylor and Francis, CRC, Boca Raton, FL.

Giordano, F., W. Fox and S. Horton (2013). *A First Course in Mathematical Modeling*, 5th ed. Cengage Publishers, Boston, MA.

Zill, D. (2008). *A First Course in Differential Equations: With Modeling Applications*. Cengage Publishers, Boston, MA.

8

Iterative Techniques in Matrix Algebra

In this chapter, we illustrate real-world applications from engineering and the economic sciences that we use numerical methods with linear algebra to solve. We will use iterative methods in this chapter. We will present Gaussian elimination, matrix inversion, eigenvalues and eigenvectors, and the Jacobi and Gauss–Seidel methods. We will present these methods as needed in our scenarios.

First, we discuss the Gauss–Seidel and Jacobi iterative methods.

8.1 The Gauss–Seidel and Jacobi Methods are Both Iterative Methods in Numerical Analysis

Gauss–Seidel Iterative Method

The Gauss–Seidel method is an iterative method to solve a square system of n linear equations. Let $Ax = b$ be a square system of n linear equations, where we define

$$A = \begin{bmatrix} a_{11} & \cdots & a_{1n} \\ \vdots & \ddots & \vdots \\ a_{n1} & \cdots & a_{nn} \end{bmatrix}, x = \begin{bmatrix} x_1 \\ \cdots \\ x_n \end{bmatrix}, b = \begin{bmatrix} b_1 \\ \cdots \\ b_n \end{bmatrix},$$

where matrices **A** and **b** are known and **x** is not known. **L** is the lower triangular matrix of **A** plus **D**, the diagonal matrix of **A**. **U** is the upper triangular matrix of **A**. The Gauss–Seidel method starts at any guess for **x** and iterates to successive approximations using x(new) = **L**$^{-1}$(b − **U**x(old)) or more simply **x(new) = Tx(old) + C**. Furthermore, we define

T = L^{-1}U and C = L^{-1}b.

We created a few other matrices as shown earlier.

Now, let's do an example.

$$A = \begin{bmatrix} 16 & 3 \\ 7 & -11 \end{bmatrix}, b = \begin{bmatrix} 11 \\ 13 \end{bmatrix}, \text{and } x_0 = \begin{bmatrix} 1 \\ 1 \end{bmatrix}$$

DOI: 10.1201/9781032703671-8

$$L = \begin{bmatrix} 16 & 0 \\ 7 & -11 \end{bmatrix}$$

$$U = \begin{bmatrix} 0 & 3 \\ 0 & 0 \end{bmatrix},$$

$$\text{So } L^{-1} = \begin{bmatrix} 0.0625 & 0 \\ 0.0398 & -0.0909 \end{bmatrix}.$$

Therefore,

$$\mathbf{T} = \begin{bmatrix} 0 & -0.1875 \\ 0 & -0.1194 \end{bmatrix}$$

$$\mathbf{C} = \begin{bmatrix} 0.6875 \\ -0.7439 \end{bmatrix}.$$

We compute the new **x**, successively as shown in Figure 8.1

Next, we call the Maple Numerical Analysis package with gaussseidel shown in Figure 8.2

Our appromxiate solution for **x** is 0.8121828486, −0.6649745509.

```
Let's repeat this example using Python.
Code:
import numpy as np
from numpy.linalg import inv
L=[[16, 0], [7, -11]]
print(L)
U=[[0, 3],[0, 0 ]]
x=[[1],[1]]
```

> $x1 := \langle\langle 1, 1 \rangle\rangle;$

$$x1 := \begin{bmatrix} 1 \\ 1 \end{bmatrix}$$

> $NX := evalf(MatrixMatrixMultiply(T1, x1) + C1);$

$$NX := \begin{bmatrix} 0.500000000000000 \\ -0.863636363618182 \end{bmatrix}$$

> $NX1 := evalf(MatrixMatrixMultiply(T1, NX) + C1)$

$$NX1 := \begin{bmatrix} 0.849431818178409 \\ -0.641270661174897 \end{bmatrix}$$

> $NX2 := evalf(MatrixMatrixMultiply(T1, NX1) + C1)$

$$NX2 := \begin{bmatrix} 0.807738248970293 \\ -0.667802932485109 \end{bmatrix}$$

FIGURE 8.1
Screenshot for iterative points.

> $b1 := \text{Vector}([11, 13]);$

$$b1 := \begin{bmatrix} 11 \\ 13 \end{bmatrix}$$

> $evalf(\text{IterativeApproximate}(A1, b1, \text{initialapprox} = \text{Vector}([1, 1]), \text{tolerance} = 10^{-5},$
> $\text{maxiterations} = 40, \text{stoppingcriterion} = \text{relative}(\infty), \text{method} = \text{gaussseidel}, \text{output}$
> $= \text{approximates}))$

$$\left[\begin{bmatrix} 1. \\ 1. \end{bmatrix}, \begin{bmatrix} 0.5000000000 \\ -0.8636363636 \end{bmatrix}, \begin{bmatrix} 0.8494318182 \\ -0.6412706612 \end{bmatrix}, \begin{bmatrix} 0.8077382490 \\ -0.6678029325 \end{bmatrix}, \begin{bmatrix} 0.8127130498 \\ -0.6646371501 \end{bmatrix}, \right.$$

$$\left. \begin{bmatrix} 0.8121194656 \\ -0.6650148855 \end{bmatrix}, \begin{bmatrix} 0.8121902910 \\ -0.6649698148 \end{bmatrix}, \begin{bmatrix} 0.8121818403 \\ -0.6649751926 \end{bmatrix}, \begin{bmatrix} 0.8121828486 \\ -0.6649745509 \end{bmatrix} \right]$$

FIGURE 8.2
Numerical analysis iterations from the Gauss–Seidel model.

```
b=[[11],[13]]
inl=(inv(L))
print(inl)
T=-inl.dot(U)
print(T)
C=inl.dot(b)
print(C)
for i in range(1,10):
 x=T.dot(x)+C
 i=i+1
 print(x)
Output
For 10 iterations, we have
[[ 0.5 ]
 [-0.86363636]]
[[ 0.84943182]
 [-0.64127066]]
[[ 0.80773825]
 [-0.66780293]]
[[ 0.81271305]
 [-0.66463715]]
[[ 0.81211947]
 [-0.66501489]]
[[ 0.81219029]
 [-0.66496981]]
[[ 0.81218184]
 [-0.66497519]]
[[ 0.81218285]
 [-0.66497455]]
[[ 0.81218273]
 [-0.66497463]]
Our approximate solution is 0.81218273, -0.66497463.
```

Jacobi Method

In numerical linear algebra, the **Jacobi method** (aka the **Jacobi iteration method**) is an iterative algorithm for determining the solutions of a strictly diagonally dominant system of linear equations. Each diagonal element is solved for, and an approximate value is substituted back into the update matrix. The process is then iterated until it converges.

The Jacobi method is an iterative method to solve a square system of n linear equation. Let $Ax = b$ be a square system of n linear equation, where we define

$$A = \begin{bmatrix} a_{11} & \cdots & a_{1n} \\ \vdots & \ddots & \vdots \\ a_{n1} & \cdots & a_{nn} \end{bmatrix}, x = \begin{bmatrix} x_1 \\ \cdots \\ x_n \end{bmatrix}, b = \begin{bmatrix} b_1 \\ \cdots \\ b_n \end{bmatrix},$$

where matrices **A** and **b** are known and **x** is not known but is a guess. The Jacobi method starts at any guess for **x** and iterates to successive approximations using

$$\mathbf{x(new)} = \mathbf{D^{-1}(b - (L + U))\ x(old)} = \mathbf{Tx(old) + C,}$$

where

$$\mathbf{T = -D^{-1}(L + U)\ and\ C = D^{-1}b.}$$

We created a few other matrices as shown earlier.
Now, let's do an example.

Example 1. Jacobi Method

$$A = \begin{bmatrix} 2 & 1 \\ 5 & 7 \end{bmatrix}, b = \begin{bmatrix} 11 \\ 13 \end{bmatrix}, \text{and } x_0 = \begin{bmatrix} 1 \\ 1 \end{bmatrix}$$

$$D = \begin{bmatrix} 2 & 0 \\ 0 & 7 \end{bmatrix} \text{ and } D^{-1} = \begin{bmatrix} 1/5 & 0 \\ 0 & 1/7 \end{bmatrix}$$

$$L = \begin{bmatrix} 0 & 0 \\ 5 & 0 \end{bmatrix} \text{ and }$$

$$U = \begin{bmatrix} 0 & 1 \\ 0 & 0 \end{bmatrix},$$

$$\text{So } T = \begin{bmatrix} 0 & -1/2 \\ -5/7 & 0 \end{bmatrix}$$

$$C = \begin{bmatrix} 11/2 \\ 13/7 \end{bmatrix}.$$

> $evalf\left(IterativeApproximate(A, b, initialapprox = Vector([1, 1]), tolerance = 10^{-5},\right.$
> $maxiterations = 40, stoppingcriterion = relative(\infty), method = jacobi, output$
> $\left. = approximates)\right);$

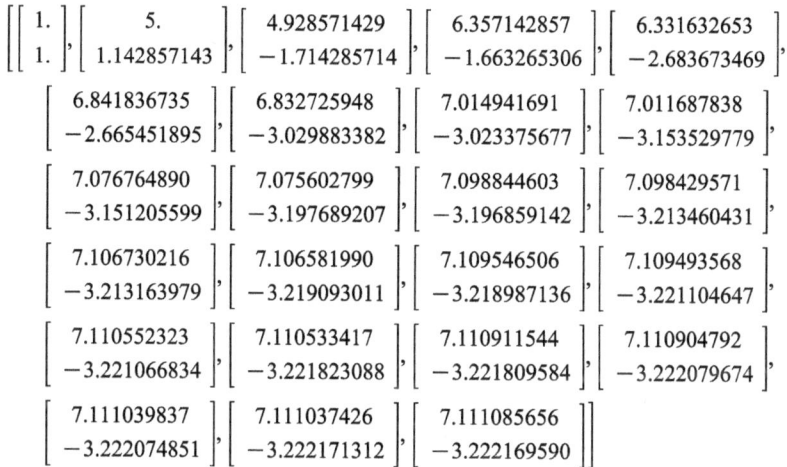

$$\left[\begin{bmatrix} 1. \\ 1. \end{bmatrix}, \begin{bmatrix} 5. \\ 1.142857143 \end{bmatrix}, \begin{bmatrix} 4.928571429 \\ -1.714285714 \end{bmatrix}, \begin{bmatrix} 6.357142857 \\ -1.663265306 \end{bmatrix}, \begin{bmatrix} 6.331632653 \\ -2.683673469 \end{bmatrix},\right.$$

$$\begin{bmatrix} 6.841836735 \\ -2.665451895 \end{bmatrix}, \begin{bmatrix} 6.832725948 \\ -3.029883382 \end{bmatrix}, \begin{bmatrix} 7.014941691 \\ -3.023375677 \end{bmatrix}, \begin{bmatrix} 7.011687838 \\ -3.153529779 \end{bmatrix},$$

$$\begin{bmatrix} 7.076764890 \\ -3.151205599 \end{bmatrix}, \begin{bmatrix} 7.075602799 \\ -3.197689207 \end{bmatrix}, \begin{bmatrix} 7.098844603 \\ -3.196859142 \end{bmatrix}, \begin{bmatrix} 7.098429571 \\ -3.213460431 \end{bmatrix},$$

$$\begin{bmatrix} 7.106730216 \\ -3.213163979 \end{bmatrix}, \begin{bmatrix} 7.106581990 \\ -3.219093011 \end{bmatrix}, \begin{bmatrix} 7.109546506 \\ -3.218987136 \end{bmatrix}, \begin{bmatrix} 7.109493568 \\ -3.221104647 \end{bmatrix},$$

$$\begin{bmatrix} 7.110552323 \\ -3.221066834 \end{bmatrix}, \begin{bmatrix} 7.110533417 \\ -3.221823088 \end{bmatrix}, \begin{bmatrix} 7.110911544 \\ -3.221809584 \end{bmatrix}, \begin{bmatrix} 7.110904792 \\ -3.222079674 \end{bmatrix},$$

$$\left.\begin{bmatrix} 7.111039837 \\ -3.222074851 \end{bmatrix}, \begin{bmatrix} 7.111037426 \\ -3.222171312 \end{bmatrix}, \begin{bmatrix} 7.111085656 \\ -3.222169590 \end{bmatrix}\right]$$

FIGURE 8.3
Screenshot of Jacobi iterative solution from Maple.

We compute the new **x**, successively:

$$x(new) = T \ x(old) + C$$

$$x(new) = \begin{bmatrix} 0 & -1/2 \\ -5/7 & 0 \end{bmatrix}\begin{bmatrix} 1 \\ 1 \end{bmatrix} + \begin{bmatrix} 11/2 \\ 13/7 \end{bmatrix} = \begin{bmatrix} 5 \\ 8/7 \end{bmatrix} \approx \begin{bmatrix} 5 \\ 1.423 \end{bmatrix}$$

Now we repeat, letting **x(old) = x(new)**.

$$x(new) = T \ x(old) + C$$

$$x(new) = \begin{bmatrix} 0 & -1/2 \\ -5/7 & 0 \end{bmatrix}\begin{bmatrix} 5 \\ 8/7 \end{bmatrix} + \begin{bmatrix} 11/2 \\ 13/7 \end{bmatrix} = \begin{bmatrix} 69/14 \\ -12/7 \end{bmatrix} \approx \begin{bmatrix} 4.983 \\ -1.7143 \end{bmatrix}$$

We now use the Numerical Analysis package in Maple to complete our iterations shown in Figure 8.3 in the Jacobi method from the Numerical Analysis package in Maple.

Our approximate solution is vector for x is 7.111085656, −3.222169590.

Using Python
Code
```
from pprint import pprint
from numpy import array, zeros, diag, diagflat, dot
def jacobi(A,b,N=25,x=None):
```

```
"""Solves the equation Ax=b via the Jacobi iterative
method."""
# Create an initial guess if needed
if x is None:
x = zeros(len(A[0]))
# Create a vector of the diagonal elements of A
# and subtract them from A
D = diag(A)
R = A - diagflat(D)
# Iterate for N times
for i in range(N):
x = (b - dot(R,x)) / D
return x
A = array([[2.0,1.0],[5.0,7.0]])
b = array([11.0,13.0])
guess = array([1.0,1.0])
sol = jacobi(A,b,N=25,x=guess)
print( "A:")
pprint(A)
print( "b:")
pprint(b)
print( "x:")
pprint(sol)
Our solution output is
A:
array([[2., 1.],
[5., 7.]])
b:
array([11., 13.])
x:
array([ 7.11110202, -3.22220342])
```

8.2 A Bridge Too Far

Trusses are lightweight structures capable of carrying heavy loads. In civil engineering bridge design, the individual members of the truss are connected with rotatable pin joints that permit forces to be transferred from one member of the truss to another. The accompanying Figure 8.4 shows a truss that is held stationary at the lower left endpoint 1, is permitted to move horizontally at the lower right endpoint 4, and has pin joints at 1, 2, 3, and 4 as shown. A load of 10 kilonewtons (kN) is placed at joint 3, and the forces on the members of the truss have magnitude given by $f_1, f_2, f_3, f_4,$ and f_5, as shown. The stationary support member has both a horizontal force F_1 and a vertical force F_2, but the movable support member has only the vertical force F_3.

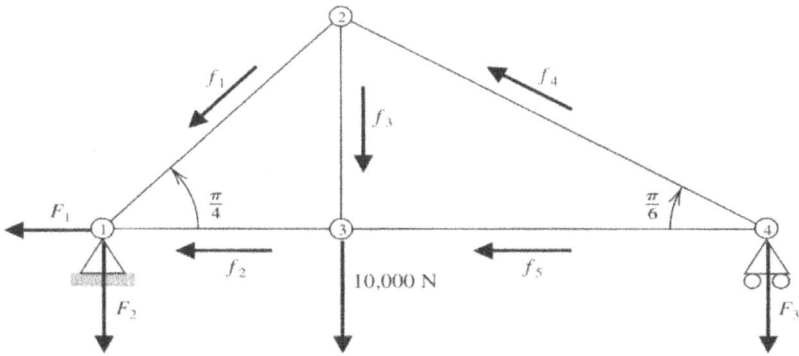

FIGURE 8.4
A bridge too far.

If the truss is in static equilibrium, the forces at each joint must add to the zero vector, so the sum of the horizontal and vertical components of the forces at each joint must be zero. This produces the system of linear equations shown in the accompanying table. According to Newton's law for equilibrium of force, $\Sigma F = 0$. We will examine each joint and calculate both the horizontal and vertical components of the applied forces. Let's begin with joint 1. We find that f_1 and f_2 are forces into joint 1 and that F_1 and F_2 are forces pulling away from joint 1. Force f_1 is acting from an angle of $\pi/4$ radians. It has two components, a horizontal (x) component and a vertical (y) component. Thus, the forces acting on joint 1 are in equilibrium when both the horizontal and vertical components are zero, or

$$-F_1 + \frac{\sqrt{2}}{2} f_1 + f_2 \text{ and } \frac{\sqrt{2}}{2} f_1 - f_2 = 0.$$

We complete the analysis at each joint and place the force vector analysis in the following table. You should verify that these are correct.

Joint	Horizontal Component	Vertical Component
1	$-F_1 + \dfrac{\sqrt{2}}{2} f_1 + f_2 = 0$	$\dfrac{\sqrt{2}}{2} f_1 - F_2 = 0$
2	$-\dfrac{\sqrt{2}}{2} f_1 + \dfrac{\sqrt{3}}{2} f_4 = 0$	$-\dfrac{\sqrt{2}}{2} f_1 - f_3 + \dfrac{1}{2} f_4 = 0$
3	$-f_2 + f_5 = 0$	$f_3 - 10,000 = 0$
4	$-\dfrac{\sqrt{3}}{2} f_4 - f_5 = 0$	$-\dfrac{1}{2} f_4 - F_3 = 0$

We can write this as a system of equations with eight equations and eight unknowns. We will put this in matrix notation, $\mathbf{Ax} = \mathbf{b}$.

-1	0	0	$\frac{\sqrt{2}}{2}$	1	0	0	0
0	-1	0	$\frac{\sqrt{2}}{2}$	0	0	0	0
0	0	0	$-\frac{\sqrt{2}}{2}$	0	0	$\frac{\sqrt{3}}{2}$	0
0	0	0	$-\frac{\sqrt{2}}{2}$	0	-1	0.5	0
0	0	0	0	-1	0	0	1
0	0	0	0	0	1	0	0
0	0	0	0	0	0	$-\frac{\sqrt{3}}{2}$	-1
0	0	-1	0	0	0	0.5	0

$$
\begin{aligned}
F_1 &= 0 \\
F_2 & 0 \\
F_2 & 0 \\
f_1 & 0 \\
f_2 & 0 \\
f_3 & 10{,}000 \\
f_4 & 0 \\
f_5 & 0
\end{aligned}
$$

We desire to interchange rows in an attempt to have an **A** matrix that is more diagonally dominant. To that end, we move the third row to the seventh row, the seventh row to the eighth row, and the eighth row to the third row to obtain the following system:

-1	0	0	$\frac{\sqrt{2}}{2}$	1	0	0	0
0	-1	0	$\frac{\sqrt{2}}{2}$	0	0	0	0
0	0	-1	0	0	0	0.5	0
0	0	0	$-\frac{\sqrt{2}}{2}$	0	-1	0.5	0
0	0	0	0	-1	0	0	1
0	0	0	0	0	1	0	0
0	0	0	$-\frac{\sqrt{2}}{2}$	0	0	$\frac{\sqrt{3}}{2}$	0
0	0	0	0	0	0	$-\frac{\sqrt{3}}{2}$	-1

$$
\begin{aligned}
F_1 &= 0 \\
F_2 & 0 \\
F_2 & 0 \\
f_1 & 0 \\
f_2 & 0 \\
f_3 & 10{,}000 \\
f_4 & 0 \\
f_5 & 0
\end{aligned}
$$

Since this matrix is sparse, we decided to solve it using iterative methods. We approximate the solution of the system to within 10^{-4} using an initial approximation vector of all 0s and the Gauss–Seidel method.

We use the GAUSS-SEIDEL METHOD FOR LINEAR SYSTEMS. The solution vector is

−.00007000,　−23660.26900000,　−13660.26905000,　−33460.66303000, 23660.26900000,　10000.00000000,　−27320.53815000,　23660.26905000, using 38 iterations with tolerance $1.0000000000e^{-04}$ in infinity norm.

We interpret the values as shown in Table 8.1.

More questions about this are contained in the exercises and projects for this chapter.

We did run the Numerical Analysis package for both Gauss–Seidel and Jacobi methods and got different results as you can see.

```
>evalf( IterativeApproximate(B, X, initialapprox = Vector( [0,
0, 0, 0, 0, 0, 0, 0] ), tolerance = 10⁻⁴, maxiterations = 100,
stoppingcriterion = relative( ∞ ), method = gaussseidel) )
```

TABLE 8.1

Forces and Their Values

Force	Values in Newtons
F_1	−0.00007
F_2	−23,660.269
F_3	−13,660.26905
f_1	−33,460.66303
f_2	23,660.269
f_3	10,000.00
f_4	−27,320.53815
f_5	23,660.26905

$$
\begin{bmatrix}
-2.639920000 \\
-23656.64784 \\
-13658.17200 \\
-33457.70771 \\
23656.64784 \\
10000.00000 \\
-27318.10394 \\
23658.17200
\end{bmatrix}
$$

```
>evalf( iterativeApproximate( B, X, initialapprox = Vector
( [0, 0, 0, 0, 0, 0, 0, 0] ), tolerance = 10⁻⁴, maxiterations
= 100, stoppingcriterion = relative( ∞ ), method = jacobi) )
```

$$
\begin{bmatrix}
-4.164080000 \\
-23658.17200 \\
-13658.17200 \\
-33457.70771 \\
23656.64784 \\
10000.00000 \\
-27318.10394 \\
23656.64784
\end{bmatrix}
$$

8.3 The Leontief Input–Output Economic Model

Leontief explained his input–output model in the April 1965 issue of *Scientific American*. Leontief organized the 1958 American economy into an 81×81 matrix. The 81 sectors of the economy, such as steel, agriculture, manufacturing, transportation, and utilities, each represented resources that rely on input from the output of other resources. For example, the production of clothing requires inputs from manufacturing, transportation, agriculture, and other manufacturing. The following is a brief example of the Leontief model and its numerical solution.

Let's consider a production model to produce 1 unit of output of

- petroleum requires 0.2 units of transportation, 0.4 unit of chemicals, and 0.1 unit of itself.
- textiles require 0.4 units of petroleum, 0.1 unit of textiles, 0.15 units of transportation, 0.3 unit of chemicals, and 0.35 units of manufacturing.

- transportation requires 0.6 unit of petroleum, 0.1 unit of itself, and 0.25 units of chemicals.
- chemical requires 0.2 units of petroleum, 0.1 unit of textiles, 0.3 units of transportation, 0.1 unit of manufacturing, and 0.2 units of itself.
- manufacturing requires 0.1 units of petroleum, 0.3 units of transportation, and 0.2 units of itself.

Units are usually measured in dollars. The technology matrix that represents this model is shown in Table 8.2.

If the economy produces 900 million dollars of petroleum, 300 million dollars of textiles, 850 million dollars of transportation, 800 million dollars of chemicals, and 750 million dollars of manufacturing, how much of this production is internally consumed by the economy? We begin by setting up these five equations and five unknowns system in matrix form: $\mathbf{Ax = b}$, shown in figure 8.5.

The Leontief exchange input–output model, or production equation, is

$x =$	$Cx +$	d
Amount Produced	Intermediate Demand	Final Demand

We put x into a matrix by multiplying it by the identity matrix, $\mathbf{I}$. Thus, $\mathbf{Ix - Cx = d}$.

TABLE 8.2

Data for Leontief Model

	Petroleum	Textiles	Transportation	Chemicals	Manufacturing
Petroleum	0.1	0.4	0.6	0.2	0.1
Textiles	0.0	0.1	0.0	0.1	0.0
Transportation	0.2	0.15	0.1	0.3	0.3
Chemicals	0.4	0.3	0.25	0.2	0.0
Manufacturing	0.0	0.35	0.0	0.1	0.2

$$\begin{bmatrix} 0.1 & 0.4 & 0.6 & 0.2 & 0.1 \\ 0 & 0.1 & 0 & 0.1 & 0 \\ 0.2 & 0.15 & 0.1 & 0.3 & 0.3 \\ 0.4 & 0.3 & 0.25 & 0.2 & 0 \\ 0 & 0.35 & 0 & 0.1 & 0.2 \end{bmatrix} \begin{bmatrix} x_1 \\ x_2 \\ x_3 \\ x_4 \\ x_5 \end{bmatrix} = \begin{bmatrix} 900 \\ 300 \\ 850 \\ 800 \\ 750 \end{bmatrix}$$

FIGURE 8.5
Leontief matrix.

To be in equilibrium (steady state) we need to set up $(\mathbf{I} - \mathbf{C})\mathbf{x} = \mathbf{d}$, where $\mathbf{I}$ is the identity matrix. So we subtract our matrix from the 5×5 identity matrix and obtain the new system of equations that we will need to solve:

$$\begin{bmatrix} 0.9 & -0.4 & -0.6 & -0.2 & -0.1 \\ 0 & 0.9 & 0 & -0.1 & 0 \\ -0.2 & -0.15 & 0.9 & -0.3 & -0.3 \\ -0.4 & -0.3 & -0.25 & 0.8 & 0 \\ 0 & -0.35 & 0 & -0.1 & 0.8 \end{bmatrix} \begin{bmatrix} x_1 \\ x_2 \\ x_3 \\ x_4 \\ x_5 \end{bmatrix} = \begin{bmatrix} 900 \\ 300 \\ 850 \\ 800 \\ 750 \end{bmatrix}.$$

We have chosen to solve this system with only an iterative method, but we present the direct method's results.

We use Maple to input the augmented matrix and solve the following:

```
> C:=<<.9,0,-.2,-.4,0>|<-.4,.9,-.15,-.3,-.35>|<-.6,0,.9,-
.25,0>|<-.2,-.1,-.3,.8,-.1>|<-.1,0,-.3,0,.8>|<900,300,850,
800,750>>;
```

$$C := \begin{bmatrix} 0.9 & -0.4 & -0.6 & -0.2 & -0.1 & 900 \\ 0 & 0.9 & 0 & -0.1 & 0 & 300 \\ -0.2 & -0.15 & 0.9 & -0.3 & -0.3 & 850 \\ -0.4 & -0.3 & -0.25 & 0.8 & 0 & 800 \\ 0 & -0.35 & 0 & -0.1 & 0.8 & 750 \end{bmatrix}$$

```
CC:=ReducedRowEchelonForm(C);
```

$$CC := \begin{bmatrix} 1 & 0 & 0 & 0 & 0 & 6944.2 \\ 0 & 1 & 0 & 0 & 0 & 1070.0 \\ 0 & 0 & 1 & 0 & 0 & 5620.6 \\ 0 & 0 & 0 & 1 & 0 & 6629.7 \\ 0 & 0 & 0 & 0 & 1 & 2234.4 \end{bmatrix}$$

We interpret the unique solution as we need to produce the following amounts of petroleum, textiles, transportation, chemicals, and manufacturing:

$$x_1 = 6944.2,\ x_2 = 1070,\ x_3 = 5620.6,\ x_4 = 6629.7,\ \text{and}\ x_5 = 2234.4.$$

We also use Gauss–Seidel and Jacobi methods for comparison in Table 8.3. Notice the differences in the methods used.

```
IterativeApproximate( C1, B, initialapprox = Vector( [0., 0.,
0., 0., 0] ), tolerance = 10⁻³, maxiterations = 100,
stoppingcriterion = relative( ∞ ), method = jacobi)
```

TABLE 8.3

Leontief Solution by Method Chosen

Variable	Direct Method	Jacobi Method	Gauss–Seidel Method
x_1	6944.2	6917.414206	6932.338102
x_2	1070	1066.45697	1068.534157
x_3	5620.6	5599.719474	5612.590107
x_4	6629.7	6603.845935	6620.803769
x_5	2234.4	2228.498681	2232.584165

$$\begin{bmatrix} 6917.414206 \\ 1066.456970 \\ 5599.719474 \\ 6603.845935 \\ 2228.498681 \end{bmatrix}$$

```
IterativeApproximate( C1, B, initialapprox = Vector( [0., 0.,
0., 0., 0] ), tolerance = 10⁻³, maxiterations = 100,
stoppingcriterion = relative( ∞ ), method = gaussseidel)
```

$$\begin{bmatrix} 6932.338102 \\ 1068.534157 \\ 5612.590107 \\ 6620.803769 \\ 2232.584165 \end{bmatrix}$$

8.4 Markov Chains with Eigenvalues and Eigenvectors (Optional)

We suggest students have a background in basic linear algebra and systems of discrete dynamical systems for this section.

Consider the following scenario. In Southern California, a new car rental company desires three rental/return locations, location 1 (the airport in Los Angeles, LAX), location 2 (downtown Los Angeles), and location 3 (San Diego). The car rental company will have a fleet of 1000 cars. Historical information from other companies indicated the following information:

- For cars rented at LAX, 80% are returned to LAX, 10% are returned downtown Los Angeles, and 10% are returned to San Diego.

- For cars rented downtown, 30% are returned to LAX, 20% are returned back to the downtown Los Angeles location, and 50% are returned to San Diego.
- For cars rented in San Diego, 20% are returned to LAX, 60% to downtown Los Angeles, and only 20% returned to San Diego.

The company desires to know how large the car lots need to beat each potential site based on this historical data. We write this out as a discrete dynamical system (DDS) first. We define the following:

$c_1(n)$ = the number of cars at location 1 (LAX) after time n

$c_2(n)$ = the number of cars at location 2 (downtown) after time n

$c_3(n)$ = the number of cars at location 3 (San Diego) after time n

Recall from Chapter 2 we use the paradigm of

"Future = Present + Change" to model our DDS.

$$c_1(n + 1) = 0.8\, c_1(n) + 0.3\, c_2(n) + 0.2 c_3(n)$$
$$c_2(n + 1) = 0.1\, c_1(n) + 0.2\, c_2(n) + 0.6 c_3(n)$$
$$c_3(n + 1) = 0.1\, c_1(n) + 0.5\, c_2(n) + 0.2 c_3(n)$$

The initial conditions are $c_1(0) = 1$, $c_2(0) = 0$, $c_3(0) = 0$ measured in percentages, where $1 = 100\%$.

We will solve this problem as a system of DDSs using eigenvalues and eigenvectors. We will use the power method for finding the largest (most dominant) eigenvalue and its associated eigenvector. In discrete dynamical systems in matrix form, the solution for any value k is

$$A(k) = c_1\, (\lambda_1)^k A_1 + c_2\, (\lambda_2)^k A_2 + c_3\, (\lambda_3)^k A_3 + \ldots. + c_n (1_n)^k A_n,$$

where $\lambda_1, \lambda_2, \lambda_3, \ldots, \lambda_n$ are the eigenvalues values of R and $A_1, A_2, A_3, \ldots, A_n$ are their corresponding eigenvectors. To get the long-term steady state, we take the limit as k goes to infinity. This makes only the dominant eigenvalue and its associated eigenvector as essential.

As we will show, we only need the dominant eigenvalue and its associated eigenvector since the other eigenvalues in a Markov chain will be between $[0, 1)$.

Python Power Method

In the power method, the basic computation is summarized as

$$u_k = \frac{Au_{k-1}}{\|Au_{k-1}\|} \text{ and } \lim_{k \to \infty} u_k = \varnothing_1.$$

This equation can be written as

$$Au_{k-1} = \lambda_1 u_{k-1} \to \lambda_1 = \frac{\|Au_{k-1}\|}{\|u_{k-1}\|}$$

The first equation is more or less the basic computation of the power method.
A is the matrix.

The $u(k-1)$ is the initial guess eigenvector.

A and $u(k-1)$ are multiplied using matrix algebra, and then the resulting vector is divided by the maximum value ($\|A\ u(k-1)\|$) of that resulting vector. This maximum value is the corresponding eigenvalue. The term $\|A\ u(k-1)\|$ is the norm of the product of the matrix, A and vector $u(k-1)$.

Here is our Python code:

```
import numpy as numpy
A = numpy.array([[0.8,.3,.2],
[.1,.2,.6],
[0.1,.5,.2]],float)
u = numpy.array([[1],[1],[1]], float)
n = 100
eigenvalue = 0
for i in range(n):
u = A@u
eigenvalue = numpy.max(u)
u = u/eigenvalue
print('eigenvector: \n', u)
print('\neigenvalue =', eigenvalue)
```

This is the code for the power method where A is a 3 × 3 matrix and u is the initial trial 3 × 1 eigenvector.

  n is the number of iterations;

  eigenvalue is the eigenvalue initially set to 0.

The next line is the for the loop that will iterate 100 times. The block of code inside the loop is the formula of the power method:

  First A is multiplied (@) by u, and that becomes a new 3 × 1 vector.

  The eigenvalue is the maximum number of that vector.

  Then the new eigenvector is set by dividing itself by the eigenvalue.

  This process will repeat 100 times.

  The two print statements print out these results:

```
[[0.8 0.3 0.2]
 [0.1 0.2 0.6]
 [0.1 0.5 0.2]]
eigenvector:
 [[1. ]
```

```
    [0.41176471]
    [0.38235294]]
neigenvalue = 1.0
```

The magnitude of the vector is 1.147057.

We divide each element of the vector by this magnitude value of 1.147057 to obtain the eigenvector that we will use in our model, [0.8717948, .3589744, 0.333333].

The other eigenvalues and eigenvectors are listed here with this eigenvalue of 1 and eigenvector (0.8717948, 0.3589744, 0.333333):

```
E-value:  [ 1.  0.5472136 -0.3472136]
E-vector [[ 0.87179487 0.81607429 0.07844683]
  [ 0.35897436 -0.43077627 -0.74305903]
  [ 0.33333333 -0.38529801 0.6646122 ]]
```

As a reminder the characteristic polynomial is found by setting the determinant of $A - lI = 0$.

After finding each of the eigenvalues, l_i, we substitute into $(A - \lambda_i I)x_i = 0$ for the eigenvectors. We will normalize the eigenvectors when we use them.

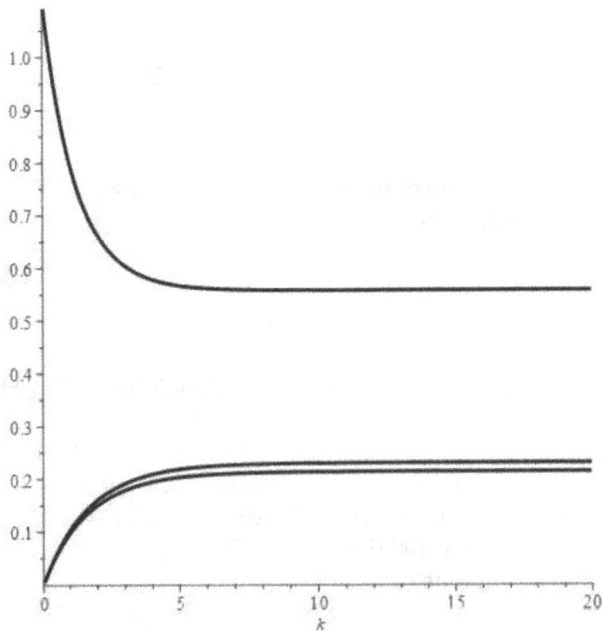

FIGURE 8.6
Long-term behavior of rental cars.

Putting the λ_i and their corresponding vectors x_i back into the DDS back into our DDS solution model, we get the following:

```
A₁(k)=c1*(1ᵏ)*.81717948+c2*(.54721359549958ᵏ)*(.81607428416248)
+c3(-.347213595499958ᵏ)(0.078446828865)
A₂(k)=c1*(1ᵏ)*.358974358974359+c2*(.54721359549958ᵏ)*
(-.430776274376376)+c3(-.347213595499958ᵏ)(-.7434059028252262)
A₃(k)  =  c1*(1ᵏ)*.33333333333334+c2*(.54721359549958ᵏ)
(-.385298012029871ᵏ)+c3*(-.347213595499958ᵏ)(-.664612199386798)
```

Solving for $c1$, $c2$, and $c3$, we get $c1 = 0.6393475128$, $c2 = 0.5429468546$, and $c3 = -0.0058971059$.

We only care about the eigenvector with the dominant eigenvalues in our long-term behavior. We take the limit as $k \to \infty$; we can see that all eigenvalue terms less than 1 raised to the k power will go to zero. This makes the steady state solution 0.5573770494, 0.2295081968, and 0.213147542. We provide a graphical representation for these steady-state solutions. We plot the long-term behavior of rental cars in Figure 8.6.

If we started with 10,000 cars then the recommended lots sizes should be approximately 5574 at location 1, 2295 at location 2, and 2131 at location 3. As a Markov chain, we could raise the matrix, A, to a large power (we used A^{50}). The steady-state solution is 0.5573770494, 0.2295081968, and 0.213147542 as we found earlier.

Exercises

1. A city is served by two cable TV companies, BestDarnTV (BDT) and CableCoolCast (CCC).

   - Due to their aggressive sales tactics, each year 40% of BDT switch to CCCC; the other 60% of BDT customers stay with BDT.
   - However, 30% of the CCC customers switch to BDT.

   The two states are BDT and CCC.

   a. Express the information above as a matrix which displays the probabilities of going from one state into another state.
   b. Write as a system of discrete dynamical systems.
   c. If the solution is with λ_1 and λ_2 being eigenvalues and $\mathbf{A}_1$ and $\mathbf{A}_2$ being corresponding eigenvectors, $A(k) = c_1(\lambda_1)_k\mathbf{A}_1 + c_2(\lambda_2)_k\mathbf{A}_2$
      Solve for the eigenvalues and eigenvectors for this matrix.
   d. Determine the long-term behavior (steady-state probabilities).

8.5 Cubic Splines with Matrices

Natural Cubic Splines as a System of Equations

In modelling with natural cubic splines, we want to fit a third-order polynomial between every pair of data points. Let's assume that we have six data pairs:

t	7	14	21	28	35	42
y	125	275	800	1200	1700	1650

We will need five third-order equations.

$$[7-14], S_1 = a_3x^3 + a_2x^2 + a_1x + a_0$$
$$[14-21], S_2 = b_3x^3 + b_2x^2 + b_1x + b_0$$
$$[21-28], S_3 = c_3x^3 + c_2x^2 + c_1x + c_0$$
$$[28-35], S_4 = d_3x^3 + d_2x^2 + d_1x + d_0$$
$$[35-42], S_5 = e_3x^3 + e_2x^2 + e_1x + e_0$$

We note that there are 20 unknowns ($a_3, a_2, \ldots, e_1, e_0$) and that we need 20 equations to uniquely solve for the these unknowns.

By substituting in the (x, y) data pairs, we obtain 10 equations. We still lack 10 equations.

We force both the first derivative (slope) and second derivatives (concavity) at the interior points to match. For $i > 1$ and less than n,

$$\frac{dS_{i+1}}{dx} = \frac{dS_{i+2}}{dx}$$
$$\frac{d^2S_{i+1}}{dx^2} = \frac{d^2S_{i+2}}{dx^2}$$

Since there are four interior points this gives eight more equations. Matching the derivatives ensures the smoothness of the curves. The last two equations concern the endpoints. Under natural cubic spline, we want the slope at the endpoints to be constants that forces the second derivatives to equal zero. This yields two more equations, and we have 20 equations. Note that if we had clamped cubic splines, then the first derivatives at the two endpoints would equal specific constants, and again, we would have 20 equations.

The 20 equations follow:

(21) $343\ a3 + 49\ a2 + 7\ a1 + a0 = 125$
(22) $2744\ a3 + 196\ a2 + 14\ a1 + a0 = 275$
(23) $2744\ b3 + 196\ b2 + 14\ b1 + b0 = 275$
(24) $9621\ b3 + 441\ b2 + 21\ b1 + b0 = 800$
(25) $9621\ c3 + 441\ c2 + 21\ c1 + c0 = 800$
(26) $21952\ c3 + 784\ c2 + 28\ c1 + c0 = 1200$
(27) $21952\ d3 + 784\ d2 + 28\ d1 + d0 = 1200$
(28) $42875\ d3 + 1225\ d2 + 35\ d1 + d0 = 1700$
(29) $42875\ e3 + 1225\ e2 + 35\ e1 + e0 = 1700$
(30) $74088\ e3 + 1764\ e2 + 42\ e1 + e0 = 1650$
(31) $588\ a3 + 28\ a2 + a1 = 588\ b3 + 28\ b2 + b1$
(32) $1323\ b3 + 42\ b2 + b1 = 1323\ c3 + 42\ c2 + c1$
(33) $2352\ c3 + 56\ c2 + c1 = 2352\ d3 + 56\ d2 + d1$
(34) $3675\ d3 + 70\ d2 + d1 = 3675\ e3 + 70\ e2 + e1$
(35) $84\ a3 + 2\ a2 = 84\ b3 + 2\ b2$
(36) $126\ b3 + 2\ b2 = 126\ c3 + 2\ c2$
(37) $168\ c3 + 2\ c2 = 168\ d3 + 2\ d2$
(38) $210\ d3 + 2\ d2 = 210\ e3 + 2\ e2$
(39) $42\ a3 + 2\ a2 = 0$
(40) $252\ e3 + 2\ e2 = 0$

We use Maple for our spline model.

```
> digits := 50;

digits := 50
> interface(rtablesize = infinity);
∞
```

$B :=$

343	49	7	1	0	0	0	0	0	0	0	0	0	0	0	0	0	0	0	0	125
2744	196	14	1	0	0	0	0	0	0	0	0	0	0	0	0	0	0	0	0	275
0	0	0	0	2744	196	14	1	0	0	0	0	0	0	0	0	0	0	0	0	275
0	0	0	0	9261	441	21	1	0	0	0	0	0	0	0	0	0	0	0	0	800
0	0	0	0	0	0	0	0	9261	441	21	1	0	0	0	0	0	0	0	0	800
0	0	0	0	0	0	0	0	21952	784	28	1	0	0	0	0	0	0	0	0	1200
0	0	0	0	0	0	0	0	0	0	0	0	21952	784	28	1	0	0	0	0	1200
0	0	0	0	0	0	0	0	0	0	0	0	42875	1225	35	1	0	0	0	0	1700
0	0	0	0	0	0	0	0	0	0	0	0	0	0	0	0	42875	1225	35	1	1700
0	0	0	0	0	0	0	0	0	0	0	0	0	0	0	0	74088	1764	42	1	1650
588	28	1	0	-588	-28	-1	0	0	0	0	0	0	0	0	0	0	0	0	0	0
0	0	0	0	1323	42	1	-1323	-42	-1	0	0	0	0	0	0	0	0	0	0	0
0	0	0	0	0	0	0	0	2352	56	1	0	-2352	-56	-1	0	0	0	0	0	0
0	0	0	0	0	0	0	0	0	0	0	0	3765	70	1	0	-3765	-70	-1	0	0
84	2	0	0	-84	-2	0	0	0	0	0	0	0	0	0	0	0	0	0	0	0
0	0	0	0	126	2	0	0	-126	-2	0	0	0	0	0	0	0	0	0	0	0
0	0	0	0	0	0	0	0	168	2	0	0	-168	-2	0	0	0	0	0	0	0
0	0	0	0	0	0	0	0	0	0	0	0	210	2	0	0	-210	-2	0	0	0
42	2	0	0	0	0	0	0	0	0	0	0	0	0	0	0	0	0	0	0	0
0	0	0	0	0	0	0	0	0	0	0	0	0	0	0	0	252	2	0	0	0

```
C:=convert(ReducedRowEchelonForm(B),float);
```

$$
B := \begin{bmatrix}
1 & 0 & 0 & 0 & 0 & 0 & 0 & 0 & 0 & 0 & 0 & 0 & 0 & 0 & 0 & 0 & 0 & 0 & 0 & 0 & 0.3285032886 \\
0 & 1 & 0 & 0 & 0 & 0 & 0 & 0 & 0 & 0 & 0 & 0 & 0 & 0 & 0 & 0 & 0 & 0 & 0 & 0 & -6.898569060 \\
0 & 0 & 1 & 0 & 0 & 0 & 0 & 0 & 0 & 0 & 0 & 0 & 0 & 0 & 0 & 0 & 0 & 0 & 0 & 0 & 53.62189371 \\
0 & 0 & 0 & 1 & 0 & 0 & 0 & 0 & 0 & 0 & 0 & 0 & 0 & 0 & 0 & 0 & 0 & 0 & 0 & 0 & -25. \\
0 & 0 & 0 & 0 & 1 & 0 & 0 & 0 & 0 & 0 & 0 & 0 & 0 & 0 & 0 & 0 & 0 & 0 & 0 & 0 & -0.549221982 \\
0 & 0 & 0 & 0 & 0 & 1 & 0 & 0 & 0 & 0 & 0 & 0 & 0 & 0 & 0 & 0 & 0 & 0 & 0 & 0 & 29.96589232 \\
0 & 0 & 0 & 0 & 0 & 0 & 1 & 0 & 0 & 0 & 0 & 0 & 0 & 0 & 0 & 0 & 0 & 0 & 0 & 0 & -462.4805656 \\
0 & 0 & 0 & 0 & 0 & 0 & 0 & 1 & 0 & 0 & 0 & 0 & 0 & 0 & 0 & 0 & 0 & 0 & 0 & 0 & 2383.478143 \\
0 & 0 & 0 & 0 & 0 & 0 & 0 & 0 & 1 & 0 & 0 & 0 & 0 & 0 & 0 & 0 & 0 & 0 & 0 & 0 & 0.410658693 \\
0 & 0 & 0 & 0 & 0 & 0 & 0 & 0 & 0 & 1 & 0 & 0 & 0 & 0 & 0 & 0 & 0 & 0 & 0 & 0 & -30.50659023 \\
0 & 0 & 0 & 0 & 0 & 0 & 0 & 0 & 0 & 0 & 1 & 0 & 0 & 0 & 0 & 0 & 0 & 0 & 0 & 0 & 807.4415678 \\
0 & 0 & 0 & 0 & 0 & 0 & 0 & 0 & 0 & 0 & 0 & 1 & 0 & 0 & 0 & 0 & 0 & 0 & 0 & 0 & -6505.97679 \\
0 & 0 & 0 & 0 & 0 & 0 & 0 & 0 & 0 & 0 & 0 & 0 & 1 & 0 & 0 & 0 & 0 & 0 & 0 & 0 & -0.4373436113 \\
0 & 0 & 0 & 0 & 0 & 0 & 0 & 0 & 0 & 0 & 0 & 0 & 0 & 1 & 0 & 0 & 0 & 0 & 0 & 0 & 40.7333735 \\
0 & 0 & 0 & 0 & 0 & 0 & 0 & 0 & 0 & 0 & 0 & 0 & 0 & 0 & 1 & 0 & 0 & 0 & 0 & 0 & -1187.277417 \\
0 & 0 & 0 & 0 & 0 & 0 & 0 & 0 & 0 & 0 & 0 & 0 & 0 & 0 & 0 & 1 & 0 & 0 & 0 & 0 & 12111.4004 \\
0 & 0 & 0 & 0 & 0 & 0 & 0 & 0 & 0 & 0 & 0 & 0 & 0 & 0 & 0 & 0 & 1 & 0 & 0 & 0 & 0.247496114 \\
0 & 0 & 0 & 0 & 0 & 0 & 0 & 0 & 0 & 0 & 0 & 0 & 0 & 0 & 0 & 0 & 0 & 1 & 0 & 0 & -31.18451036 \\
0 & 0 & 0 & 0 & 0 & 0 & 0 & 0 & 0 & 0 & 0 & 0 & 0 & 0 & 0 & 0 & 0 & 0 & 1 & 0 & 1290.479268 \\
0 & 0 & 0 & 0 & 0 & 0 & 0 & 0 & 0 & 0 & 0 & 0 & 0 & 0 & 0 & 0 & 0 & 0 & 0 & 1 & -15877.14509
\end{bmatrix}
$$

```
> xdata := [7, 14, 21, 28, 35, 42];
xdata :=[7, 14, 21, 28, 35, 42]
> ydata := [125, 275, 800, 1200, 1700, 1650];
ydata :=[125, 275, 800, 1200, 1700, 1650]
> p1 := pointplot( { seq( [xdata[i], ydata[i], i = 1 ..6) } ) :
```

$>$ $p2 := plot\left(C[1,21]\cdot x^3 + C[3,21]\cdot x + C[4,21], x = 7..14\right):$

$>$ $p3 := plot\left(C[5,21]\cdot x^3 + C[6,21]\cdot x^2 + C[7,21]\cdot x, +C[8,21], x = 14..21\right):$

$>$ $p4 := plot\left(C[9,21]\cdot x^3 + C[10,21]\cdot x^2 + C[11,21]\cdot x, +C[12,21], x = 21..28\right):$

$>$ $p5 := plot\left(C[13,21]\cdot x^3 + C[14,21]\cdot x^2 + C[15,21]\cdot x + C[16,21], x = 28..35\right):$

$>$ $p6 := plot\left(C[17,21]\cdot x^3 + C[18,21]\cdot x^2 + C[19,21]\cdot x + C[20,21], x = 35..42\right):$

$>$ $display\left(\{p1, p2, p3, p4, p5, p6\}\right);$

Figure 8.7 shows the cubic spline data and the cubic spline model.

We note that in general, with N pairs of data, we will need $N-1$ cubic equations. To set up the system of equations, we will need

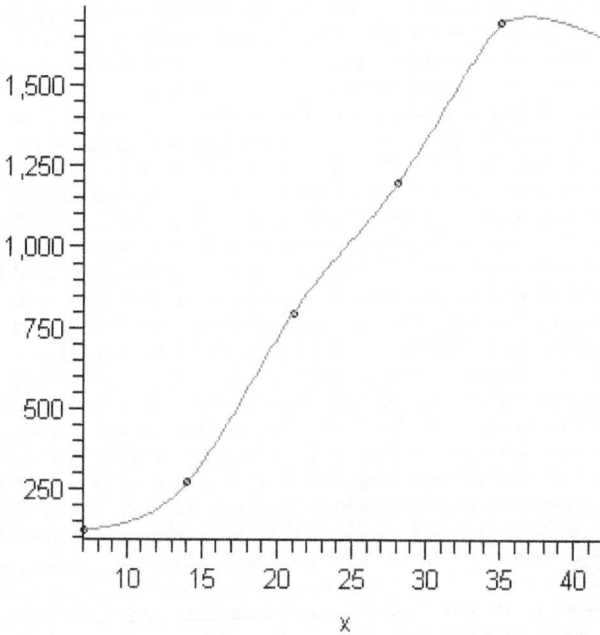

FIGURE 8.7
Cubic spline model.

1. $2N - 2$ cubic equations (2 for each successive pairs of points).
2. for each interior point $(N - 2)$ points, we obtain 2 equations: One equation is the first derivative being set equal at each interior point, and the second is the second derivatives being set equal at each interior point. This yields another $2(N - 2)$ equations.
3. finally, if we are need clamped cubic splines then we use the fact that the first derivatives at the two endpoints equal specific constants to obtain the last two equations. If we need natural cubic splines, we assume the slopes are unknown but constants, then we make the second derivatives equal to zero at the two endpoints to obtain the last two equations.

Exercises

1. A Bridge Too Far (revisited). Set up a model and solve if the angles at joint 1 and joint 4 are both $\frac{\pi}{4}$.
2. A Bridge Too Far (revisited). Set up a model and solve if the angles at joint 1 and joint 4 are both $\frac{\pi}{6}$.

3. A Bridge Too Far (revisited). Set up a model and solve if the angles at joint 1 and joint 4 are both $\frac{\pi}{3}$.

4. Consider the Leontief model in Example 2.

   a. Determine the solution with the following technology matrix.

	Petroleum	Textiles	Transportation	Chemicals	Manufacturing
Petroleum	0.2	0.3	0.6	0.2	0.1
Textiles	0.0	0.2	0.0	0.1	0.0
Transportation	0.2	0.15	0.2	0.3	0.3
Chemicals	0.4	0.35	0.25	0.2	0.0
Manufacturing	0.0	0.35	0.0	0.1	0.2

   b. If the economy now produces 1000 million dollars of petroleum, 400 million dollars of textiles, 950 million dollars of transportation, 750 million dollars of chemicals, and 950 million dollars of manufacturing, how much of this production is internally consumed by the economy?

4. Use least squares and fit the model $y = kx - b$ using the following data using the following normal equations:

$$K\Sigma x - nb = \Sigma y$$
$$k\Sigma x^2 - b\Sigma x = Sxy,$$

where n is the number of data pairs.

x	0.5	1.0	1.5	2.0	2.5
y	0.7	3.4	7.2	12.4	20.1

5. Use least squares and fit the model $W = kL^3$.

   Data Available:

Length, L	12.5	12.625	12.625	14.125	14.5	14.5	17.27	17.75
Weight, W	17	16	17	23	26	27	43	49

6. Use cubic spline interpolation to obtain the cubic equations for the following data:

   a.

X	1	2	3
Y	9	27	48

b.

X	0.5	1.0	1.5	2.0	2.5
Y	0.7	3.4	7.2	12.4	20.1

Projects

1. The Bridge Truss. In general, you are asked to explain the model of the earlier bridge situation and to compare and contrast different solution techniques for all the forces on the various nodes.

   a. Explain in words the meaning and derivation of each of the 16 equations in the table for the forces acting on the bridge. What assumptions are inherent in this model?

   b. Write these equations in matrix form. You may want to rearrange the equations so that all of the following solution techniques will minimize numerical errors and determine an accurate solution. In other words, you want your matrix A to be as strictly diagonally dominant as possible.

   c. Determine numerically (assume all numbers given are exact) the following and describe the meaning of each calculation. (1) d e t (A); (2) $\|A\|$ ∞; (3) A ^(−1), if it exists; (4) spectral radius of A; (5) $\|A\|$ 2; and (6) K (A), the condition number of A.

   d. Solve the system to within $10^{\wedge}(-4)$ by Gaussian elimination with back substitution using Burden and Faires's (1997) Algorithm 6.1.

   e. Approximate the solution of the system to within $10^{\wedge}(-4)$ using an initial approximation vector of all 1s and the Jacobi method. Compute the residual vector.

   f. Approximate the solution of the system to within $10^{\wedge}(-4)$ using an initial approximation vector of all 1s and the Gauss–Seidel method. Compute the residual vector.

   g. Compute one iterative refinement on each answer for parts e and f to improve the approximations in parts e and f. Show that the refinement did or did not improve these approximations and give reasons why this did or did not work.

   h. Compare and contrast the efficiencies of parts d, e, and f.

2. Resolve Problem 1 when the initial angles for joint 1 and joint 4 are π/3 and π/4, respectively.

3. A third-world country has six main industries: mining, tourism, an electric plant, a water sewerage plant, a railroad, and a Coca-Cola plant. To mine $1 of ore, the mining company must purchase $0.25 of electricity to run its equipment and $0.25 of transportation to ship its goods. To produce $1 of electricity, the plant requires $0.65 of ore for fuel, $0.05 of its own electricity to run equipment, $0.05

for transportation, and $0.10 of water. To provide $1 of transportation, the railroad requires $0.55 of ore for fuel, $0.10 of electricity for its auxiliary equipment, and $0.05 of water. To produce $1 of tourism, the country requires $0.50 for transportation, $0.05 for water, and $0.10 for Coca-Cola. To produce $1 of water, $0.65 of electricity is required and $0.01 of its own water. To produce $1 of Coca-Cola requires $0.25 of electricity, $0.25 of transportation, and $0.25 of water. The outside weekly demand for mining is $50,000 in ore, the electric plant receives orders for $25,000 of electricity, the water plant receives orders for $10,000 in water products, and Coca-Cola has orders for $75,000 of cola products. Determine how much of each of the industries must produce in a week to exactly satisfy the total demand. Use a Leontief model.

4. Consider a DDS representing students who either eat at the college dining hall, the college cafe, or out on the town. The transition matrix A is defined as follows: A = [[.2 .1 .7][.6 .4 .2] [.2 .5 .1]]. If 10,000 students are originally scheduled to be at this college, determine the steady state (or long-term behavior) of this system. Make a recommendation to the college dining and cafe facilities about what they can expect during the next year.

5. One question that manufacturers must decide is how much of something to produce. Considering that an economy consists of manufacturers of many things, you can see how this problem can get really complex. Wassily Leontief's work on this problem earned him the Nobel Prize. What we will discuss now comes from his work, hence the name Leontief Models. Here is an example of such a model.

First, we divide an economy into certain sectors. In reality, there are hundreds of sectors, but to keep things simple, we will say for this example that there are three: manufacturing (M), electronics (E), and agriculture. We must decide how many units of each sector to produce. We can put this into a <u>production matrix</u>:

$$X = \begin{matrix} M \\ E \\ A \end{matrix} \begin{bmatrix} ?? \\ ?? \\ ?? \end{bmatrix}.$$

Now, let's say that the public wants 100 units of manufacturing, 200 units of electronics, and 300 units of agriculture. We can put this into an <u>(external) demand matrix</u>:

$$D = \begin{matrix} M \\ E \\ A \end{matrix} \begin{bmatrix} 100 \\ 200 \\ 300 \end{bmatrix}.$$

Now, one might say that if this is what the people want, then this is what should be produced (i.e., $X = D$). The problem, however, is that the production of certain resources actually *uses up* resources as well. In other words, it takes stuff to make stuff. How much stuff it takes to make stuff can be expressed by an <u>input–output matrix</u>.

```
Output
Input
```

$$T_E = \begin{bmatrix} .1 & .2 & .3 \\ .2 & .1 & .3 \\ .1 & .1 & .2 \end{bmatrix}$$

This matrix says that the production of 1 unit of manufacturing uses up 0.1 units of manufacturing, 0.2 units of electronics, and 0.1 units of agriculture. The production of 1 unit of electronics uses up 0.2 units of manufacturing, 0.1 units of electronics, and 0.1 units of agriculture. Finally, the production of 1 unit of agriculture uses up 0.3 units of manufacturing, 0.3 units of electronics, and 0.1 units of agriculture. So, not only must we account for what the people want, but we must also make up for what is used up in the process of making what the people want. This is called <u>internal demand,</u> and it is given by the matrix product TX. Hence, what we produce needs to satisfy both internal demand and external demand. That is,

$$X = TX + D.$$

6. <u>Wassily Leontief</u> (1906–1999) was a Russian-born American economist who, aside from developing highly sophisticated economic theories, also enjoyed trout fishing, ballet and fine wines. He won the 1973 Nobel Prize for economics for his work in creating mathematical models to describe various economic phenomena. In the remainder of this problem, we will look at a very simple special case of his work called a closed exchange model. Here is the premise:

Suppose in a faraway land of Eigenbazistan, in a small country town called Matrixville, there lived a farmer, a tailor, a carpenter, a coal miner, and Slacker Bob. The farmer produced food; the tailor, clothes; the carpenter, housing; the coal miner supplied energy; and Slacker Bob made high-quality 100 proof moonshine, half of which he drank himself. Let us make the following assumptions:

- Everyone buys from and sells to the central pool (i.e., there is no outside supply and demand).
- Everything produced is consumed.

For these reasons, this is called a *closed* exchange model. Next we must specify what fraction of each of the goods is consumed by each person in our town. Here is a table containing this information:

	Food	Clothes	Housing	Energy	High-Quality 100 Proof Moonshine
Farmer	0.25	0.15	0.25	0.18	0.20
Tailor	0.15	0.28	0.18	0.17	0.05
Carpenter	0.22	0.19	0.22	0.22	0.10
Coal miner	0.20	0.15	0.20	0.28	0.15
Slacker Bob	0.18	0.23	0.15	0.15	0.50

So for example, the carpenter consumes 22% of all food, 19% of all clothes, 22% of all housing, 22% of all energy, and 10% of all high-quality 100 proof moonshine.

If $I - T$ is invertible, this equation can be solved for X:

$$X - TX = D$$
$$(I - T)X = D$$
$$(I - T)^{-1}[(I - T)X] = (I - T)^{-1}D$$
$$[(I - T)^{-1}(I - T)]X = (I - T)^{-1}D$$
$$IX = (I - T)^{-1}D$$
$$X = (I - T)^{-1}D.$$

In this example, $I - T = \begin{bmatrix} 1 & 0 & 0 \\ 0 & 1 & 0 \\ 0 & 0 & 1 \end{bmatrix} - \begin{bmatrix} .1 & .2 & .3 \\ .2 & .1 & .3 \\ .1 & .1 & .2 \end{bmatrix} = \begin{bmatrix} .9 & -.2 & -.3 \\ -.2 & .9 & -.3 \\ -.1 & -.1 & .8 \end{bmatrix}.$

The inverse of this matrix is (approximately)

$$(I - T)^{-1} \approx \begin{bmatrix} 1.255 & .345 & .6 \\ .345 & 1.255 & .6 \\ .2 & .2 & 1.4 \end{bmatrix}.$$

And so

$$X = (I - T)^{-1}D = \begin{bmatrix} 1.255 & .345 & .6 \\ .345 & 1.255 & .6 \\ .2 & .2 & 1.4 \end{bmatrix}\begin{bmatrix} 100 \\ 200 \\ 300 \end{bmatrix} \approx \begin{bmatrix} 375 \\ 465 \\ 480 \end{bmatrix}.$$

That is, we want to produce 375 units of manufacturing, 465 units of electronics, and 480 units of agriculture.

The internal demand is now easy to calculate. While we could do it by finding the matrix product TX, we can just say it is total production minus external demand, that is, $X-D$.

$$\text{internal demand} = X - D = \begin{bmatrix} 375 \\ 465 \\ 480 \end{bmatrix} - \begin{bmatrix} 100 \\ 200 \\ 300 \end{bmatrix} = \begin{bmatrix} 275 \\ 265 \\ 180 \end{bmatrix}$$

In reality, the numbers in T will be much smaller, and so the internal demand is usually very small compared to the external demand.

References and Further Readings

Burden, R. and D. Faires (1997). *Numerical Analysis*. Brooks-Cole, Pacific Grove, CA.

Fox, W. P. (2018). *Mathematical Modeling for Business Analytics*. Taylor and Francis Publishers, Boca Raton, FL.

Fox, W. P. and R. Burks (2021). *Advanced Mathematical Modeling*. Taylor and Francis, CRC Press, Boca Raton, FL.

Fox, W. P. and R. Burks (2022). *Mathematical Modeling Under Change, Uncertainty, and Machine Learning*. Taylor and Francis, CRC, Boca Raton, FL.

Giordano, F., W. Fox and S. Horton (2013). *A First Course in Mathematical Modeling*, 5th ed. Cengage Publishers, Boston, MA.

9

Modelling with Single-Variable Unconstrained Optimization and Numerical Methods

9.1 Introduction

Consider an oil-drilling rig that is 8.5 miles offshore. The drilling rig is to be connected by an underwater pipe to a pumping station. The pumping station is connected by land-based pipe to a refinery, which is 14.7 miles down the shoreline from the drilling rig (see Figure 9.1). The underwater pipe costs $31,575 per mile, and land-based pipe costs $13,342 per mile. You are to determine where to place the pumping station to minimize the cost of the pipe.

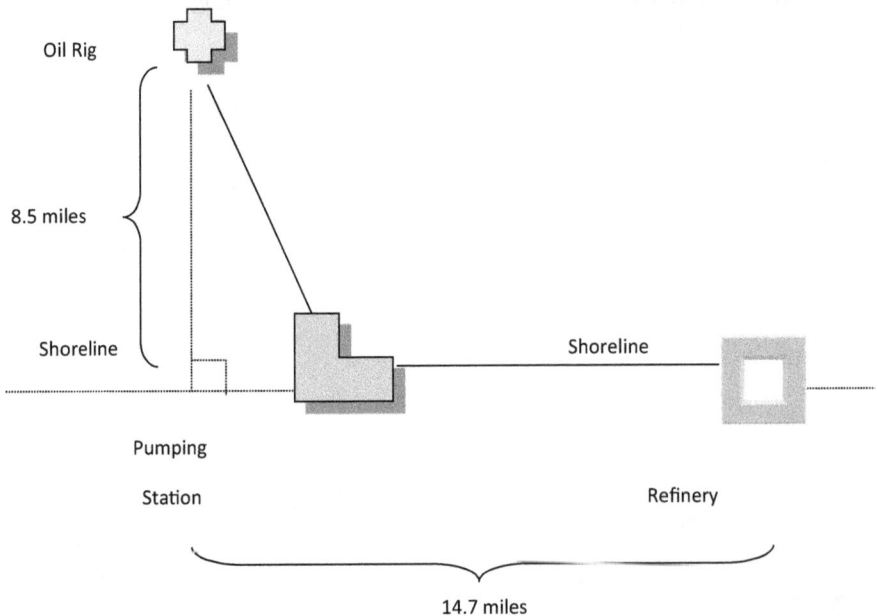

FIGURE 9.1
Location of oil pumping station.

DOI: 10.1201/9781032703671-9

In this chapter, we will discuss models that require single-variable calculus to solve. We will review the calculus concepts for optimization and then apply them to the application. We will use Maple to assist us.

9.2 Single-Variable Optimization and Basic Theory

We want to solve problems of the form:

$$\max \text{ (or min) } f(x)$$
$$x \in (a, b)$$

If $a = -\infty$ and $b = \infty$, then we are looking at R^2—the xy plane. If either a, b, or both a and b are restricted, then we must consider possible end points in our solution. We will examine three cases.

Case 1. Points where $a < x < b$ and $f'(x) = 0$

Case 2. Points where $f'(x)$ does not exist

Case 3. Endpoints a and b of the interval $[a, b]$

Additionally, we will define points where $f'(x) = 0$ as critical points or stationary points. Here are some additional definitions and theorems from calculus that might be useful.

Definition: A function f has a maximum (global) at a point c, if $f(c) \geq f(x)$ for all x in the domain of f, and a function f has a minimum (global) at a point c, if $f(c) \leq f(x)$ for all x in the domain of f.

Extreme Value Theorem: If f is continuous on a closed interval $[a, b]$, then f has both a global maximum and a global minimum over the interval.

We recall from your study of calculus the analysis of the first derivative test.

If $f'(x) > 0$ to the left of x^* and $f'(x) < 0$ to the right of x^*, then x^* is a local maximum, and we find $f(x)$ increasing to the left and decreasing to the right of point x^*.

Also, recall the second derivative test. If $f'(x_0) = 0$, then we compute $f''(x_0)$.

If $f''(x_0) < 0$, then $f(x_0)$ is a local maximum.

If $f''(x_0) > 0$, then $f(x_0)$ is a local minimum.

If $f''(x_0) = 0$, then $f(x_0)$ might be an inflection point.

Theorem 9.1 If $f'(x_0) = 0$ and $f''(x_0) < 0$, then $f(x_0)$ is a local maximum.
If $f'(x_0) = 0$ and $f''(x_0) > 0$, then $f(x_0)$ is a local minimum.

Theorem 9.2 If $f''(x_0) = 0$, and

1. if the first nonzero derivative at x_0 occurs at an odd-order derivative, then $f(x_0)$ is neither a local maximum nor a local minimum.

2. if the first nonzero derivative is positive and occurs at an even-order derivative, then $f(x_0)$ is a local minimum.

3. if the first nonzero derivative is negative and occurs at an even-order derivative, then $f(x_0)$ is a local maximum.

Examples: Case 2. Points where $f'(x)$ does not exist.

If $f(x)$ does not have a derivative at x_0, then $f(x_0)$ might be a local maximum, a local minimum, or neither. In this case, we test points near x_0 and evaluate the function at those neighboring points where $x_1 < x_0 < x_2$.

Relationship between $f(x0)$ and close neighbors x_0	Classification
$f(x_0) > f(x_1), f(x_0) < f(x_2)$	Not a local extrema
$f(x_0) < f(x_1), f(x_0) > f(x_2)$	Not a local extrema
$f(x_0) \geq f(x_1), f(x_0) \geq f(x_2)$	Local maximum
$f(x_0) \leq f(x_1), f(x_0) \leq f(x_2)$	Local minimum

Case 3. Endpoints a and b.
From the following figure, we see that

1. if $f'(a) > 0$, then $f(a)$ is a local minimum.

2. if $f'(a) < 0$, then $f(a)$ is a local maximum.

3. if $f'(b) > 0$, then $f(b)$ is a local minimum.

4. if $f'(b) < 0$, then $f(b)$ is a local maximum.

If both $f'(a) = f'(b) = 0$, then draw a sketch and test neighboring points to determine if $f(a)$ and/or $f(b)$ is/are extrema.

Example 1. Find the Local Minimum of $f(x) = 1.5x^2$

Solution:

$$f(x) = 1.5x^2$$

The first derivative is equal to 0 at $x = 0$. The second derivative is positive at $x = 0$, so the critical point 0 yields a local minimum of the function.

Example 2. Min $x3$ on the Interval $-1 \leq x \leq 3$

We find $f'(x) = 0$ at $x = 0$. The second derivative test yields $f''(0) = 0$, so we have an inflection point at $x = 0$. We also note the $f(0)=0$. Next, we test the endpoints.

$$f(-1) = -1 \text{ and } f''(-1) = 6 > 0$$

$f(3) = 27$ and $f''(3) = 18 >$. Therefore, $f(-1) = -1$ is the minimum on the interval $[-1,3]$.

9.3 Models with Basic Applications of Max-Min Theory (Calculus Review)

Example 3. Chemical Company

A chemical manufacturing company sells sulfuric acid at a price of $100 per unit. If the daily total production cost in dollars for x units is

$$C(x) = 100000 + 50x + 0.0025x^2,$$

and the daily production is at most 7000 units. How many units of sulfuric acid should the manufacturer produce to maximize daily profits?

Solution: Profit = Revenue − Cost

$$P = 100x - (100000 + 50x + 0.0025x^2)$$
$$P = -100000 + 50x - 0.0025x^2, \text{ for } 0 \le x \le 7000$$
$$\frac{dP}{dx} = 50 - .005x = 0$$

$$X^* = 10,000 \text{ units with } P(10000) = \$740,000$$

$\frac{d^2P}{dx^2} = -.005$, since $P'' < 0$, we have a local maximum at $x = 10,000$.

We must also check the endpoints.

$x = 0, P = -\$100,000$
$x = 7000, P = \$217,500$

Our solution is $x = 7000$ and $P = \$217,500$ at the endpoint since x^* is outside of the domain of production even though its mathematical results are better: $x^* = 10,000$ units with $P(10,000) = \$240,000$.

In Maple, we will use commands that we have previously learned:

```
diff or Diff - Differentiation or Partial
Differentiation
Calling Sequence
      diff(f, x1, ..., xj)
      diff(f, [x1$n])
      diff(f, x1$n, [x2$n, x3], ... xi, [xj$m])
solve - solve one or more equations
Calling Sequence
      solve(equations, variables)
> cost:=100*x-(10000+50*x+0.0025*x^2);
cost = 50x - 10000 - 0.0025 x²
> dc:=diff(cost,x);
dc := 50 - 0.0050 x
> ans:=solve(dc=0,x);
ans := 10000.
> ddc:=diff(dc,x);
```

```
ddc := - 0.0050
> subs(x=ans,cost);
240000.0000
> plot(cost, x=0..12000);
```

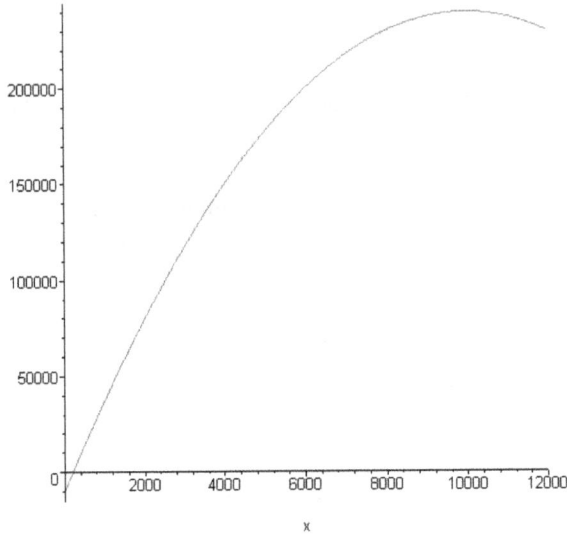

Checking the endpoints:

```
> subs(x=0, cost);
-10000.
> subs(x=7000,cost);
217500.0000
```

The endpoint solution is selected because the critical point at $x = 10{,}000$ is outside the domain.

Example 4. SP6 Computer Development

A company spends \$200 in variable costs to produce an SP6 computer, plus a fixed cost of \$5000 if any SP6 computers are produced. If the company spends x dollars on advertising their new SP6 computer, it can sell $x^{1/2}$ at \$500 per computer. How many SP6 computers should the company produce to maximize profits?

Solution: Maximize Profit = Revenue − Cost

$$\text{Cost} = \text{fixed} + \text{variable} + \text{advertising costs} = 5000 + 200 \cdot x^{1/2} + x$$

$$\text{Revenue} = 500 \cdot x^{1/2}$$

$$\text{Maximize } P = 500 \times x^{1/2} - (5000 + 200 \cdot x^{1/2} + x)$$

$$\frac{dP}{dx} = \frac{d(500 \cdot x^{1/2} - (5000 + 200 \cdot x^{1/2} + x))}{dx} = -\frac{-150 + \sqrt{x}}{\sqrt{x}} = \frac{150}{\sqrt{x}} - 1$$

$$\text{Set } \frac{dP}{dx} = 0, \text{ yields } \frac{150}{\sqrt{x}} - 1 = 0, \ x = 22{,}500$$

$$P(x) = 500 \times x^{1/2} - (5000 + 200 \cdot x^{1/2} + x)$$

$\dfrac{d^2P}{dx^2} = \dfrac{-75}{x^{3/2}} < 0$ for all $x > 0$. Therefore, we have found a maximum.

The following Maple commands achieve the solution:

```
> restart;
> profit:=500*sqrt(x)-(500+200*sqrt(x)+x);
```
$Profit := 300\sqrt{x} - 500 - x$
```
> dp:=diff(profit,x);
```
$dp := \dfrac{150}{\sqrt{x}} - 1$
```
> ans:=solve(dp=0,x);
ans := 22500
> evalf(subs(x=ans,profit));
22000.00000
> evalf(subs(x=ans,diff(dp,x)));
-0.00002222222222
Since p"<0 we found maximum.
> plot(profit, x=0..30000, title=`Profit`,
thickness=3);
```

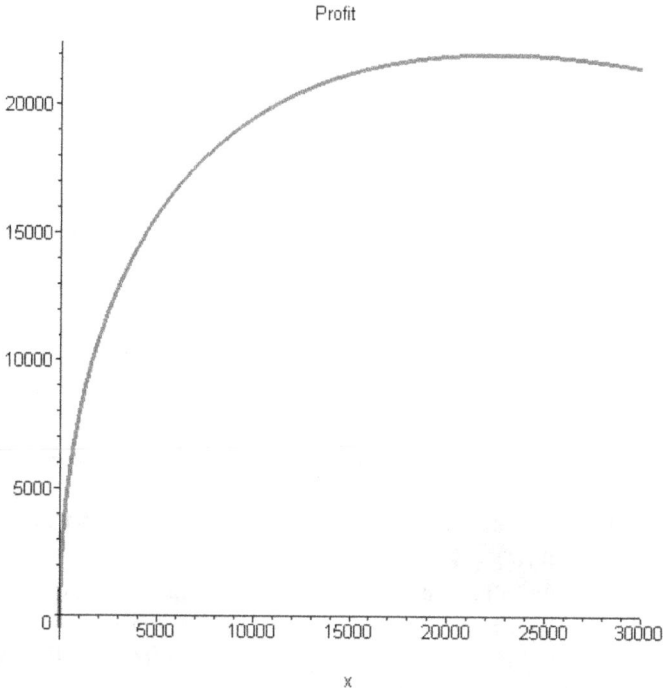

Profit

Exercises

1. Each morning during rush hour, 10,000 people travel from New Jersey to New York City. If a person takes the subway, the trip lasts 40 minutes. If x thousand people drive to New York City, it takes $20 + 5x$ minutes to make the trip.

   a. Formulate the problem with the objective to minimize the average travel time per person.

   Let x = number of people (in 1000s) that drive to New York City from New Jersey.

   b. Find the optimal number of people that drive so that the average time per person is minimized.

2. Use calculus to find the optimal solution to

   MAX $f(x) = 2x^3 - 1$

   $-1 \leq x \leq 1$

3. Given the following plot of the function: $f(x) = 0.5x^3 - 8x^2$

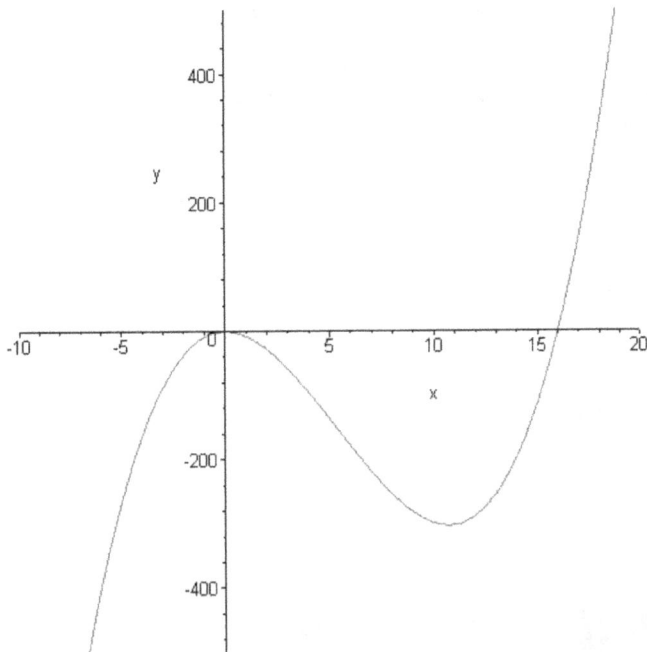

   a. Using analytical techniques (i.e., calculus), find and classify all *extrema* for $f(x)$.

   b. Apply the definition of convexity to show that the function $f(x)$ is concave over the interval between $x_1 = -6$ and $x_2 = 5$, using $c = 0.5$. (We're fixing the value of c here to keep the algebra from getting ugly.)

c. Confirm your response to part b using the second derivative test to show that $f(x)$ is concave over this entire interval.

d. Over what interval is the function in part c convex? concave?

e. Why is knowing the concavity important in optimization?

4. Find the optimal solution to the bounded single variable optimization problem:

(You may use calculus here)

MAX $f(x) = 2x^3 - 2$

s.t. $-1 \le x \le 1$

5. Dr. E. N. Throat has been taking X-rays of the trachea contracting during coughing. He has found that the trachea appears to contract by 33% (1/3) of its normal size. He has asked the department of mathematics to confirm or deny his claim. You perform some initial research, and you find that under reasonable assumptions about the elasticity of the tracheal wall and about how air near the wall is slowed by friction, that the average flow of velocity v can be modeled by the equation:

$$v = c\,(r_0 - r)r^2 \text{ cm/s, between } r_0/2 \le r \le r_0,$$

where c is a positive constant (let $c = 1$), r_0 is the resting radius of the trachea in centimeters.

Find the value of r that maximizes v and then support or deny the claim.

9.4 Applied Single-Variable Optimization Models

Oil-Rig Location Problem

Consider an oil-drilling rig that is 8.5 miles off shore. The drilling rig is to be connected by underwater pipe to a pumping station. The pumping station is connected by land-based pipe to a refinery, which is 14.7 miles down the shoreline from the drilling rig (see Figure 9.2). The underwater pipe costs $31,575 per mile, and land-based pipe costs $13,342 per mile. You are to determine where to place the pumping station to minimize cost of the pipe.

Problem Identification: Find a relationship between the location of the pumping station and cost of the installation of the pipe.

Assumptions: First, we assume no cost saving for the pipe if we purchase in larger lot sizes. We further assume no additional costs are incurred in preparing the terrain to lay the pipe.

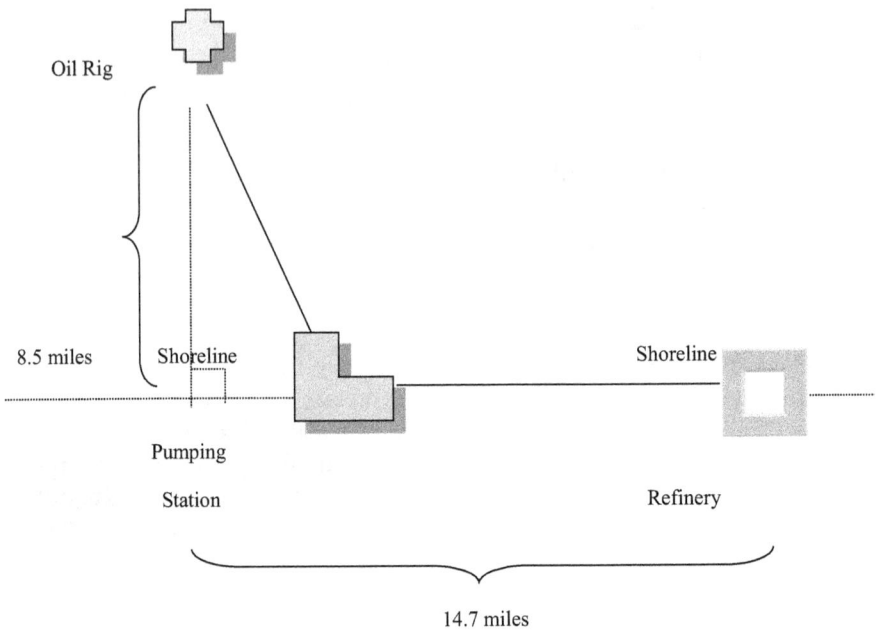

FIGURE 9.2
Oil refinery.

Variables:

x = the location of the pumping station along the horizontal distance from $x = 0$ to $x = 14.7$ miles.

TC = total cost of the pipe for both underwater and on shore piping.

Model Construction:

We use Pythagorean's theorem for the underwater distance of the pipe that is the hypotenuse of the right triangle with height 8.5 miles and base $= x$. The hypotenuse is

$\sqrt{8.5^2 + x^2}$. The length of the pipe on shore is 14.7 − x.

Total cost = 31,575 $\sqrt{8.5^2 + x^2}$ + 13342 (14.7−x).

```
>
> TC:= 31575*sqrt(8.5^2+x^2)+13342*(14.7-x);
```

$$TC := 31575\sqrt{75.25 + x^2} + 196127.4 - 13342x$$

```
> cp:=diff(TC,x);
```

$$cp := \frac{31575x}{\sqrt{72.25 + x^2}} - 13342$$

```
> xstar:=solve(cp=0,x);
```

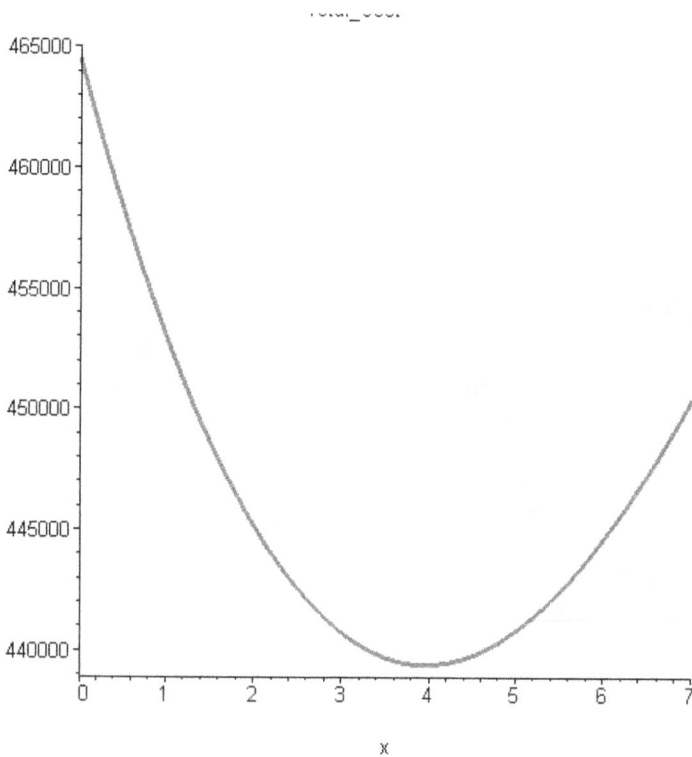

FIGURE 9.3
Plot of total cost.

```
xstar := 3.962829844
> TC_at_xstar:=subs(x=xstar,TC);
TC_at_xstar:=439377.6878
> plot(TC,x=0..7, thickness=3, title=`Total_Cost`);
```

Thus, if the pumping station is located at 14.7 − 3.963 = 10.737 miles from the refinery, we will minimize the total cost at a cost of $439,377.69. The figure is shown as Figure 9.3

9.5 Single-Variable Numerical Search Techniques

The basic approach of most numerical methods in optimization is to produce a sequence of improved approximations to the optimal solution according to a specific scheme. We will examine both elimination (dichotomous, golden section, and Fibonacci) and interpolation methods (Newton's and bisection).

In numerical methods of optimization, a procedure is used in obtaining values of the objective function at various combinations of the decision variables and conclusions are then drawn regarding the optimal solution. The elimination methods can be used to find an optimal solution for even discontinuous functions. An important relationship (assumption) must be made to use these elimination methods. The function must be unimodal. A unimodal function is one that has only one peak (maximum) or one valley (minimum). This can be stated mathematically as follows:

A function $f(x)$ is unimodal if (1) $x_2 < x^*$ implies that $f(x_2) < f(x_1)$, and (2) $x_1 > x^*$ implies that $f(x_1) < f(x_2)$, where x^* is a minimum and $x_1 < x_2$.

A function $f(x)$ is unimodal if (1) $x_2 > x^*$ implies that $f(x_2) > f(x_1)$, and (2) $x_1 < x^*$ implies that $f(x_1) > f(x_2)$, where x^* is a maximum and $x_1 < x_2$.

Some examples of unimodal functions are shown in Figure 9.4. Thus, as seen unimodal functions may or may not be differentiable.

Thus, a function can be a non-differentiable (corners) or even a discontinuous function. If a function is known to be unimodal in a given interval, then the optimum (maximum or minimum) can be found as a smaller interval.

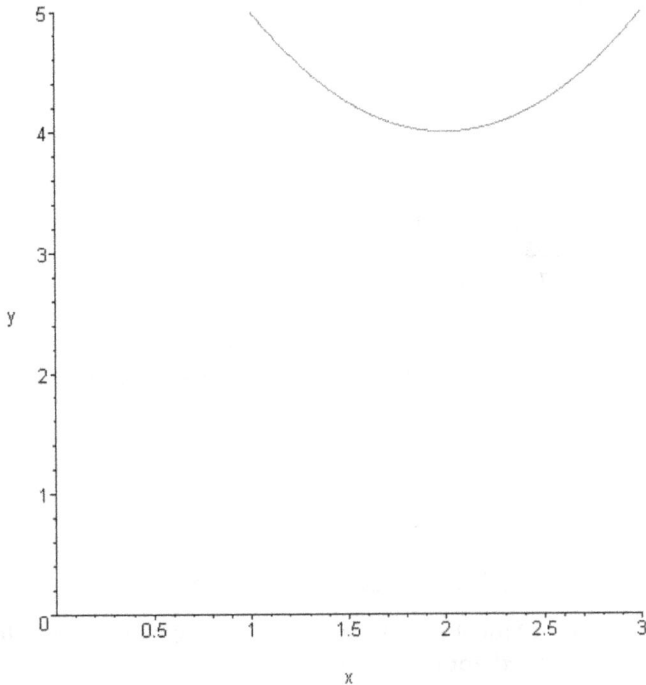

FIGURE 9.4a
Unimodel function.

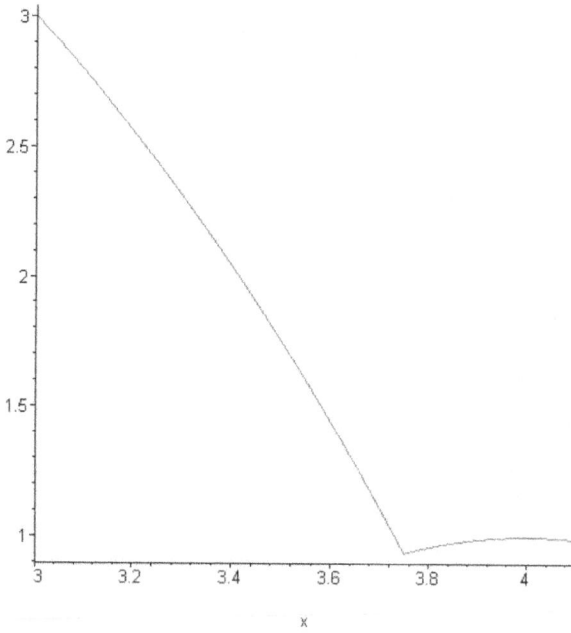

FIGURE 9.4b
Unimodal function.

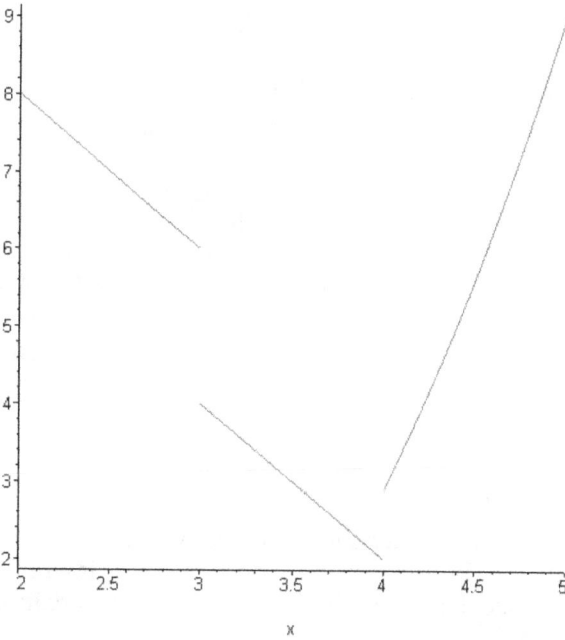

FIGURE 9.4c
Unimodal function.

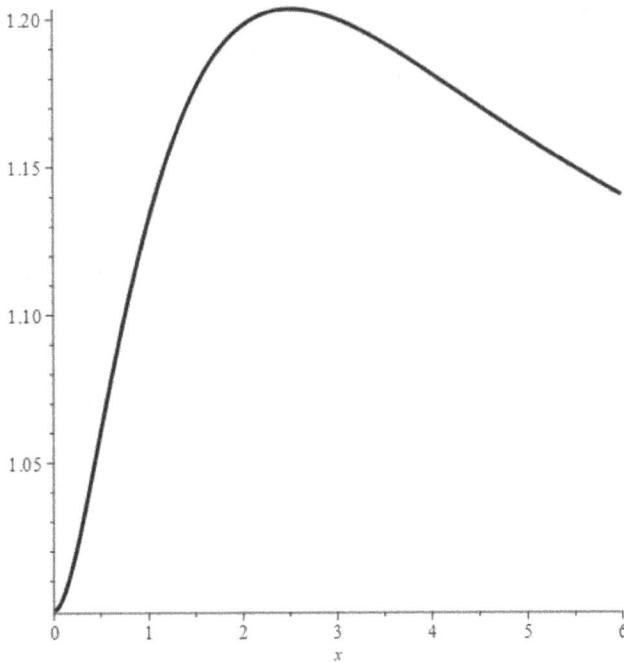

FIGURE 9.5
Plot of $L(X)$ from $t = 0$ to $t = 6$ years.

In this section, we will learn many techniques for numerical searches. For the elimination methods, we accept an interval answer. If a single value is required, then we usually evaluate the function at each endpoint of the final interval and the midpoint of the final interval and take the optimum of those three values to approximate our single value.

We are provided a function for the lifetime in years of an object being sent to the space station. We need to find its maximum value and at what time period it is achieved, The lifetime function, as expressed by the engineering section is, $L(X) = x - e^{-x} + (1/1 + x)$. We find we cannot apply classical calculus methods.

We plot $L(x)$ as seen in Figure 9.5 that shows the maximum and at least an approximate value.

Unrestricted Search

The method called unrestricted search is used when the optimum needs to be found, but we have no known interval of uncertainty. This will involve a search with a fixed step size. This method is not very computationally effective.

1. Start with a guess, say, x_1.
2. Find $f_1 = f(x_1)$.

3. Assume a step size, S; find $x_2 = x_1 + S$.

4. Find $f_2 = f(x_2)$.

5. For a minimization problem, if $f_2 < f_1$, then the solution interval cannot possibly lie in $x < x_1$. So we find points $x_3, x_4, \ldots, x_n$. This is continued until an increase of the function is found.

6. The search is terminated at x_i and x_{i-1}.

7. If $f_2 > f_1$, then the search goes in the reverse direction.

8. If $f_1 = f_2$, then either x_1 or x_2 is optimum.

9. If the search proceeds too slowly, an acceleration step size, a constant c, times S, cS, can be used.

Example 5. Find the Minimum of the Function

	$-x/2$	$x \le 2$
$f(x) =$		
	$x - 3$	$x > 2$

Using an unrestricted search method with an initial guess as 1.0, and a step size $= 0.4$.

$$x_1 = 1 \text{ and } f(x_1) = f_1 = -.5$$
$$x_2 = x_1 + S = 1 + .4 = 1.4, f(1.4) = f_2 = -.7$$
$$f_1 > f_2, \text{ so}$$
$$x_3 = 1.8, f_3 = -.9, f_2 > f_3$$
$$x_4 = 2.2, f_4 = -.8, f_4 > f_3. \text{ Stop}$$

Thus, the optimum (minimum) must lie between 1.8 and 2.2. If a single value is required, we evaluate

$$f(1.8) = -.9$$
$$f(2.2) = -.8$$
$$f((1.8 + 2.2)/2) = f(2) = -1.$$

Since $f(2)$ is the smallest value of $f(x)$, we will use $x = 2$, as our approximation to the minimum yielding a value of $f(2) = -1$.

Dichotomous Search

This search method is a simultaneous search method in which all the experiments are conducted before any judgment is made concerning the location of the optimum point.

The following is the dichotomous algorithm:

1. Initialize a distinguishable constant $2e > 0$ (e is a very small number, like 0.01)

2. Select a length of uncertainty for the final interval, $t > 0$ (t is also small)

3. Calculate the number of iterations required, N, using
   $0.5^n = t/(b - a)$.
4. Let $k = 1$.

Main Steps

1. If $(b - a) < t$, then stop because $(b - a)$ in the final interval.
   If $(b - a) > t$, then let

$$x_1 = \frac{(a+b)}{2} - e, x_2 = \frac{(a+b)}{2} + e .$$

2. Perform comparisons of function values at these points

Minimization Problem

If f(x1)<f(x2)	If f(x1)>f(x2)
a=a	a=x1
b=x2	b=b
k=k+1	k=k+1
Return to Main Step 1	Return to Main Step 1

For maximization problems, just use $-f$ for f in the algorithm.

Example 6. Maximize $L(X) = x - e^{-x} + (1/1 + x)$ over the Interval [0, 6] Using a Dichotomous Search.

We will replace f with $-f$.

Let $t = 0.2$ and $e = 0.01$.
Find n using $0.5^n = t/(b - a)$
$0.5^n = 0.2/(6 - (0)) = 0.2/6$
$0.5^n = (0.2/6)$
$n \ln(0.5) = \ln(0.2/6)$
$n = 4.906$ or 5 (rounding up)

We provide a screenshot of our solution in Figure 9.6.

Golden Section Search

A golden section search is a search procedure that utilizes the *golden ratio*. To better understand the *golden ratio*, consider a line segment over the interval that is divided into two separate regions as shown in Figure 9.7. These segments are divided into the *golden ratio* if the length of the whole line is to the length of the larger part as the length of the larger part is to the length of the smaller part of the line. Symbolically, this can be written as

Maximize the function $f(x)=1-exp(x)+1/(1+x)$ over the interval $[0,6]$.

```
> f := x->1-exp(-x)+(1/(1+x));
```

$$f := x \mapsto 1 - e^{-x} + \frac{1}{1+x}$$

```
> DICHOTOMOUS (f,0,6,.001,.001);
```

The interval [a,b] is [0.00, 6.00]and user specified tolerance level is 0.00100.
The first 2 experimental endpoints are x1= 2.999 and x2 = 3.001.

Iteration	x(1)	x(2)	f(x1)	f(x2)	Interval
1	2.9990	3.0010	1.2002	1.2002	[0.0000, 6.0000]
2	1.4995	1.5015	1.1768	1.1770	[0.0000, 3.0010]
3	2.2492	2.2512	1.2023	1.2023	[1.4995, 3.0010]
4	2.6241	2.6261	1.2034	1.2034	[2.2492, 3.0010]
5	2.4367	2.4387	1.2035	1.2035	[2.2492, 2.6261]
6	2.5304	2.5324	1.2036	1.2036	[2.4367, 2.6261]
7	2.4835	2.4855	1.2036	1.2036	[2.4367, 2.5324]
8	2.5070	2.5090	1.2036	1.2036	[2.4835, 2.5324]
9	2.5187	2.5207	1.2036	1.2036	[2.5070, 2.5324]
10	2.5128	2.5148	1.2036	1.2036	[2.5070, 2.5207]
11	2.5099	2.5119	1.2036	1.2036	[2.5070, 2.5148]
12	2.5114	2.5134	1.2036	1.2036	[2.5099, 2.5148]
13	2.5121	2.5141	1.2036	1.2036	[2.5114, 2.5148]
14	2.5117	2.5137	1.2036	1.2036	[2.5114, 2.5141]

The midpoint of the final interval is 2.512919 and f(midpoint) = 1.204.

The maximum of the function is 1.204 and the x value = 2.511370

FIGURE 9.6
Screenshot of solution by dichotomous search.

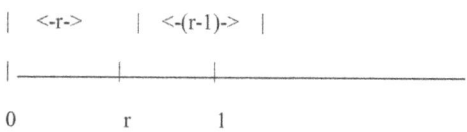

FIGURE 9.7
Line segment illustration.

$$\frac{1}{r} = \frac{r}{(r-1)}.$$

Algebraic manipulation of the golden ratio relationship yields $r^2 + r - 1 = 0$. Solving this function for its roots (using the quadratic formula) gives us two real solutions:

$$r_1 = \frac{\sqrt{5}-1}{2}, r_2 = \frac{-\sqrt{5}-1}{2}.$$

Only the positive root, r_1, satisfies the requirement of residing on the given line segment. The numerical value of r_1 is 0.618. This value is known as the *golden ratio*. This ratio has, among its properties, being the limiting value for the ratio of the consecutive Fibonacci sequences, which we will see in the next method. It is noted here because there is also a Fibonacci search method that could be used in lieu of the golden section method.

In order to use the golden section search procedure, we must ensure that certain assumptions hold. These key assumptions include

1. the function must be unimodal over a specified interval,
2. the function must have an optimal solution over a known interval of uncertainty, and
3. we must accept an interval solution since the exact optimal cannot be found by this method.

Only an interval solution, known as the final interval of uncertainty, can be found using this technique. The length of this final interval is

controllable by the user and can be made arbitrarily small by the selection of a *tolerance value*. The final interval is guaranteed to be less than this tolerance level.

Line search procedures use an initial interval of uncertainty to iterate to the final interval of uncertainty. The procedure is based, as shown earlier, on solving for the unique positive root of the quadratic equation, $r^2 + r = 1$. The positive root form using the quadratic formula with $a = 1$, $b = 1$, and $c = -1$ is

$$r = \frac{-b \pm \sqrt{b^2 - 4ac}}{2a} = 0.618.$$

Finding the Maximum of a Function over an Interval with the Golden Section Search

This search procedure to find a maximum is iterative, requiring evaluations of $f(x)$ at experimental points x_1 and x_2, where $x_1 = b - r(b - a)$ and $x_2 = a + r(b - a)$. These experimental points will lie between the original interval $[a, b]$. These experimental points are used to help determine the new interval of search. If $f(x_1) < f(x_2)$, then the new interval is $[x_1, b]$ and if $f(x_1) > f(x_2)$, then the new interval is $[a, x_2]$. The iterations continue in this manner until the final interval length is less than our imposed tolerance. Our final interval contains the optimum solution. It is the size of this final interval that determines our accuracy in finding the approximate optimum solution. The number of iterations required to achieve this accepted interval length can be found as the smallest integer greater than k, where k equals [1,4]:

$$k = \frac{\ln(\frac{tolerance}{(b-a)})}{\ln(0.618)} \, .$$

Often we are required to provide a point solution instead of the interval solution. When this occurs, the method of selecting a point is to evaluate the function, $f(x)$, at the endpoints of the final interval and at the midpoint of this final interval. For maximization problems, we select the value of x that yields the largest $f(x)$ solution. For minimization problems, we select the value of x that yields the smallest $f(x)$ solution.

The algorithm used is shown in Figure 9.8.

Examples

```
For our problem to maximize L(x).
GOLD(f,0,6,.01);
The interval [a,b] is [ 0.00, 6.00] and user-
specified tolerance level is 0.01000.
The first 2 experimental endpoints are x1= 2.292 and
x2 = 3.708.
```

Iteration	x(1)	x(2)	f(x1)	f(x2)	Interval
1	2.2920	3.7080	1.2027	1.1879	[0.0000, 3.7080]
	0.0000	3.7080			
2	1.4165	2.2920	1.1713	1.2027	[1.4165, 3.7080]
	1.4165	3.7080			
3	2.2920	2.8326	1.2027	1.2021	[1.4165, 2.8326]
	1.4165	2.8326			
4	1.9574	2.2920	1.1969	1.2027	[1.9574, 2.8326]
	1.9574	2.8326			
5	2.2920	2.4983	1.2027	1.2036	[2.2920, 2.8326]
	2.2920	2.8326			
6	2.4983	2.6261	1.2036	1.2034	[2.2920, 2.6261]
	2.2920	2.6261			
7	2.4196	2.4983	1.2035	1.2036	[2.4196, 2.6261]
	2.4196	2.6261			
8	2.4983	2.5472	1.2036	1.2036	[2.4196, 2.5472]
	2.4196	2.5472			
9	2.4684	2.4983	1.2036	1.2036	[2.4684, 2.5472]
	2.4684	2.5472			
10	2.4983	2.5171	1.2036	1.2036	[2.4983, .5472]
	2.4983	2.5472			
11	2.5171	2.5285	1.2036	1.2036	[2.4983, 2.5285]
	2.4983	2.5285			

To find a maximum solution to given a function, *f(x)*, on the interval *[a, b]* where the function, *f(x)*, is unimodal.

INPUT:　　　endpoints a, b; tolerance, t

OUTPUT:　　final interval *[a_i, b_i], f(midpoint)*

Step 1.　　　Initialize the tolerance, $t > 0$.

Step 2.　　　Set r=0.618 and define the test points

$$x_1 = a + (1-r)(b-a)$$

$$x_2 = a + r(b-a)$$

Step 3. Calculate $f(x_1)$ and $f(x_2)$

Step 4.　　　Compare $f(x_1)$ and $f(x_2)$

　　　　a. If $f(x_1) \leq f(x_2)$, then the new interval is $[x_1, b]$:

　　　　　　a becomes the previous x_1

　　　　　　b does not change

　　　　　　x_1 becomes the previous x_2

　　　　　　Find the new x_2 using the formula in Step 2.

　　　　b. If $f(x_1) > f(x_2)$, then the new interval is $[a, x_2]$:

　　　　　　a does not change

　　　　　　b becomes the previous x_2

　　　　　　x_2 becomes the previous x_1

Find the new x_1 using the formula in Step 2.

Step 5. If the length of the new interval from Step 4 is less than the tolerance

　　　　specified, then stop. Otherwise go back to Step 3.

Step 6. Estimate x^* as the midpoint of the final interval and compute, $f(x^*)$, the

　　　　estimated maximum of the function.

STOP

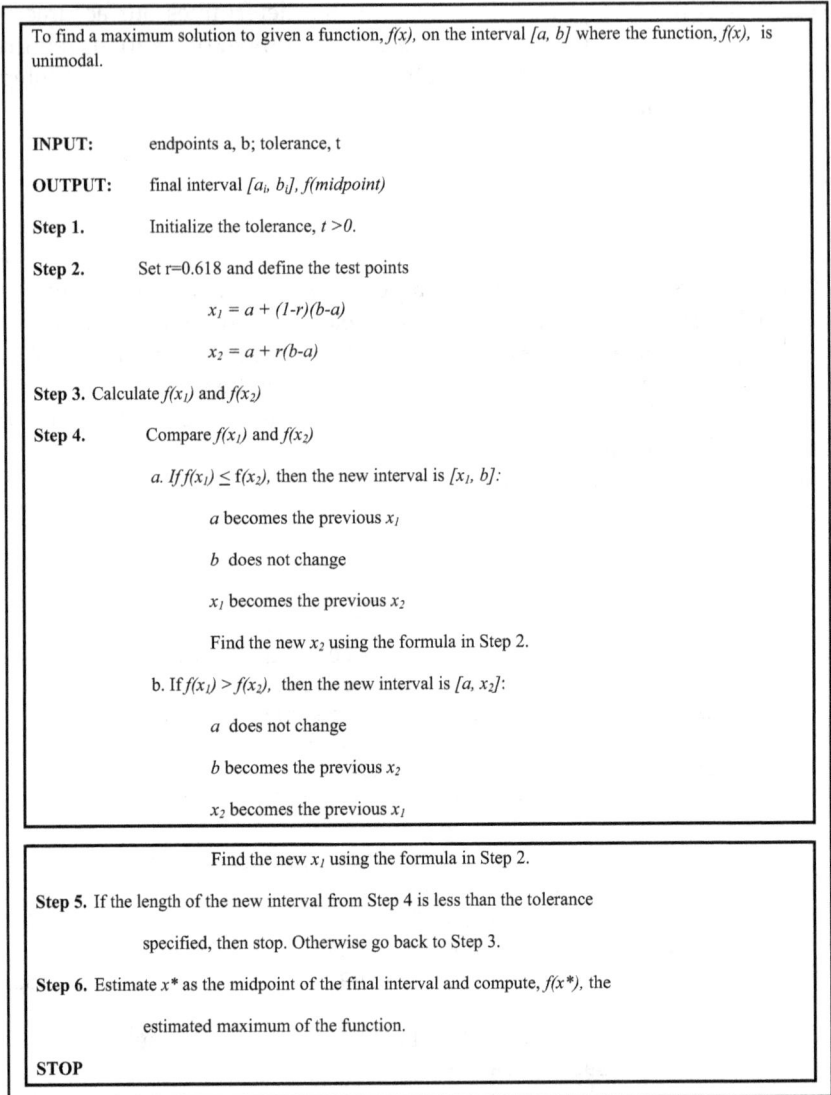

FIGURE 9.8
Golden section algorithm.

12	2.5099	2.5171	1.2036	1.2036	[2.4983, 2.5171]
	2.4983	2.5171			
13	2.5055	2.5099	1.2036	1.2036	[2.5055, 2.5171]
	2.5055	2.5171			
14	2.5099	2.5127	1.2036	1.2036	[2.5099, 2.5171]
	2.5099	2.5171			

```
The midpoint of the final interval is 2.513483 and
f(midpoint) = 1.204.
The maximum of the function is 1.204 and the x value
= 2.513483
```

Although a golden section search can be used with any unimodal function to find the maximum (or minimum) over a specified interval, its main advantage comes when normal calculus procedures fail as we just showed. Consider the following example.

Given data for x and y and the fact we need to minimize the sum of the absolute deviation.

x	1	4	9
y	2	5	8

Maximize f(x) = -|2-x|-|5-4x|-|8-9x| over the interval 0 ≤ x ≤ 3.

In calculus, absolute values are not differentiable because they have corner points. Thus, taking the first derivative and setting it equal to zero is not an option. Another method needs to be used to find the solution. We use the golden section to solve this problem and other examples.

Example 7. Maximizing a Function That Does Not Have a Derivative

Maximize f(x) = -|2-x|-|5-4x|-|8-9x| over the
interval 0 ≤ x ≤ 3.
> f:= x->-(abs(2-x)+abs(5-4*x)+abs(8-9*x));

$$f := x \rightarrow -|2-x|-|5-4x|-|8-9x|$$

```
> GOLD(f,0,3,.1);
The interval [a,b] is [ 0.00,  3.00]and user-specified
tolerance level is 0.10000.
The first 2 experimental endpoints are x1= 1.146 and
x2 = 1.854.
```

Iter	x(1)	x(2)	f(x1)	f(x2)	Interval
1	1.1460	1.8540	-3.5840	-11.2480	[0.0000, 1.8540]
2	0.7082	1.1460	-5.0848	-3.5840	[0.7082, 1.8540]
3	1.1460	1.4163	-3.5840	-5.9958	[0.7082, 1.4163]
4	0.9787	1.1460	-2.9149	-3.5840	[0.7082, 1.1460]
5	0.8755	0.9787	-2.7436	-2.9149	[0.7082, 0.9787]
6	0.8116	0.8755	-3.6382	-2.7436	[0.8116, 0.9787]
7	0.8755	0.9149	-2.7436	-2.6594	[0.8755, 0.9787]
8	0.9149	0.9393	-2.6594	-2.7571	[0.8755, 0.9393]

```
The midpoint of the final interval is 0.907364 and
f(midpoint) = -2.629.
The maximum of the function is -2.629 and the x
value = 0.907364
```

In this example, we want a specific point as our solution. The midpoint yields the maximum value of $f(x)$. Thus, we will use 0.907364 as the value of x that maximizes this function.

Example 8. Maximizing a Transcendental Function

```
Maximize the function f(x)=1-exp(-x)+1/(1+x) over the
interval [0,20].
> f:= x->1-exp(-x)+(1/(1+x));
```

$$f := x \rightarrow 1 - e^{-x} + \frac{1}{1+x}$$

```
> GOLD(f,0,20,.001);
The interval [a,b] is [ 0.00, 20.00] and user-
specified tolerance level is 0.00100.
The first 2 experimental endpoints are x1= 7.640 and
x2 = 12.360.
```

Iteration	x(1)	x(2)	f(x1)	f(x2)	Interval
1	7.6400	12.3600	1.1153	1.0748	[0.0000, 12.3600]
2	4.7215	7.6400	1.1659	1.1153	[0.0000, 7.6400]
3	2.9185	4.7215	1.2012	1.1659	[0.0000, 4.7215]
4	1.8036	2.9185	1.1920	1.2012	[1.8036, 4.7215]
5	2.9185	3.6069	1.2012	1.1899	[1.8036, 3.6069]
6	2.4925	2.9185	1.2036	1.2012	[1.8036, 2.9185]
7	2.2295	2.4925	1.2021	1.2036	[2.2295, 2.9185]
8	2.4925	2.6553	1.2036	1.2033	[2.2295, 2.6553]
9	2.3921	2.4925	1.2034	1.2036	[2.3921, 2.6553]
10	2.4925	2.5548	1.2036	1.2036	[2.3921, 2.5548]
11	2.4543	2.4925	1.2036	1.2036	[2.4543, 2.5548]
12	2.4925	2.5164	1.2036	1.2036	[2.4925, 2.5548]
13	2.5164	2.5310	1.2036	1.2036	[2.4925, 2.5310]
14	2.5072	2.5164	1.2036	1.2036	[2.5072, 2.5310]
15	2.5164	2.5219	1.2036	1.2036	[2.5072, 2.5219]
16	2.5128	2.5164	1.2036	1.2036	[2.5072, 2.5164]
17	2.5107	2.5128	1.2036	1.2036	[2.5107, 2.5164]
18	2.5128	2.5142	1.2036	1.2036	[2.5107, 2.5142]
19	2.5120	2.5128	1.2036	1.2036	[2.5120, 2.5142]
20	2.5128	2.5134	1.2036	1.2036	[2.5120, 2.5134]
21	2.5125	2.5128	1.2036	1.2036	[2.5125, 2.5134]

```
The midpoint of the final interval is 2.512961 and
f(midpoint) = 1.204.
The maximum of the function is 1.204 and the x value =
2.512961.
```

Again, assuming that we desire a specific numerical value as the solution, our solution is $x = 2.512705$ with $f(2.512705) = 1.204$.

Golden Section in Python

We solve our refinery location problem here using the following code.
   First, get a plot to ensure unimodal and get a search domain.

```
import numpy as np
import math
from matplotlib import pyplot as plt
plt.rcParams["figure.figsize"] = [7.50, 3.50]
plt.rcParams["figure.autolayout"] = True
def f(x):
    return 31575*np.sqrt(72.25+x**2)+13342*(14.7-x);
x = np.linspace(0, 10, 100,endpoint=True)
plt.plot(x, f(x), color='black')
plt.show()
```

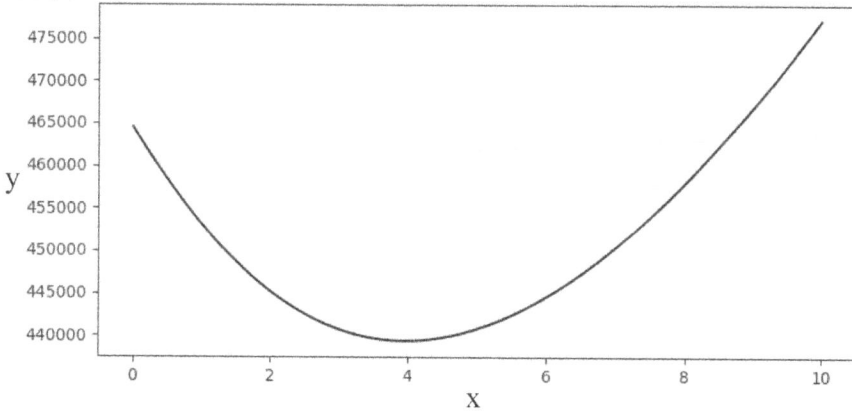

Code

```
# Golden-section search method
import math
def cal_f(_x): # Location
    return -((31575*math.sqrt(8.5**2+_x**2)+13342*(14.7-_x)))
def cal_d(_xu, _xl):
    return R * (_xu - _xl)
def check_e(_xu, _xl, _xopt, _e):
    _ea = (1 - R) * abs((_xu - _xl) / _xopt)
    print("Ea = {} %".format(_ea*100))
    if _ea < _e:
        return 1
    return 0
def search(_xu, _xl, _e, _n):
    _x = [0, 0]
    _f = [0, 0]
    _result = 0
```

```
    for i in range(0, _n):
        _d = cal_d(_xu, _xl)
        _x[0] = _xl + _d
        _x[1] = _xu - _d
        _f[0] = cal_f(_x[0])
        _f[1] = cal_f(_x[1])
        if _f[0] > _f[1]:
            _xl = _x[1]
            _result = _x[0]
            if check_e(_xu, _xl, _x[0], _e):
                break
        else:
            _xu = _x[0]
            _result = _x[1]
            if check_e(_xu, _xl, _x[1], _e):
                break
        if i % 1 == 0:
            print("Iteration {}: xL = {}\t xU = {}\t x1 = {}\t
fx1 = {}\t x2 = {}\t fx2 = {}\t d = {}\t"
                .format(i, round(_xl, 5), round(_xu, 5),
round(_x[0], 5), round(_f[0], 5), round(_x[1], 5),
                        round(_f[1], 5), round(_d, 5)))
    return _result
R = (math.sqrt(5) - 1) * 0.5
xU = 4
xL = 2
eS = 1 / 100
n = 100
result = search(xU, xL, eS, n)
print("x = {}, fx = {}".format(result, cal_f(result)))
Output
```

```
Ea = 14.589803375031543 %
Iteration 0: xL = 2.76393    xU = 4 x1 = 3.23607    fx1 = -440131.80613
x2 = 2.76393 fx2 = -441470.97645 d = 1.23607

Ea = 8.271182329550232 %
Iteration 1: xL = 3.23607    xU = 4 x1 = 3.52786    fx1 = -439644.41916
x2 = 3.23607 fx2 = -440131.80613 d = 0.76393

Ea = 4.863267791677183 %
Iteration 2: xL = 3.52786    xU = 4 x1 = 3.7082     fx1 = -439468.36888
x2 = 3.52786 fx2 = -439644.41916 d = 0.47214

Ea = 2.9179606750063103 %
Iteration 3: xL = 3.7082     xU = 4 x1 = 3.81966    fx1 = -439406.21501
x2 = 3.7082 fx2 = -439468.36888 d = 0.2918

Ea = 1.7714525201697713 %
Iteration 4: xL = 3.81966    xU = 4 x1 = 3.88854    fx1 = -439385.34432
x2 = 3.81966 fx2 = -439406.21501 d = 0.18034
```

```
Ea = 1.0829614118923137 %
Iteration 5: xL = 3.88854   xU = 4 x1 = 3.93112   fx1 = -439379.08055
x2 = 3.88854 fx2 = -439385.34432 d = 0.11146

Ea = 0.6648570266933098 %
x = 3.9574275274955837, fx = -439377.7281521616
```

Fibonacci Search

A Fibonacci search is a search procedure that utilizes the ratio of Fibonacci numbers to set up experimental points in a sequence. The Fibonacci numbers are a sequence that follows the rules:

$$F_0 = 1$$
$$F_1 = 1$$
$$F_i = F_{i-1} + F_{i-2}$$

This generates the sequence {1, 1, 2, 3, 5, 8, 13, 21, 34, 55, 89, . . .}.

The limiting value for the ratio of the consecutive Fibonacci sequences is the golden ratio 0.618. It is noted here because the golden section search method could be used in lieu of the Fibonacci method. However, the Fibonacci search converges faster than the golden section method.

In order to use the Fibonacci search procedure, we must ensure that certain assumptions hold. These key assumptions include that

1. the function must be unimodal over a specified interval,
2. the function must have an optimal solution over a known interval of uncertainty, and
3. we must accept an interval solution since the exact optimal cannot be found by this method.

Only an interval solution, known as the final interval of uncertainty, can be found using this technique. The length of this final interval is controllable by the user and can be made arbitrarily small by the selection of a *tolerance value*. The final interval is guaranteed to be less than this tolerance level.

Line search procedures use an initial interval of uncertainty to iterate to the final interval of uncertainty.

Finding the Maximum of a Function over an Interval with the Fibonacci Method

This search procedure to find a maximum is iterative, requiring evaluations of $f(x)$ at experimental points x_1 and x_2, where $x_1 = a + (F_{n-2}/F_n)(b - a)$ and $x_2 = a + (F_{n-1}/F_n)(b - a)$. These experimental points will lie between the original interval $[a, b]$. These experimental points are used to help determine the new interval of search. If $f(x_1) < f(x_2)$, then the new interval is $[x_1, b]$ and if $f(x_1) > f(x_2)$, then the new interval is $[a, x_2]$. The iterations continue in this manner until the final interval length is less than our imposed tolerance. Our final interval contains the optimum solution. It is the size of this final interval that determines our accuracy in finding the approximate optimum solution. The number of iterations required to achieve this accepted interval length can be found as the smallest Fibonacci number from the sequence that satisfies the inequality

$$F_k > \frac{(b-a)}{tolerance}.$$

Often we are required to provide a point solution instead of the interval solution. When this occurs the method of selecting a point is to evaluate the function, $f(x)$, at the endpoints of the final interval and at the midpoint of this final interval. For maximization problems, we select the value of x that yields the largest $f(x)$ solution. For minimization problems, we select the value of x that yields the smallest $f(x)$ solution.

The algorithm used is shown in Figure 9.9.

Although Fibonacci can be used with any unimodal function to find the maximum (or minimum) over a specified interval, its main advantage comes when normal calculus procedures fail. Consider the following example:

Maximize $f(x) = -|2 - x| - |5 - 4x| - |8 - 9x|$ over the interval $0 \le x \le 3$.

In calculus, absolute values are not differentiable because they have corner points. Thus, taking the first derivative and setting it equal to zero is not an option. Another method needs to be used to find the solution. We use the Fibonacci to solve this problem and other examples.

Example 9. Maximizing a Function That Does Not Have a Derivative

```
Maximize f(x) = -|2-x|-|5-4x|-|8-9x|  over the interval
0 ≤ x ≤ 3.
```

```
> f:= x->-(abs(2-x)+abs(5-4*x)+abs(8-9*x));
```

$$f := x \to -|2 - x| - |5 - 4x| - |8 - 9x|$$

```
> FIBSearch(f,0,3,.1);
```

```
The interval [a,b] is [ 0.00,  3.00]and user-specified
tolerance level is 0.10000.
The first 2 experimental endpoints are x1= 1.147 and
x2 = 1.853.
```

To find a maximum solution to a given function, $f(x)$, on the interval $[a, b]$, where the function, $f(x)$, is unimodal.

INPUT: endpoints a, b; tolerance, t, Fibonacci sequence

OUTPUT: final interval $[a_i, b_i]$, $f(midpoint)$

Step 1.   Initialize the tolerance, $t > 0$.

Step 2. Set $F_n > (b-a)/t$ as the smallest F_n and define the test points

$$x_1 = a + (F_{n-2}/F_n)(b-a)$$

$$x_2 = a + (F_{n-1}/F_n)(b-a)$$

Step 3. Calculate $f(x_1)$ and $f(x_2)$

Step 4. Compare $f(x_1)$ and $f(x_2)$

      a. If $f(x_1) \leq f(x_2)$, then the new interval is $[x_1, b]$:

           a becomes the previous x_1

           b does not change

           x_1 becomes the previous x_2

           $n = n - 1$

           Find the new x_2 using the formula in Step 2.

      b. If $f(x_1) > f(x_2)$, then the new interval is $[a, x_2]$:

           a does not change

           b becomes the previous x_2

           x_2 becomes the previous x_1

           $n = n - 1$

           Find the new x_1 using the formula in Step 2.

Step 5. If the length of the new interval from Step 4 is less than the tolerance specified, then stop. Otherwise go back to Step 3.

Step 6. Estimate x^* as the midpoint of the final interval and compute, $f(x^*)$, the estimated maximum of the function.

STOP

FIGURE 9.9
Fibonacci algorithm.

```
Iteration  x(1)     x(2)      f(x1)     f(x2)     Interval
   1       1.1471  1.8529   -3.5882  -11.2353  [0.0000, 1.8529]
   2       0.7059  1.1471   -5.1176   -3.5882  [0.7059, 1.8529]
   3       1.1471  1.4118   -3.5882   -5.9412  [0.7059, 1.4118]
   4       0.9706  1.1471   -2.8824   -3.5882  [0.7059, 1.1471]
   5       0.8824  0.9706   -2.6471   -2.8824  [0.7059, 0.9706]
   6       0.7941  0.8824   -3.8824   -2.6471  [0.7941, 0.9706]
```

The midpoint of the final interval is 0.882353 and
f(midpoint) = -2.647.
The maximum of the function is -2.882 and the x value =
0.970588.

In this example, we want a specific point as our solution. The midpoint
yields the maximum value of $f(x)$. Thus, we will use $x = 0.970588$ with
$f(0.970588) = -2.882$ as our solution.

Example 10. Maximizing a Transcendental Function

Maximize the function $f(x) = 1 - \exp(-x) + 1/(1 + x)$ over the interval
[0, 20].

```
> f:=(x)->1-exp(-x)+1/(1+x);
```

$$f := x \rightarrow 1 - e^{-x} + \frac{1}{1+x}$$

```
> FIBSearch(f,0,20,.1);
```

The interval [a,b] is [0.00, 20.00]and user-specified
tolerance level is 0.10000.
The first 2 experimental endpoints are x1= 7.639 and
x2 = 12.361.

```
Iteration  x(1)     x(2)      f(x1)    f(x2)     Interval
   1       7.6395  12.3605  1.1153  1.0748  [0.0000, 12.3605]
   2       4.7210   7.6395  1.1659  1.1153  [0.0000,  7.6395]
   3       2.9185   4.7210  1.2012  1.1659  [0.0000,  4.7210]
   4       1.8026   2.9185  1.1919  1.2012  [1.8026,  4.7210]
   5       2.9185   3.6052  1.2012  1.1900  [1.8026,  3.6052]
   6       2.4893   2.9185  1.2036  1.2012  [1.8026,  2.9185]
   7       2.2318   2.4893  1.2021  1.2036  [2.2318,  2.9185]
   8       2.4893   2.6609  1.2036  1.2033  [2.2318,  2.6609]
   9       2.4034   2.4893  1.2034  1.2036  [2.4034,  2.6609]
  10       2.4893   2.5751  1.2036  1.2036  [2.4034,  2.5751]
```

The midpoint of the final interval is 2.489270 and
f(midpoint) = 1.204.
The maximum of the function is 1.204 and the x value =
2.403433

Again, assuming that we desire a specific numerical value as the solution, our solution is $x = 2.489270$ with $f(2.489270) = 1.204$.

9.6 Interpolation with Derivatives: Newton's Method for Nonlinear Optimization

Finding the Critical Points (Roots) of a Function

Newton's method has been adapted to solve nonlinear optimization problems. For a function of a single variable, the adaptation is straightforward. Newton's method is applied to the *derivative* of the function we wish to optimize, for the function's critical points occur where the derivative's roots are found. When finding the critical points of the function, Newton's method is based on the derivative of the quadratic approximation of the function $f(x)$ at the point x_k:

$$q(x) = f(x_k) + f'(x_k)(x - x_k) + \tfrac{1}{2} f''(x_k)(x - x_k)^2.$$

The result, $q'(x)$, is a linear approximation of $f'(x)$ at the point x_k. Setting $q'(x) = 0$ and solving for x yields the formula

$$x_{k+1} = x_k - \frac{f'(x_k)}{f''(x_k)},$$

where $x_{k+1} \equiv x$.

Newton's method can be terminated when $|x_{k+1} - x_k| < \varepsilon$, where epsilon is a prespecified scalar tolerance or when $|f'(x)| < \varepsilon$.

In order to use Newton's method to find the critical points of a function, the function's first and second derivatives must exist in the neighborhood of interest. Also note that when the second derivative at x_k is zero, the point x_{k+1} cannot be computed.

It is important to first master the computations required in the algorithm. It is also noted that Newton's method finds only the approximate critical value, it does not know whether it is finding a maximum or a minimum. The sign of the second derivative may be used to determine if we have a maximum or a minimum.

The Basic Application

Consider any simple polynomial, such as $f(x) = 5x - x^2$, whose critical point can easily be found using calculus by taking the first derivative and setting it equal to zero. We find that the critical point $x = 2.5$ yields a maximum of the function. Applying Newton's method to find critical points requires finding $f'(x)$ and $f''(x)$ and then using a computation device to perform the iterations:

$f'(x) = 5 - 2x$ and $f''(x) = -2$ Newton's method uses $x_{k+1} = x_k - \dfrac{f'(x_k)}{f''(x_k)}$ or

$$x_{k+1} = x_k - \frac{(5 - 2x)}{(-2)}.$$

Starting at $x_0 = 1$ yields:

TABLE 9.1

Newton's Method Iterations

k	x_k	$f'(x)$	$f''(x)$	x_{k+1}	$\|x_k - x_{k+1}\|$
0	1	−3	−2	2.5	1.5
1	2.5	0	−2	2.5	0

Starting at other values also yields $x = 2.5$. Since this simple quadratic function has a derivative that is a linear function, the linear approximation of the derivative will be exact regardless of the starting point, and the answer will be confirmed at the second iteration. Newton's method produces the critical values of $f'(x)$ without regard to the point x_k being a maximum or a minimum. We know we have found a maximum by looking at the entries in the table for $f''(x)$. Since $f''(x)$ at $x = 2.5$ is −2, which is less than or equal to 0.

Note that the slope of the linear approximation of the function at the point x_k is precisely the slope of the function at that point, so the linear approximation is tangent to the function at the point x_k as shown in Figure 9.10.

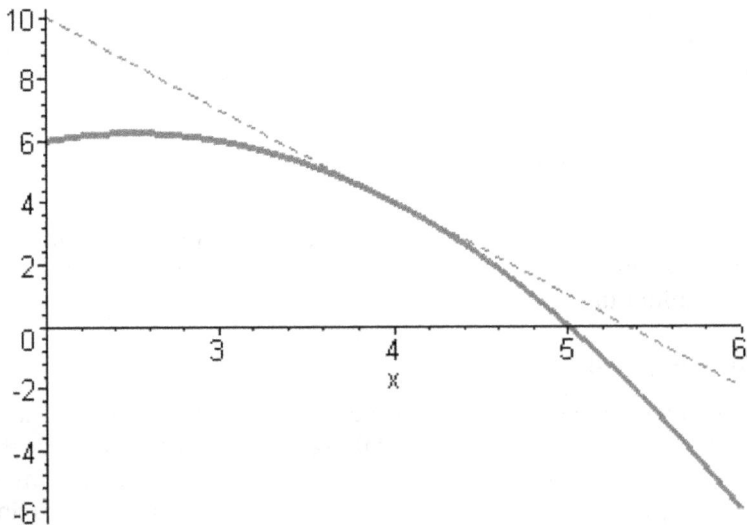

FIGURE 9.10
A graph of the function $f(x) = 5x - x^2$ and its linear approximation at the point $x_k = 4$.

Example 11. Minimize $f(x) = x^2 + 2x$

We will start our guess with $x = 4$ and use a stopping criteria of epsilon = 0.01.

$$f(x) = x^2 + 2x$$
$$f'(x) = 2x + 2$$
$$f''(x) = 2$$

$$x_2 = x_1 - f'(x_1)/f''(x_1)$$
$$x_2 = 4 - (10/2) = 4 - 5 = -1$$

$$x_3 = x_2 - f'(x_2)/f''(x_2)$$
$$x_2 = -1 - (0/2) = -1$$

Stop

$$|f'(x)| = 0 < 0.01 \text{ or } |x_2 - x_1| = 0 < 0.01$$

Example 12. Maximize $f(x) = -2x^3 + 10x - 10$

We will begin at $x = 1$ and use a stopping criteria of epsilon = 0.01

$$f'(x) = -6x^2 + 10$$
$$f''(x) = 12x$$
At $x = 1, f'(1) = 4$ and $f''(1) = -12$.
$$x_2 = 1 - (-4/12) = 1.33333$$

Neither $|f'(x)|$ nor $|x_{k+1} - x_k|$ are less than 0.01, so we continue. We summarize in the following table.

k	X	$f'(x)$	$f''(x)$	x (k + 1)
1	1	4	−12	1.333333
2	1.333333	−0.66667	−16	1.291667
3	1.291667	−0.01042	−15.5	1.290995
4	1.290995	−2.7E-06	−15.4919	1.290994

Since $f'(x) = |-2.7\text{E-}06| < 0.01$, we stop. Our critical point is $x = 1.290994$. Since $f''(x) < 0$, then we have found a maximum.

The Bisection Method with Derivatives

To utilize bisection method with derivatives correctly there are certain properties that must hold for the given function:

Let f' be a function that has opposite sign values at each end of some specified interval. Then, by the *intermediate-value property* (IVP) of continuous functions (mentioned in many basic college algebra texts), we are guaranteed to find a root between the endpoints of the given interval. Specifically, the IVP states that given two points (x_1, y_1) and (x_2, y_2), with $y_1 \neq y_2$ on a graph of the continuous function f, the function f takes on every value between y_1 and y_2. Thus, with values having opposite signs, then there must be a value for which $f'(x) = 0$.

The algorithm is listed in Figure 9.11.

Example 13. Minimize $f(x) = x^2 + 2x$, $-3 \leq x \leq 6$. Let epsilon = 0.2.

The number of required observations is found by solving the formula $(0.5)^n = t/(b - a)$.
   We find that $n = 6$ observations.

$$f'(x) = 2x + 2$$
$$k = 1$$

Step 1. Find two values a and b, where $f'(a)$ and $f'(b)$ have opposite signs.

Step 2. Fix a tolerance for the final interval of the solution $[a_f, b_f]$ so that

$|b_f - a_f| < $ tolerance.

Step 3. Find the midpoint, $m_i = (b_i + a_i)/2$

Step 4. Compute $f'(a_i), f'(b_i)$, and $f'(m_i)$.

Step 5. Determine if $f'(a_i) * f'(m_i) < 0$.

   If true, then

   $a_i = a_i$

   $b_i = m_i$

   Otherwise,

   $a_i = m_i$

   $b_i = b_i$

Step 6. If $f'(m_i) \neq 0$ and the new $|b_i - a_i| > $ tolerance then using the new interval $[a_i, b_i]$

   go back to Step 3 and repeat the process. Otherwise, **STOP**

FIGURE 9.11
The bisection with derivatives algorithm.

1. $x_{mp} = 0.5(6 - 3) = 1.5, f'(1.5) = 5, f'(1.5) > 0,$
   so
   $a = -3,$
   $b = 1.5,$ and
   $x_{mp} = -0.75.$

2. $f'(-0.75) = 0.5 > 0,$
   so
   $a = -3,$
   $b = -0.75,$ and
   $x_{mp} = -1.875.$

3. $f'(-1.875) = -1.75 < 0,$
   so
   $a = -1.875,$
   $b = -0.75,$ and
   $x_{mp} = -1.3125.$

4. $f'(-1.3125) = -0.625 < 0,$
   so
   $a = -1.3125,$
   $b = -0.75,$ and
   $x_{mp} = -1.03125.$

5. $f'(-1.03125) = -0.0625 < 0,$
   so
   $a = -1.30125,$
   $b = -0.75,$ and
   $x_{mp} = -0.89025.$

6. $f'(-0.89025) = 0.2195 > 0,$
   $a = 1.030125$ and
   $b = -0.89025.$

The final interval is less than 0.2 so we stop. Our solution is between $[-1.030125, -0.89025]$.

Just as previously discussed, if we need a single value we can evaluate the derivative of the function, f', at a, at b, and at the midpoint selecting the value that is closest to $f'(x) = 0$ from those points.

$$f'(-1.030125) = -0.0625$$
$$f'(-0.89025) = 0.2195$$
$$f'(-0.96075) = 0.0705$$

$x = -1.030125$ with $f'(x) = -0.0625$ is best. The exact value, via single variable calculus, for the maximization of f is the solution $x = -1$.

Exercises

Use the golden section method, Fibonacci's method, Newton's method, and bisection method to solve the following:

1. Maximize $f(x) = -x^2 - 2x$ on the closed interval $[-2, 1]$. Using a tolerance for the final interval of **0.6**. Hint (Start Newton's method at $x = -0.5$.)

2. Maximize $f(x) = -x^2 - 3x$ on the closed interval $[-3, 1]$. Using a tolerance for the final interval of **0.6**. (Hint start Newton's method at $x = 1$.)

3. Minimize $f(x) = x^2 + 2x>$ on the closed interval $[-3, 1]$. Using a tolerance for the final interval of **0.5**. (Start Newton's at $x = -3$.)

4. Minimize $f(x) = -x + e^x$ over the interval $[-1, 3]$ using a tolerance of 0.1. (Start Newton's method at $x = -1$.)

5. List at least two assumptions required by both the golden section and Fibonacci's search methods.

6. Consider minimizing $f(x) = -x + e^x$ over the interval $[-1, 3]$. Assume your final interval yielded a solution within the tolerance of $[-0.80, 0.25]$. Report a single best value of x to minimize $f(x)$ over the interval.

Projects

1. We are considering buying the new E-phone system, which is computer-compatible with the new E-computer system. The company is concerned with how often it will replace the machines and the cost involved. The company's research and development team estimates that when the E-phone is t years old, it will allow the user to earn revenue at a rate of e^{-t} per year. After t years of use, the E-phone can be sold to a third-world company for $1/(1 + t)$ dollars. Maintenance costs after t years are estimated as $0.01\,t$ dollars. Build a model for the company and determine how long they should keep the E-phone system before replacing it. Make a recommendation to the CEO as to what you would do based on your modelling.

2. Previously we discussed the least squares method to fit the parameters to a proposed model. Another choice to fit the model is minimize the sum of the absolute deviations between the proposed model and the data. For example, consider the model using $W = kL^3$ and the data available.

Data Available:

Length	12.5	12.625	12.625	14.125	14.5	14.5	17.27	17.75
Weight	17	16	17	23	26	27	43	49

A model that we could use to find the slope is a search method. Find the value of k that minimizes the function S.

Minimize $S = |17 - k * 12.5^3| + |16 - k * 12.625^3 + \ldots + |49 - k * 17.75^3|$

Use any of our numerical techniques and find the value of k.

3. In the following figure, the profit function is given for a sulfuric acid alkylation reactor as a function of feed rate and catalyst concentration. Plot the profit function as a function of feed rate for a constant catalyst concentration of 95%. Place six golden section experiments on the interval giving their location, the corresponding value of the profit function and the length and location of the final interval of uncertainty. Determine the optimal solution to a tolerance of 0.25, 0.1, 0.01?

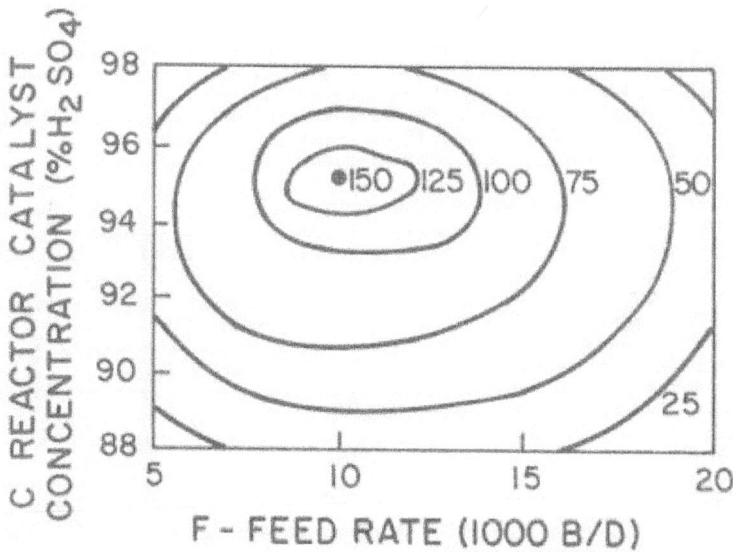

Profit Function for the Operation of a Sulfuric Acid Alkylation Reactor as a Function of Catalyst Concentration and Feed Rate.

4. An economic analysis of a proposed facility is being conducted in order to select an operating life such that the maximum uniform annual income is achieved. A short life results in high annual amortization costs, but the maintenance costs become excessive for a long life. The annual income after deducting all operating costs, except maintenance costs, is $180,000. The installed cost of the facility, C, is $500,000 borrowed at 10% interest compounded annually. The maintenance charges on an annual basis are evaluated using the product of the gradient present-worth factor and the capital-recovery factor. In the gradient present-worth method, there are no maintenance charges the first year, a cost M for the second year, 2M for the third year, 3M for the fourth year, and so on. The second-year cost, M, for the problem is $10,000. The annual profit is given by the following equation:

$$P = 180{,}000 - \{[(i+1)^N - 1 - iN]/i[(1+i)^N - 1]\}M - \{i(1+i)^N/[(1+i)^N - 1]\}C,$$

where i is the interest rate and N is the number of years. Determine the number of years, N, that give the maximum uniform annual income, P. For your convenience the following table gives the values of the coefficients of M and C for $i = 0.1$ as a function of N.

Coefficient of			Coefficient of		
Year	M	C	Year	M	C
1	0	1.10	11	4.05	0.154
2	0.476	0.576	12	4.39	0.147
3	0.909	0.403	13	4.69	0.141
4	1.30	0.317	14	5.00	0.136
5	1.80	0.264	15	5.28	0.131
6	2.21	0.230	16	5.54	0.128
7	2.63	0.205	17	5.80	0.125
8	2.98	0.188	18	6.05	0.122
9	3.38	0.174	19	6.29	0.120
10	3.71	0.163	20	6.51	0.117

5. It is proposed to recover the waste heat from exhaust gases leaving a furnace (flow rate, $m = 60{,}000$ lb/hr; heat capacity, $c_p = 0.25$ BTU/lb°F) at a temperature of $T_{in} = 500$°F by installing a heat exchanger (overall heat transfer coefficient, $U = 4.0$ BTU/hr, ft^2, °F) to produce steam at $T_s = 220$°F from saturated liquid water at 220°F. The value of heat in the form of steam is $p = \$0.75$ per million BTUs, and the installed cost of the heat exchanger is $c = \$5.00$ per ft^2 of gas-side area. The life

of the installation is $n = 5$ years, and the interest rate is $i = 8.0\%$. The following equation gives the net profit P for the 5 year period from the sale of the steam and the cost of the heat exchanger. The exhaust gas temperature T_{out} can be between the upper and lower limits of 500 °F and 220 °F.

$$P = pqn - cA(1 + i)n,$$

where

$$q = mc_p(T_{in} - T_{out}) = UADT_{LM}$$

$$\Delta T_{LM} = \frac{(T_{in} - T_s) - (T_{out} - T_s)}{\ln\left(\dfrac{T_{in} - T_s}{T_{out} - T_s}\right)}$$

a. Derive the following equation for this design.

$$P = 91{,}137 - 492.75T_{out} + 27550\ln(T_{out} - 220)$$

b. Use a Fibonacci search with seven experiments to locate the optimal outlet temperature T_{out} to maximize the profit P on the interval of T_{out} from 220 °F to 500 °F. Find the largest value of the profit and the size and location of the final interval of uncertainty for the fractional resolution based on the initial interval of $e = 0.01$.

c. Use another numerical method and compare your results.

Further Readings

Bazarra, M., C. Shetty and H. D. Scherali (1993). *Nonlinear Programming: Theory and Applications*. Wiley, New York.

Fox, W. P. (1992, January–March). Teaching Nonlinear Programming with Minitab. *COED Journal*, II(1): 80–84.

Fox, W. P. (1993). Using Microcomputers in Undergraduate Nonlinear Optimization. *Collegiate Microcomputer*, XI(3): 214–218.

Fox, W. P. (2021). *Nonlinear Optimization*. CRC Press, Boca Raton, FL.

Fox, W. P. and R. Burks (2021). *Advanced Mathematical Modeling*. Taylor and Francis, CRC Press, Boca Raton, FL.

Fox, W. P. and R. Burks (2022). *Mathematical Modeling Under Change, Uncertainty, and Machine Learning*. Taylor and Francis, CRC, Boca Raton, FL.

Fox, W. P., F. Giordano and M. Weir (2003). *A First Course in Mathematical Modeling*, 3rd ed. Brooks/Cole, Monterey, CA.

Fox, W. P. and M. Witherspoon (2001). Single Variable Optimization When Calculus Fails: Golden Section Search Methods in Nonlinear Optimization Using MAPLE. *COED*, XI(2): 50–56.

Phillips, D. T., A. Ravindran and J. Solberg (1976). *Operations Research*. John Wiley and Sons, New York.

Rao, S. S. (1979). *Optimization: Theory and Applications*. Wiley Eastern Limited, New Delhi, India.

Winston, W. (2002). *Introduction to Mathematical Programming: Applications and Algorithms*, 4th ed. Duxbury Press, ITP, Belmont, CA.

10

Multivariable Numerical Search Methods

10.1 Introduction

Consider the following problem. A small company is planning to install a central computer with cable links to five departments. According to the floor plan, the peripheral computers for the five departments will be situated using a 100 by 100 grid (measured in feet) as shown by the circles in Figure 10.1. The company wishes to locate the central computer so that the minimal amount of cable will be used to link the central computer to the five peripheral computers (Fox et al., 2002).

The five peripherals are located at integer-coordinate positions $(15, 60)$, $(25, 90)$, $(60, 75)$, $(75, 60)$, and $(80, 25)$. Cable may be strung over the ceiling panels in a straight line from a point above any peripheral to a point above the central computer, and it is not necessary to consider lengths of cable from a peripheral computer itself to a point above the ceiling panel located immediately over that computer. That is, we work only with lengths of cable strung over the ceiling panels. The central computer will be located at coordinates (m, n), where m and n are integers in the grid representing the office space. A solution to this minimization problem can be sought using concepts from multivariable calculus, nonlinear programming, computer programming, and integer programming. We will show how numerical methods work to solve this problem.

Background Theory

In the previous sections we discussed analytical and numerical techniques to solve the unconstrained nonlinear programming problem (NLP):

$$\text{Max } z = f(x_1, x_2, x_3, \dots, x_n) \text{ over } R^n. \tag{10.1}$$

In many problems, it is quite difficult to find the stationary points (critical points) and use them to determine the nature of the stationary point. In this chapter, we will discuss several numerical techniques to either maximize or minimize a multivariable function as expressed in Equation 10.1.

DOI: 10.1201/9781032703671-10

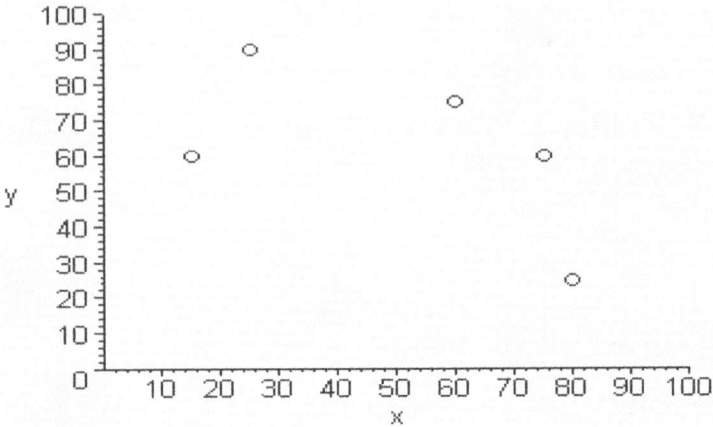

FIGURE 10.1
A grid showing locations of the five peripheral computers.

10.2 Gradient Search Methods

Suppose we want to solve the following unconstrained NLP:

$$\text{Max } z = f(x_1, x_2, x_3, \ldots, x_n). \quad (10.2)$$

In calculus, if Equation 10.1 is a concave function, then the optimal solution (if there is one) will occur at a stationary point x^* having the following property:

$$\frac{\partial f(x^*)}{\partial x_1} = \frac{\partial f(x^*)}{\partial x_2} = \ldots = \frac{\partial f(x^*)}{\partial x_n} = 0.$$

In many problems, it is not an easy task to find the stationary point. Thus, the method of steepest ascent (maximization problems) and the method of steepest descent (minimization problems) offers an alternative to finding an approximate stationary point. We will continue to discuss the gradient method for the steepest ascent.

Given a function, like the one in Figure 10.2, assume that we want to find the maximum point of the function. If we started at the bottom of the hill, then we might proceed by finding the gradient. The gradient is the vector of the partial derivatives that points "up the hill." We define the gradient vector as follows:

$$\nabla f(x) = [\frac{\partial f(x^*)}{\partial x_1}, \frac{\partial f(x^*)}{\partial x_2}, \ldots, \frac{\partial f(x^*)}{\partial x_n}].$$

If we were lucky, the gradient would point all the way to the top of the surface, but the contours of functions do not always cooperate and rarely do.

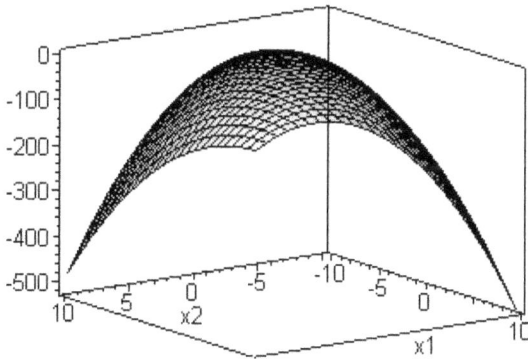

FIGURE 10.2
The function $Z = f(x_1, x2) = 2x_1x_2 + 2x_2 - x_1^2 - 2x_2^2$.

Thus, the gradient "points up hill," but for how far? We need to find the distance along the gradient for which to travel that maximizes the height of the function in that direction. From that new point, we re-compute a new gradient vector to find a new direction that "points up hill." We continue this method until we get to the top of the hill.

From a starting point, we move in the direction of the gradient as long as we continue to increase the value of *f*. At that point, we move in the direction of a newly calculated gradient as far as we can so long as it continues to increase *f*. This continues until we achieve our maximum value within some specific tolerance (or margin of acceptable error). Figure 10.3 displays an algorithm for the method of steepest ascent using the gradient. For descent, just substitute *−f* for *f* in the algorithm.

Example 1. Computer Placement with Gradient Search

For our computer placement problem, our function follows:
The model is to minimize the distance *d*, where

$$d = \sqrt{(x - X_i)^2 + (y - Y_i)^2}$$

$$f(x,y) = \sum_{i=1}^{5}\sqrt{(x - X_i)^2 + (y - Y_i)^2}$$

$$fx := (x,y) \rightarrow \frac{1}{2}\frac{2x - 30}{\sqrt{x^2 - 30x + 3825 + y^2 - 120y}} + \frac{1}{2}\frac{2x - 50}{\sqrt{x^2 - 50x + 8725 + y^2 - 180y}}$$

$$+ \frac{1}{2}\frac{2x - 120}{\sqrt{x^2 - 120x + 9225 + y^2 - 150y}} + \frac{1}{2}\frac{2x - 150}{\sqrt{x^2 - 150x + 9225 + y^2 - 120y}}$$

$$+ \frac{1}{2}\frac{2x - 160}{\sqrt{x^2 - 160x + 7025 + y^2 - 50y}}$$

To find a maximum solution to a given multivariable unconstrained function, $f(x)$

INPUT: starting point x_0; tolerance, t

OUTPUT: Approximate x^*, and evaluate $f(x^*)$

Step 1.   Initialize the tolerance, $t > 0$.

Step 2.   Set $x=x_0$ and define the gradient at that point.

$$\nabla f(x_0)$$

Step 3.   Calculate the maximum of the new function $f(x_i + t_i \ \nabla f(x_i))$,

where $t_i \geq 0$, by finding the value of t_i.

Step 4.   Find the new x_i point by substituting t_i into

$$x_{i+1} = x_i + t_i \ \nabla f(x_i)$$

Step 5. If the length (magnitude) of **x**, defined by

$$\| \mathbf{x} \| = (x_1^2 + x_2^2 + ... + x_n^2)^{\frac{1}{2}},$$

(difference between 2 successive points) is less than the tolerance specified or if the **absolute magnitude of gradient is less than our tolerance** (derivative approximately zero), , then **continue.**

Otherwise, go back to Step 3.

Step 6. Use x^* as the approximate stationary point and compute $f(x^*)$, the estimated maximum of the function.

STOP

FIGURE 10.3
Steepest ascent algorithm.

$$fy := (x,y) \rightarrow \frac{1}{2}\frac{2x-120}{\sqrt{x^2-30x+3825+y^2-120y}} + \frac{1}{2}\frac{2x-180}{\sqrt{x^2-50x+8725+y^2-180y}}$$

$$+ \frac{1}{2}\frac{2x-150}{\sqrt{x^2-120x+9225+y^2-150y}} + \frac{1}{2}\frac{2x-120}{\sqrt{x^2-150x+9225+y^2-120y}}$$

$$+ \frac{1}{2}\frac{2x-50}{\sqrt{x^2-160x+7025+y^2-50y}}$$

With our steepest ascent algorithm, we find the computer placement should be at (56.8299, 68.0631) and f(56.8299, 68.0631) =157.6635 in four iterations.

```
> f:=(sqrt((x1-15)^2+(x2-60)^2)+sqrt((x1-25)^2+(x2-90)^2)+sqrt(
  (x1-60)^2+(x2-75)^2)+sqrt((x1-80)^2+(x2-25)^2)+sqrt((x2-60)^2+
  (x1-75)^2));
```

$$f := \sqrt{x1^2 - 30x1 + 3825 + x2^2 - 120x2} + \sqrt{x1^2 - 50x1 + 8725 + x2^2 - 180x2} \qquad (1)$$

$$+ \sqrt{x1^2 - 120x1 + 9225 + x2^2 - 150x2} + \sqrt{x1^2 - 160x1 + 7025 + x2^2 - 50x2}$$

$$+ \sqrt{x2^2 - 120x2 + 9225 + x1^2 - 150x1}$$

```
> UP(200,.005,0,3,f);
```


Initial Condition: (0.0000, 3.0000)

Iter Length	Gradient Vector G	magnitude G	x[k]	Step
1 -18.6620	(-2.9312, -3.5666)	4.6166	(0.0000, 3.0000)	
2 -6.6799	(-.3001, .2466)	.3884	(54.7025, 69.5607)	
3 -9.2843	(-.0133, -.0161)	.0209	(56.7069, 67.9134)	
4	(.0018, -.0015)	.0023		

Approximate Solution:	(56.8299, 68.0631)
Maximum Functional Value:	157.6635
Number gradient evaluations:	5
Number function evaluations:	4

☑ Editable Maple Default Profile E:\Maple files from old computer\MAPLE NLP Me

FIGURE 10.4
Maple screenshot of algorithm solution to the computer placement example.

Example 2. Maximize $f(x_1, x_2) = 2x_1x_2 + 2x_2 - x_1^2 - 2x_2^2$

The gradient of $f(x_1, x_2)$, ∇f, is found using the partial derivatives. The gradient is the vector

$$[2x_2 - 2x_1, 2x_1 + 2 - 4x_2].$$

$\nabla f(0, 0) = [0, 2]$. From $(0, 0)$, we move along (up) the x_2-axis in the direction of $[0, 2]$. How far do we go? We need to maximize the function starting at the point $(0, 0)$ using the function $f(x_i + t_i \nabla f(x_i)) = f(0 + 0t, x\ 0 + 2t) = 2(2t) - 2(2t)^2 = 4t - 8t^2$.

This function can be maximized by using any of the one-dimensional search techniques that we discussed in Chapter 8 (single-variable optimization). This function can also be maximized by simple single-variable calculus:

$$\frac{df}{dt} = 0 = 4 - 16t = 0, t = 0.25$$

The new point is found by substitution into $x_{i+1} = x_i + t_i \nabla f(x_i)$.

So, $x_1 = [0 + 0(0.25), 0 + 2(0.25)], [0, 0.5]$.

The magnitude of x_1 is 0.5, which is not less than our tolerance of 0.01 (chosen arbitrarily). Since we are not optimal we continue. We now repeat the calculations from the new point $[0, 0.5]$.

Iteration 2

The gradient vector is $[2 x_2 - 2x_1, 2 x_1 + 2 - 4x_2]$.

$\nabla f(0, 0.5) = [1, 1]$. From $(0, 0.5)$, we move in the direction of $[1, 0]$. How far do we go? We need to maximize the function starting at the new point $(0, 0.5)$ using the function $f(x_i + t_i \nabla f(x_i)) = f(0 + 1t, 0.5 + 0t) = 2(t)(.5) + 2(.5) - t^2 - 2(.5)^2 = -t^2 + t + .5$.

This function can also be maximized be using any of the one-dimensional search techniques that we discussed in Chapter 9 or be maximized by simple single-variable calculus:

$$\frac{df}{dt} = 0 = -2t + 1 = 0, t = 0.50.$$

The new point is found by substitution into $x_{i+1} = x_i + t_i \nabla f(x_i)$.

So, $x_1 = [0 + 1(0.5), 0.5 + 0(0.5)], [0.5, 0.5]$.

The magnitude of x_1 is $\sqrt{.5} = 0.707$, which is not less than our tolerance of 0.01 (chosen arbitrarily). The magnitude of $\nabla f = 1$, which is also not less than 0.01. Since we are not optimal, we continue. We repeat the calculations from the new point $[0.5, 0.5]$.

We use Maple to complete the process.

```
> f:=2*x1*x2+2*x2-x1^2-2*x2^2;
f:= 2 x1 x2 + 2 x2 - x1² - 2 x2
```

```
                Initial Condition: ( 0.0000, 0.0000)
Iter  Gradient Vector G magnitude G      x[k]      Step Length
 1    ( 0.0000, 2.0000)    2.0000    (0.0000, 0.0000)   .25
 2    ( 1.0000, 0.0000)    1.0000    (0.0000,  .5000)   .50
 3    ( 0.0000, 1.0000)    1.0000    ( .5000,  .5000)   .25
 4    (  .5000, 0.0000)     .5000    ( .5000,  .7500)   .50
 5    ( 0.0000,  .5000)     .5000    ( .7500,  .7500)   .25
 6    (  .2500, 0.0000)     .2500    ( .7500,  .8750)   .50
 7    ( 0.0000,  .2500)     .2500    ( .8750,  .8750)   .25
 8    (  .1250, 0.0000)     .1250    ( .8750,  .9375)   .50
 9    ( 0.0000,  .1250)     .1250    ( .9375,  .9375)   .25
10    (  .0625, 0.0000)     .0625    ( .9375,  .9688)   .50
11    ( 0.0000,  .0625)     .0625    ( .9688,  .9688)   .25
12    (  .0313, 0.0000)     .0313    ( .9688,  .9844)   .50
13    ( 0.0000,  .0313)     .0313    ( .9844,  .9844)   .25
14    (  .0156, 0.0000)     .0156    ( .9844,  .9922)   .50
15    ( 0.0000,  .0156)     .0156    ( .9922,  .9922)   .25
16    (  .0078,  .0000)     .0078
```

```
           Approximate Solution: ( .9922, .9961)
           Maximum Functional Value:        1.0000
           Number gradient evaluations:     17
           Number function evaluations:     16
```

The solution, via calculus, is as follows:

$$\text{Maximize } f(x_1, x_2) = 2x_1x_2 + 2x_2 - x_1^2 - 2x_2^2$$

$$\frac{\partial f}{\partial x_1} = 0 = 2x_2 - 2x_1$$

$$\frac{\partial f}{\partial x_2} = 0 = 2x_1 + 2 - 4x_2$$

Solving these two equations simultaneously yields

$$x_1 = 1$$
$$x_2 = 1$$
$$f(x_1, x_2) = 1.$$

The Hessian matrix, $\begin{bmatrix} -2 & 2 \\ 2 & -4 \end{bmatrix}$, is negative definite so the point x^* is a maximum. Note that our approximate solutions, x, (0.9922, 0.9961), and $f(x_1, x_2) =$ 1.000, are close to the exact value of x^*, (1, 1), and $f(x^*) = 1$. To get a closer approximation, we should make our tolerance smaller. A look at the contour plot confirms a hill at approximately (1, 1).

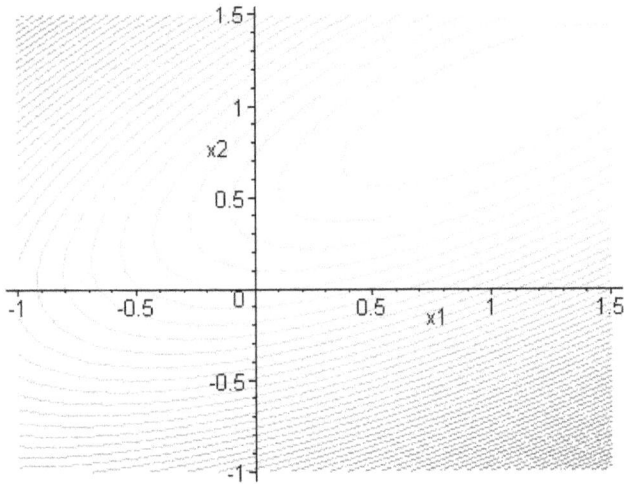

FIGURE 10.5
Contour plot of $2x_1x_2 + 2x_2 - x_1^2 - 2x_2^2$.

Example 3. Maximizing a Transcendental Multivariable Function When Calculus Fails

$$f(x_1, x_2) = 2x_1x_2 + 2x_2 - e^{x_1} - e^{x_2} + 10$$

These partial derivative equations,

$$2x_2 - e^{x1} = 0$$
$$2x_1 + 2 - e^{x2} = 0,$$

are not solvable in closed form for (x_1, x_2) without numerical methods. So, we will use the gradient method to approximate the solution. Here we have no analytical solution for comparison.

We will start at the point $(0.5, 0.5)$.
The gradient is $\nabla f = [2x_2 - e^{x1}, 2x_1 + 2 - e^{x2}]$.

$\nabla f(0, 0) = [-0.6487, 1.3513]$ with magnitude $= 1.4989$.

A new point is found by maximizing $f[0.5 - .6487t, 0.5 + 1.3513t] =$

$$2(0.5 - .6487)(0.5 + 1.3513t) + 2(0.5 + 1.3513t) - e^{0.5 - .6487t} - e^{0.5 + 1.3513t}$$
$$df/dt = 0 = 3.40520 - 3.50635324\,t + 0.6487\,e^{(0.5 - 0.6487t)} - 1.3513\,e^{(0.5 + 1.3513t)}$$
$$t = 0.2873853295 = .2874 \text{ (rounded to 4 decimals)}$$

This moves us to the next point $(0.3136, 0.8883)$. We continue the search using technology.

```
> f:=2*x1*x2+2*x2-exp(x1)-exp(x2)+10;
```

```
                 Initial Condition: ( .5000, .5000)
Iter Gradient Vector G magnitude G      x[k]        Step Length
 1   ( -.6487, 1.3513)   1.4989    ( .5000,   .5000)    .2874
 2   (  .4084,  .1961)    .4530    ( .3136,   .8883)   1.7041
 3   ( -.2993,  .6235)    .6917    ( 1.0095, 1.2224)    .2001
 4   (  .1097,  .0526)    .1216    ( .9496,  1.3472)    .7345
 5   ( -.0298,  .0621)    .0689    ( 1.0302, 1.3859)    .1868
 6   (  .0090,  .0043)    .0099    ( 1.0246, 1.3975)
```

```
        Approximate Solution: ( 1.0246, 1.3975)
           Maximum Functional Value:      8.8277
        Number gradient evaluations:         7
        Number function evaluations:         6
```

10.3 Modified Newton's Method

An alternative search method is the Newton–Raphson numerical method illustrated in two variables. This numerical method appears to do a more efficient and faster job in converging to the near-optimal solution. It is an iterative root finding technique using the partial derivatives of the function as the new system of equations. The algorithm uses Cramer's rule to find the solution of the system of equations.

Newton's method for multivariable optimization searches is based on Newton's single-variable algorithm for finding the roots and the Newton–Raphson method for finding roots of the first derivative; given a x_0, iterate $x_{n+1} = x_n - f'(x_n)/f''(x_n)$ until $|x_{n+1} - x_n|$ is less than some small tolerance. In several variables, we may use a vector $\mathbf{x}_0$, or two variables, (x_0, y_0). The algorithm is expanded to include partial derivatives with respect to each variable's dimension. In two variables (x, y), this would yield a system of equations where F is the derivative of $f(x, y)$ with respect to x and G is the derivative of $f(x, y)$ with respect to y. Thus, we need to find both $F = 0$ and $G = 0$ simultaneously.

This yields a matrix equation $\sum_{j=1}^{N} \alpha_{ij} \delta x_j = \beta_i$, where

$$\alpha_{ij} = \frac{\partial f_i}{\partial x_i}, \beta_i = -f_i.$$

The matrix equation can be solved by Lower-Upper (LU) decomposition or in the case of a 2 × 2 by Cramer's rule. The corrections are then added to the solution vector

$$x_i^{new} = x_i^{old} + \delta x_i, i = 1, \dots N$$

and iterated until it converges within a tolerance.

Modified Newton with Technology

INPUT: $x(0), y(0), N$, Tolerance

OUTPUT: $x(n), y(n)$

Step 1. For $n = 1$ to N do

Step 2. Calculate the new estimate for $x(n)$ and $y(n)$ as follows:

$$\frac{\partial F}{\partial x}(x(n-1), y(n-1)) \rightarrow q$$

$$\frac{\partial F}{\partial y}(x(n-1), y(n-1)) \rightarrow r$$

$$\frac{\partial G}{\partial x}(x(n-1), y(n-1)) \rightarrow s$$

$$\frac{\partial G}{\partial x}(x(n-1), y(n-1)) \rightarrow t$$

$$- F(x(n-1), y(n-1)) \rightarrow u$$

$$- G(x(n-1), y(n-1)) \rightarrow v$$

$$qt - rs \rightarrow D$$

$$x(n-1) + (ut - vr)/D \rightarrow x(n)$$

$$y(n-1) + (qv - su)/D \rightarrow y(n)$$

Step 3. If $((x(n)-x(n-1))^2 + (y(n)-y(n-1))^2)^{1/2} <$ tolerance,

         Then **Stop**

         Else , Go back to Step 2.

STOP

FIGURE 10.6
Pseudocode for Newton's method in two variables.

Let's repeat our examples with technology.

Example 4. Computer Placement Problem

```
> f:=-(sqrt((x1-15)^2+(x2-60)^2)+sqrt((x1-25)^2+(x2-90)^2)+sqrt
  ((x1-60)^2+(x2-75)^2)+sqrt((x1-80)^2+(x2-25)^2)+sqrt((x2-60)
  ^2+(x1-75)^2));
```

$$f := -\sqrt{x1^2 - 30x1 + 3825 + x2^2 - 120x2} - \sqrt{x1^2 - 50x1 + 8725 + x2^2 - 180x2} \quad (3.2)$$
$$- \sqrt{x1^2 - 120x1 + 9225 + x2^2 - 150x2} - \sqrt{x1^2 - 160x1 + 7025 + x2^2 - 50x2}$$
$$- \sqrt{x2^2 - 120x2 + 9225 + x1^2 - 150x1}$$

```
> Newtons(f,10,.5,55,55);
>

Hessian: [    -.077      -.004  ]
         [    -.004      -.093  ]
eigenvalues:    -.093      -.076
pos def: false
new x=  57.083    new y=  69.257

Hessian: [    -.158       .027  ]
         [     .027      -.116  ]
eigenvalues:    -.172      -.103
pos def: false
new x=  56.786    new y=  68.088

Hessian: [    -.141       .015  ]
         [     .015      -.110  ]
eigenvalues:    -.147      -.104
pos def: false
new x=  56.818    new y=  68.075


final new x=  56.818    final new y=  68.075
final fvalue is -157.663
```

FIGURE 10.7
Screenshot of solution for the computer placement problem.

Solution is (56.818, 68.075) and f(56.818, 68.075) = 157.663.
    Example 5. Maximize $f(x_1, x_2) = 2x_1x_2 + 2x_2 - x_1^2 - 2x_2^2$ starting at the point (0, 0) with a tolerance of 0.01.

Example 5. $2x^1x^2 + 2x_2 - e^{x1} - e^{x2} + 10$

$$f := 2\,x1\,x2 + 2\,x2 - e^{x1} - e^{x2} + 10$$

```
> df1:=diff(f,x1);
```

$$df1 := 2x2 - e^{x1}$$

```
> df2:=diff(f,x2);
```

$$df2 := 2\,x1 + 2 - \mathbf{e}^{x2}$$

```
> f1:=unapply(df1,x1,x2);
```

$$f1 := (x1,x2) \rightarrow 2\,x2 - \mathbf{e}^{x1}$$

```
> f2:=unapply(df2,x1,x2);
```

$$f2 := (x1, x2) \rightarrow 2\,x1 + 2 - \mathbf{e}^{x2}$$

```
> Steepest(f,f1,f2,100,.1,0,0);
pos def:
Hessian: [ -1.000     2.000 ]
         [  2.000    -1.000 ]
eigenvalues: -3.000   1.000
pos def: false
new x=    -.333    new y=    .333
Hessian: [ -.717     2.000 ]
         [  2.000   -1.396 ]
eigenvalues: -3.085   .973
pos def: false
new x=    -.269    new y=    .381
final new x= -.269 final new y= .381
final fvalue is 8.329
```

Newton's Method in Python

Comparisons of Methods

We compared these two routines and found that the Newton's method converges faster than the gradient method. This is displayed in the following table.

Function 1	Initial condition	Iterations	Feval	Gevals	Soln	Max F
Steepest ascent	(0, 0)	16	16	17	x = 0.9922 y = 0.9961	1.0
Newton's method	(0, 0)	2	2		x = 1 y = 1	1.00000
Function 2	Initial Condition	iterations	Feval	Gevals	Soln	Max F
Steepest ascent	(0, 0)	4	5	4	x= −0.26638, y = 0.3853	8.3291
Newton's method	(0, 0)	2			x = −0.269, y = 0.381	8.329

10.4 Applications

Manufacturing

The manager of a new television plant is planning to introduce two new products, a 40-inch set with a manufacturer's suggested retail price (MSRP) of $339 and a 50-inch set with a MSRP of $399. The costs to the company are $195 per 40-inch set and $225 per 50-inch set, plus an additional $400,000 in fixed costs of parts, labor, and machinery. Our research team has found that in a competitive market, in which they desire to sell their sets, the number of sales per year will affect the average selling price. It is estimated that for each type of set, the average selling price drops by 1 cent for each additional type of set sold. Furthermore, sales of the 40-inch sets will affect the sales of the 50-inch sets and vice versa. It is estimated that the average selling price for the 40-inch set will be reduced by an additional 0.3 cents for each 50-inch set sold and that the price of each 50-inch set will decrease by 0.4 cents for each 40-inch set sold. Management wants to find the optimal number of units of each type to produce to maximize profits. Our research team reminds us that Profit = Revenue − Cost.

Let's start by defining our variables and building our equations.

Let x_1 = number of 40-inch sets produced.

Let x_2 = number of 50-inch sets produced.

The cost function, $C(x_1, x_2) = 195 x_1 + 225 x_2 + 400{,}000$.

The revenue function, $R(x_1, x_2) = (339 - 0.01 * x_1 - 0.003 * x_2)x_1 + (399 - 0.004x_1 - .01x_2)x_2$.

$$P = R - C = (339 - 0.01 * x_1 - 0.003 * x_2)x_1 + (399 - 0.004x_1 - .01x_2)x_2 - (400000 + 195x_1 + 225x_2)$$

This is an unconstrained optimization problem, and we will use numerical methods.

Steepest Descent Method

```
              Initial Condition: ( 0.0000, 3.0000)
Iter  Gradient Vector G   magnitude G    x[k]              Step Length
1   (143.9790,173.9400)  225.7988 ( 0.0000, 3.0000)         37.2069
2   ( -8.4635,  7.0057)   10.9868 (5357.0082,6474.7633) 76.2007
3   (  0.6981,  0.8434)    1.0949 (4712.0833,7008.6005) 37.2069
4   ( -0.0410,  0.0340)    0.0533 (4738.0585,7039.9810) 76.2006
5   (  0.0034,  0.0041)    0.0053 (4734.9314,7042.5695)
```

```
      Approximate Solution: (4734.9314,7042.5695)
      Maximum Functional Value:       553641.0260
      Number gradient evaluations:         6
      Number function evaluations:         5
```

```
Using Newton-Raphson method
Hessian: [ -0.020 -0.007 ]
        [ -0.007 -0.020 ]
eigenvalues: -0.027 -0.013
pos def: false
new x=4735.043 new y=7042.735
Hessian: [ -0.020 -0.007 ]
        [ -0.007 -0.020 ]
eigenvalues: -0.027 -0.013
pos def: false
new x=4735.043 new y=7042.735
final new x=4735.043 final new y=7042.735
final fvalue is 553641.026
Summary:
Make 4735 of the 40-inch TVs and 7042 of the 50-inch TV.
Our profit is $553,641.02.
Python Code
import numpy as np
import numpy.linalg as la
import math
import scipy.optimize as sopt
import matplotlib.pyplot as pt
from mpl_toolkits.mplot3d import axes3d
#Here are two functions.
def f(x):
  return (339-.01*x[0]-0.003*x[1])*x[0]+(399-0.004*x[0]-.01*
  x[1]*x[1])-(400000+195*x[0]+225*x[1])
def df(x):
  return np.array([144-0.2*x[0]-.007*x[1],174-.007*x[0]-
  0.2*x[1]])
def ddf(x):
  return np.array([
             [-.02,-.007],
             [-.007,-.02]
             ])
q1=np.empty((2))
q2=np.empty((2,2))
q3=np.empty((2))
#initialize guess
N=10
x[0]=0
x[1]=0
q1=df(x)
q2=ddf(x)
q3=np.array([x[0],x[1]])
s=la.solve(ddf(x),df(x))
next_q=q3-s
for i in range(0,N):
  print(next_q)
Output
[4735.04273504 7042.73504274]
```

TV Manufacturing

The manager of a new television plant is planning to introduce three new products, a 40-inch set with a MSRP of $339, a 50-inch set with a MSRP of $399, and a 60-inch set has a MSRP of $425. The costs to the company is $195 per 40-inch set, $225 per 50-inch set, and $250 per 60-inch set plus an additional $400,000 in fixed costs of parts, labor, and machinery. Our research team has found that in a competitive market, in which they desire to sell their sets, the number of sales per year will affect the average selling price. It is estimated that for each type of set, the average selling price drops by 1 cent for each additional type of set sold. Furthermore, sales of the 40-inch sets will affect the sales of the 50-inch sets and 60-inch sets and vice versa. It is estimated that the average selling price for the 40-inch set will be reduced by an additional 0.3 cents for each 50-inch set sold, the price of each 50-inch set will decrease by 0.4 cents for each unit of 40-inch sets sold, and the price of the 60-inch set will decrease 0.25 cents of the 50-inch sets. Management wants to find the optimal number of units of each type to produce to maximize profits. Our research team reminds us that Profit = Revenue − Costs.

Let's start by defining our variable and building our equations.

Let x_1 = number of 40-inch sets produced.

Let x_2 = number of 50-inch sets produced.

Let x_3 = number of 60-inch sets produced

The cost function is $C(x_1, x_2, x_3) = 195x_1 + 225x_2 + 250 * x_3 + 400{,}000$.

The revenue function is $R(x_1, x_2, x_3) = (339 - 0.01 * x_1 - 0.003 * (x_2 + x_3)) * x_1 + (399 - 0.004(x_1 + x_2) - 0.01x_2) * x_2 + (425 - 0.0025*(x_2 + x_3) - 0.01 * x_3) * x_3$.

$$P = R - C = (339 - 0.01 * x_1 - 0.003 * (x_2 + x_3) * x_1 + (399 - 0.004(x_1 + x_2) - 0.01x_2) * x_2 + (425 - 0.0025 * (x_2 + x_3) - 0.01 * x_3) * x_3 - 195x_1 + 225x_2 + 250 * x_3 + 400{,}000$$

$$339 - 0.01 * x_1 - 0.003 * (x_2 + x_3) * x_1 + (399 - 0.004(x_1 + x_2) - 0.01x_2) * x_2 + (425 - 0.0025 * (x_2 + x_3) - 0.01 * x_3) * x_3 - 195x_1 + 225x_2 + 250 * x_3 + 400{,}000$$

```
f := (339 - 0.01 x1 - 0.003 x2 - 0.003 x3) x1 + (399 - 0.004x1
- 0.004.x3 - 0.01 x2) x2 + (425 - 0.0025x2 - 0.0025x1 - 0.01
x3) x3 - 195x1 - 225x2 - 250x3 - 400000
> df1 := diff(f,x1);
df1 := -0.02 x1 + 144 - 0.007 x2 - 0.003 x3
> df2 := diff(f,x2);
df2 := - 0.007 x1 - 0.028 x2 + 624 - 0.0025 x3
> df3 := diff(f,x3);
df3 := - 0.003 x1 - 0.0250 x3 + 675 - 0.0025 x2
> ddf1 := diff(df1, x1);
ddf1 := -0.02
> ddf1x2 := diff(df1, x2);
ddf1x2 := -0.007
> ddf1x3 := diff(df1, x3);
```

```
ddf1x3 := -0.003
> ddf2 := diff(df2, x2);
ddf2 := -0.028
> ddf2x := diff(df2, x3);
ddf2x := -0.0025
> ddf3 := diff(df3, x3);
ddf3:= -0.0250
> with(Optimization) :
> NLPSolve(f, maximize)
> [861420.311340607936, [x1 = 3638.74282639172,
x2 =5487.53277312025, x3 = 5965.89454048237]]
> df1 := diff(f, x1);
df1 := -0.02 x1 + 144 - 0.007 x2 - 0.0055 x3
> df2 := diff(f, x2);
df2 := -0.007 x1 + 0.02 x2 + 174 - 0.0065 x3
> df3 := diff(f, x3);
df3 := -0.0055 x1 - 0.0065 x2 - 0.02 x3 + 175
> ddf1 := diff(df1, x1);
ddf1 := -0.02
> ddf1x2 := diff(df1, x2);
ddf1x2 := -0.007
> ddf1x3 := diff(df1, x3);
ddf1x3 := -0.0055
> ddf2 := diff(df2, x2);
ddf2 := -0.02
> ddf2x := diff(df2, x3);
ddf2x := -0.0065
> ddf3 := diff(df3, x3);
ddf3 := -0.02


Python Code
import numpy as np
import numpy.linalg as la
import scipy.optimize as sopt
import matplotlib.pyplot as pt
guess=np.empty(3)
q1=np.empty((3))
q2=np.empty((3,3))
q3=np.empty((3))
from mpl_toolkits.mplot3d import axes3d
def f(x):
  return (339-.01*x[0]-0.003*(x[1]+x[2])*x[0])+(399-
0.004*(x[0]+x[2])-.01*x[1])*x[1]-(400000+195*x[0]+225*x[1])+
(425-0.0025*x[1]-0.0025*(x[1]+x[0])-0.01*x[2])*x[2]
def df(x):
  return np.array([[(-0.02*x[0] + 144 - 0.007*x[1] -
0.0055*x[2]), (-0.007*x[0] - 0.02*x[1] + 174 - 0.0065*x[2]),
(-0.0055*x[0] - 0.0065*x[1] - 0.02*x[2] + 175)])
def ddf(x):
```

```
    return np.array([
           [-.02,-.007,-0.0055],
           [-.007,-.02,-.0065],[-0.0055,-.0065,-.02]])
#initialize guess
N=10
x[0]=0
x[1]=0
x[2]=0
q1=df(x)
q2=ddf(x)
q3=np.array([x[0],x[1],x[2]])
print(q3)
s=la.solve(ddf(x),df(x))
next_q=q3-s
for i in range(0,N):
  print(next_q)

guesses = [np.array([1,1,1])]
x = guesses[-1]
s = la.solve(ddf(x), df(x))
next_guess = x - s
print(f(next_guess), next_guess)
guesses.append(next_guess)
Python Output
[0 0 0]
[3638.7447492   5487.52162095 5965.90066716]
[3638.7447492   5487.52162095 5965.90066716]
[3638.7447492   5487.52162095 5965.90066716]
[3638.7447492   5487.52162095 5965.90066716]
[3638.7447492   5487.52162095 5965.90066716]
[3638.7447492   5487.52162095 5965.90066716]
[3638.7447492   5487.52162095 5965.90066716]
[3638.7447492   5487.52162095 5965.90066716]
[3638.7447492   5487.52162095 5965.90066716]
[3638.7447492   5487.52162095 5965.90066716]
```

The 40-inch TV is 3639, the 50-inch TV is 5488, and the 60-inch TV is 5966 for a total profit of about \$861,420.31.

Exercises

1. Given: MAX $f(x, y) = 2xy - 2x^2 - y^2$

   Assume our tolerance for the magnitude of the gradient is 0.10.

   a. Start at the point $(x, y) = (1, 1)$. Perform two complete iterations of gradient search. For each iteration, clearly show X_n, X_{n+1}, $\nabla f(X_n)$, and t^*. Justify that we will eventually find the approximate maximum.

   b. Use Newton's method to find the maximum starting at $(x, y) = (1, 1)$. Clearly show X_n, X_{n+1}, $\nabla f(X_n)$, and H^{-1} for each iteration. Clearly indicate when the stopping criterion is achieved.

2. MAX $f(x,y) = 3xy - 4x^2 - 2y^2$

   Assume our tolerance for the magnitude of the gradient is 0.10.

   a. Start at the point $(x, y) = (1, 1)$. Perform two complete iterations of gradient search. For each iteration, clearly show X_n, X_{n+1}, $\nabla f(X_n)$, and t^*. Justify that we will eventually find an approximate maximum.

   b. Use Newton's method to find the maximum starting at $(x, y) = (1, 1)$. Clearly show X_n, X_{n+1}, $\nabla f(X_n)$, and H^{-1} for each iteration. Clearly indicate when a stopping criterion is achieved.

3. Apply the modified Newton's method (multivariable) to find the following:

   a. MAX $f(x, y) = -x^3 + 3x + 84\,y - 6y^2$

   Start at $(1, 1)$.

   Why can't we start at $(0, 0)$?

   b. MIN $f(x,y) = -4x + 4x^2 - 3y -+ y^2$

   Start at $(0, 0)$.

   c. Perform three iterations to

   MIN $f(x, y) = (x - 2)^4 + (x - 2y)^2$. Start at $(0, 0)$.

   Why is this problem not converging as quickly as part b?

4. Use the gradient search to find the approximate minimum to
   $f(x,y) = (x - 2)^2 + x + y^2$. Start at $(2.5, 1.5)$.

Projects

1. Write a computer program in Maple that uses a one-dimensional search algorithm, say, a Golden section search, instead of calculus to perform iterations of gradient search. Use your code to find the maximum of

$$f(x, y) = xy - x^2 - y^2 - 2x - 2y + 4.$$

2. Write a computer program in Maple that uses a one-dimensional search algorithm, say, a Fibonacci search, instead of calculus to perform iterations of gradient search. Use your code to find the maximum of

$$f(x,y) = xy - x^2 - y^2 - 2x - 2y + 4.$$

3. Redo the TV manufacturing of two TVs from Section 10.4 regarding the manager of a new television plant planning to introduce two new products, a 40-inch set with an MSRP of $239 and a 50-inch set with an MSRP of $299. The costs to the company are $155 per 40-inch set and $195 per 50-inch set, plus an additional $400,000 in fixed costs of parts, labor, and machinery. Our research team has found that in a competitive market, in which they desire to sell their sets, the number of sales per year will affect the average selling price. It is estimated that for each type of set, the average selling price drops by 1 cent for each additional type of set sold. Furthermore, sales of the 40-inch sets will affect the sales of the 50-inch sets and vice versa. It is estimated that the average selling price for the 40-inch set will be reduced by an additional 0.25 cents for each 50-inch set sold and that the price of each 50-inch set will decrease by 0.35 cents for each unit of 40-inch sets sold. Management wants to find the optimal number of units of each type to produce to maximize profits. Our research team reminds us that Profit = Revenue – Costs.

4. The manager of a new television plant is planning to introduce three new products, a 40-inch set with an MSRP of $239, a 50-inch set with an MSRP of $299, and a 60-inch set has an MSRP of $325. The costs to the company are $155 per 40-inch set, $195 per 50-inch set, and $200 per 60-inch set plus an additional $400,000 in fixed costs of parts, labor, and machinery. Our research team has found that in a competitive market, in which they desire to sell their sets, the number of sales per year will affect the average selling price. It is estimated that for each type of set, the average selling price drops by 1 cent for each additional type of set sold. Furthermore, sales of the 40-inch sets will affect the sales of the 50-inch sets and 60-inch sets and vice versa. It is estimated that the average selling price for the 40-inch set will be reduced by an additional 0.3 cents for each 50-inch set sold, that the price of each 50-inch set will decrease by 0.4 cents for each unit of 40-inch sets sold, and that the price of the 60-inch set will decrease 0.25 cents of the 50-inch sets. Management wants to find the optimal number of units of each type to produce to maximize profits. Our research team reminds us that Profit = Revenue – Costs.

References and Further Reading

Bazarra, M., C. Shetty and H. D. Scherali (1993). *Nonlinear Programming: Theory and Applications*. Wiley, New York.

Fox, W. P. (1992, January–March). Teaching Nonlinear Programming with Minitab. *COED Journal*, II(1): 80–84.

Fox, W. P. (1993). Using Microcomputers in Undergraduate Nonlinear Optimization. *Collegiate Microcomputer*, XI(3): 214–218.

Fox, W. P. (2021). *Nonlinear Optimization: Models and Applications*. CRC Press (Taylor and Francis Group), Boca Raton, FL.

Fox, W. P. and J. Appleget (2000, October–December). Some Fun with Newton's Method. *COED Journal*, X(4): 38–43.

Fox, W. P., F. Giordano, S. Maddox and M. Weir (1987). *Mathematical Modeling with Minitab*. Brooks/Cole, Monterey, CA.

Fox, W. P., F. Giordano and M. Weir (1997). *A First Course in Mathematical Modeling*, 2nd ed. Brooks/Cole, Monterey, CA.

Fox, W. P. and W. Richardson (2000, October). Mathematical Modeling with Least Squares Using MAPLE. *Maple Application Center*, Nonlinear Mathematics.

Fox, William P. and William H. Richardson. (2002). "Multivariable Variable Optimization When Calculus Fails: Gradient Search Methods in Nonlinear Optimization Using MAPLE", *Computers in Education Journal*, pp 2–11. Sept-Dec 2002.

Meerschaert, M. (1993). *Mathematical Modeling*. Academic Press, San Diego, CA.

Phillips, D. T., A. Ravindran and J. Solberg (1976). *Operations Research*. John Wiley and Sons, New York.

Rao, S. S. (1979). *Optimization: Theory and Applications*. Wiley Eastern Limited, New Delhi, India.

William, H., B. Flannery, S. Teukolsky and W. Vetterling (1987). *Numerical Recipes*. Cambridge University Press, New York, pp. 269–271.

Winston, W. (2002). *Introduction to Mathematical Programming: Applications and Algorithm*, 4th ed. Duxbury Press, ITP, Belmont, CA.

11

Boundary Value Problems in Ordinary Differential Equations

11.1 Introduction

Before we start off this section, we need to make it very clear that we are only going to cover a few numerical techniques of differential equation's boundary value problems (BVPs for short). The intent of this chapter is to give a brief look at the idea of boundary value problems and give enough information to allow us to do solve boundary problems using numerical methods only.

The first thing that we need to do is to define just what we mean by a BVP. With initial value problems (IVPs), Chapter 7, we had a differential equation, and we specified the value of the solution and an appropriate number of derivatives and their values at the same point, called initial conditions. For instance, for a second-order differential equation, the initial conditions are.

An example of possible initial conditions for a second-order ordinary differential equation (ODE) would be $y(t_0) = y_0$ and $y'(t_0) = y'_0$.

With BVPs, we have the differential equation, and we specify the function's boundary conditions that can be any of the following:

$$y(t_0) = y_0, y(x_1) = y_1$$
$$y'(t_0) = y_0, y'(x_1) = y_1$$
$$y'(t_0) = y_0, y(x_1) = y_1$$
$$y(t_0) = y_0, y'(x_1) = y_1$$

In our examples, we only consider boundaries in the form: $y(t_0) = y_0 y(x_1) = y_1$.

Our second-order ODE will be of the form in Equation 11.1:

$$y'' + p(x)y' + q(x)y = g(x), \qquad (11.1)$$

along with our boundary conditions: $y(t_0) = y_0 y(x_1) = y_1$.

DOI: 10.1201/9781032703671-11

The biggest difference that we're going to see here comes when we go to solve the BVP. When solving linear IVPs, a unique solution will be guaranteed under very mild conditions. We only looked at this idea for first-order IVPs, but the idea does extend to higher order IVPs. In that section of Chapter 7, we saw that all we needed to guarantee a unique solution was some basic continuity conditions. With BVPs, we will often have no solution or infinitely many solutions even for very nice differential equations that would yield a unique solution if we had initial conditions instead of boundary conditions.

Before we get into solving some of these, let's next address the question of why we're even talking about these in the first place. As we'll see in Chapter 12 in the process of solving some partial differential equations, we will run into BVPs that will need to be solved as well. In fact, a large part of the solution process there will be in dealing with the solution to the BVP. In these cases, the boundary conditions will represent things like the temperature at either end of a bar or the heat flow into/out of either end of a bar. Or maybe they will represent the location of ends of a vibrating string. So, the boundary conditions there will really be conditions on the boundary of some process.

So, with some of basic stuff out of the way, let's find some solutions to a few BVPs. Note as well that there really isn't anything new here yet. We might know how to solve the differential equation and we know how to find the constants by applying the conditions. The only difference is that here we'll be applying boundary conditions instead of initial conditions.

Again, in this chapter, we are presenting two numerical approximation methods: the shooting point method and linear finite differences method. We will discuss both first and provide basic examples before we apply the methods to applications.

In Chapter 7, we discussed initial value problems for differential equations. In this chapter we discuss methods to solve differential equations with two boundary values, where conditions are posed at different points. In first-order differential equations there is no difference between initial value and boundary value problems and their solutions.

The differential equation that we present here is of the form

$$y'' = f(x, y, y') \text{ for } a \leq x \leq b \text{ with boundary condition}$$
$$y(a) = \alpha \text{ and } y(b) = \beta, \text{ for constants } \alpha \text{ and } \beta.$$

We can find a unique solution provided that

1. the function f and its partial derivatives with respect to y and y' are continuous,
2. the partial derivatives with respect to y are positive, and
3. the partial derivatives with respect to y' is bounded.

There are several numerical methods that we will present. They are the linear shooting method and linear finite difference method.

11.2 Linear Shooting Method

If the second-order differential equation has the form

$$y'' = f(x, y, y') = p(x)\, y' + q(x)\, y + r(x),$$

then we can use the linear shooting method to numerically obtain an approximate solution.

Linear ODEs are easier to solve than nonlinear ODEs. This is because adding the solution to the homogeneous ODE (set ODE equal to 0) to the nonhomogeneous part gives the complete solution desired.

In the shooting point method, take two IVPs, such as shown in Equations 11.2 and 11.3:

$$y'' = p(x)y' -+ q(x)y =+ r(x) \text{ for } a \le x \le b, \text{ where } y(a) = \alpha \text{ and } y'(a) = 0 \quad (11.2)$$

and

$$y'' = p(x)y' + q(x)y \text{ for } a \le x \le b, \text{ where } y(a) = 0 \text{ and } y'(a) = 1, \quad (11.3)$$

both of which have unique solutions.

Let $y_1(x)$ be the solution to Equation 11.2 and $y_2(x)$ be the solution to Equation 11.3; then we obtain Equation 11.4 as

$$y(x) = y_1(x) + \frac{(\beta - y_1(\beta))}{y_2(\beta)} y_2(x), \quad (11.4)$$

as the unique solution to our original boundary problem:

$$y'' = f(x, y, y') \text{ for } a \le x \le b, \text{ with boundary condition } y(a) = \alpha \text{ and } y(b) = \beta.$$

The Euler, the improved Euler, and Runge–Kutta 4 (RK4) methods might be used to get the solutions to Equations 11.2 and 11.3 and be placed in the weighted Equation 11.4 to approximate the solution to our boundary problem.

Example 1. Shooting Point Method

Given the BVP,

$$y'' = -2/x\, y' + 2/x^2\, y + \sin(\ln(x))/x^2 \text{ for } 1 \le x \le 2, \text{ where } y(1) = 1 \text{ and } y(2) = 2$$
$$\text{(see Burden et al., 2003)}.$$

We set up two problems each requiring numerical solutions:

$$y_1'' = -2/x\,y_1' + 2/x^2\,y_1 + \sin(\ln(x))/x^2 \text{ for } 1 \le x \le 2,$$
$$\text{where } y_1(1) = 1 \text{ and } y_1'(1) = 0.$$

and

$$y_2'' = -2/x y_2' + 2/x^2 y_2 \text{ for } 1 \le x \le 2, \text{ where } y_2(1) = 0 \text{ and } y_2'(1) = 1.$$

We will use RK4 from Chapter 7 and the algorithm from Burden and Faires (2003) to estimate $y(2)$.

The exact solution is

$$
\begin{aligned}
&> ecalf\left(dsolve\left(\left[ode2, bc2\right]\right)\right); \\
&y(x) = 1.1392.7013x - \frac{0.0392070132}{x^2} - 0.6000000000 \sin\left(0.5000000000 \ln(x)\right) \\
&\cos\left(0.5000000000 \ln(x)\right) - 0.2000000000 \cos\left(0.5000000000 \ln(x)\right)^2 + 0.1000000000
\end{aligned}
$$

The result was $y(2) = 2.000000105$. The exact answer is 2.0000000000. The error with our approximation is $|2.000000105 - 2.000000000| = 1.05 \times 10^{-7}$.

First, we use the Burden and Faires program in Maple and obtain our estimates:

LINEAR SHOOTING METHOD

```
     I X(I)        W(1,I)       W(2,I)
 0 1.00000000 1.00000000 0.91762139
 1 1.10000000 1.09262916 0.93528286
 2 1.20000000 1.18708471 0.95383867
 3 1.30000000 1.28338226 0.97197732
 4 1.40000000 1.38144589 0.98909652
 5 1.50000000 1.48115938 1.00495322
 6 1.60000000 1.58239245 1.01948769
 7 1.70000000 1.68501396 1.03273244
 8 1.80000000 1.78889854 1.04476394
 9 1.90000000 1.89392951 1.05567694
10 2.00000000 2.00000000 1.06557077
```

The column for variable w values is our estimates to the solution.

We also mention that Professor Doug Meade has a Maple application for the shooting point method that is available from the Maple Applications website that we will also illustrate.

We repeat this example using Python and following Python code.

```
# Author: Carlos eduardo da Silva Lima-- Modifed by W Fox
# Solving EDO initial value problem (IVP) via scipy and 4Order
Runge-Kutta
# 4Order Runge-Kutta
import numpy as np
import matplotlib.pyplot as plt
from scipy.integrate import odeint
import math
# Initial conditions
t_initial = 1.0
t_final = 2.0
a=1
b=2
y0 = 0.0
u0 = 1.0
N = 10
h =(b-a)/N # Stepsize
# Enter the definition of the set of ordinary differential
equations
def ode(t,y,u):
    ode_1 = u
    ode_2 = -(2/t)*u+2*(y/(t**2))
    #+(np.sin(math.log(t))/t**2)
    return np.array([ode_1,ode_2])
# RK4
t = np.empty(N+1)
y = np.empty(N+1); u = np.empty(N+1)
t[0] = t_initial
y[0] = y0; u[0] = u0
for i in range(0,N,1):
k11 = h*ode(t[i],y[i],u[i])[0]
    k12 = h*ode(t[i],y[i],u[i])[1]
    k21 = h*ode(t[i]+(h/2),y[i]+(k11/2),u[i]+(k12/2))[0]
    k22 = h*ode(t[i]+(h/2),y[i]+(k11/2),u[i]+(k12/2))[1]
    k31 = h*ode(t[i]+(h/2),y[i]+(k21/2),u[i]+(k22/2))[0]
    k32 = h*ode(t[i]+(h/2),y[i]+(k21/2),u[i]+(k22/2))[1]
    k41 = h*ode(t[i]+h,y[i]+k31,u[i]+k32)[0]
    k42 = h*ode(t[i]+h,y[i]+k31,u[i]+k32)[1]
    y[i+1] = y[i] + ((k11+2*k21+2*k31+k41)/6)
    u[i+1] = u[i] + ((k12+2*k22+2*k32+k42)/6)
    t[i+1] = t[i] + h
#print(t,y,u)
# Graphics
#plt.style.use('dark_background')
#plt.figure(figsize=(7,7))
#plt.xlabel(r'$t(s)$')
#plt.ylabel(r'$y(t)$ and $u(t)$')
#plt.title(r'$\frac{d^{2}y(x)}{dt^{2}}+ + 4\frac{dy(x)}{dt} +
2y(x) = 0$ with $y(t_{0} = 0) = 1$ and $\frac{dy(0)}{dt} = 3$')
```

```
#plt.plot(t,y,'b-o',t,u,'r-o')
#plt.grid()
#y
# Initial conditions
t_initial = 1.0
t_final = 2.0
a=1
b=2
x0 = 1.0
v0 = 0
N = 10
h =(b-a)/N # Step
# Enter the definition of the set of ordinary differential
equations
def ode(t,x,v):
   ode_3 = v
   ode_4 = (-2/t)*v+2*(x/(t**2))+(np.sin(math.log(t))/t**2)
   return np.array([ode_3,ode_4])
# RK4
t = np.empty(N+1)
x = np.empty(N+1); v = np.empty(N+1)
t[0] = t_initial
x[0] = x0; v[0] = v0
for i in range(0,N,1):
k11 = h*ode(t[i],x[i],v[i])[0]
   k12 = h*ode(t[i],x[i],v[i])[1]
   k21 = h*ode(t[i]+(h/2),x[i]+(k11/2),v[i]+(k12/2))[0]
   k22 = h*ode(t[i]+(h/2),x[i]+(k11/2),v[i]+(k12/2))[1]
   k31 = h*ode(t[i]+(h/2),x[i]+(k21/2),v[i]+(k22/2))[0]
   k32 = h*ode(t[i]+(h/2),x[i]+(k21/2),v[i]+(k22/2))[1]
   k41 = h*ode(t[i]+h,x[i]+k31,v[i]+k32)[0]
   k42 = h*ode(t[i]+h,x[i]+k31,v[i]+k32)[1]
   x[i+1] = x[i] + ((k11+2*k21+2*k31+k41)/6)
   v[i+1] = v[i] + ((k12+2*k22+2*k32+k42)/6)
   t[i+1] = t[i] + h
#print(t,x,v)
# Graphics
#plt.style.use('dark_background')
#plt.figure(figsize=(7,7))
#plt.xlabel(r'$t(s)$')
#plt.ylabel(r'$y(t)$ and $u(t)$')
#plt.title(r'$\frac{d^{2}y(x)}{dt^{2}}+ + 4\frac{dy(x)}{dt} +
2y(x) = 0$ with $y(t_{0} = 0) = 1$ and $\frac{dy(0)}{dt} =
3$')
#plt.plot(t,y,'b-o',t,u,'r-o')
#plt.grid()
#x
w=np.empty(N+1)
c=(b-y[10])/x[10]
```

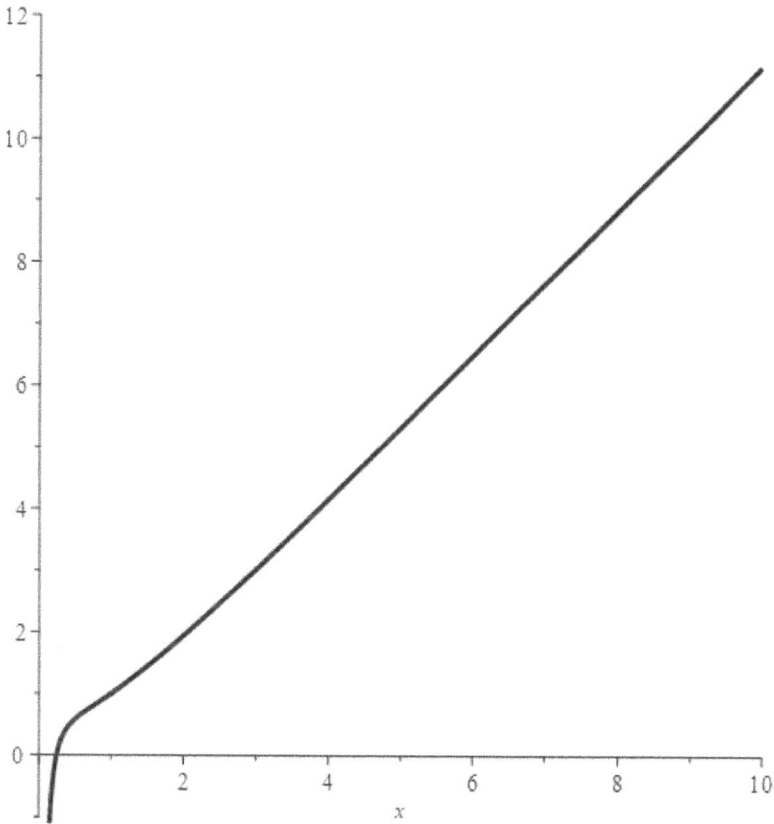

FIGURE 11.1
Plot of Example 1 solution.

```
print('c is', c)
for i in range(0,N+1):
 w[i]=x[i]+c*y[i]
print('estimates are', w)
c is 0.9671928674164761
estimates are [1. 1.09714908 1.19543808 1.29508545 1.39614833
1.4986008
 1.60237555 1.70738648 1.81354108 1.92074715 2.0289163 ]
```

The plot is shown in Figure 11.1.

Example 2. $y^{(2)}(t) + 4\,y^{(1)}(t) + 2\,y(t) = 0$, $y(1) = 4$, $y(3) = 2$

We applied the shooting point method within Maple. We see the output in Figure 11.2. We summarize in Table 11.1

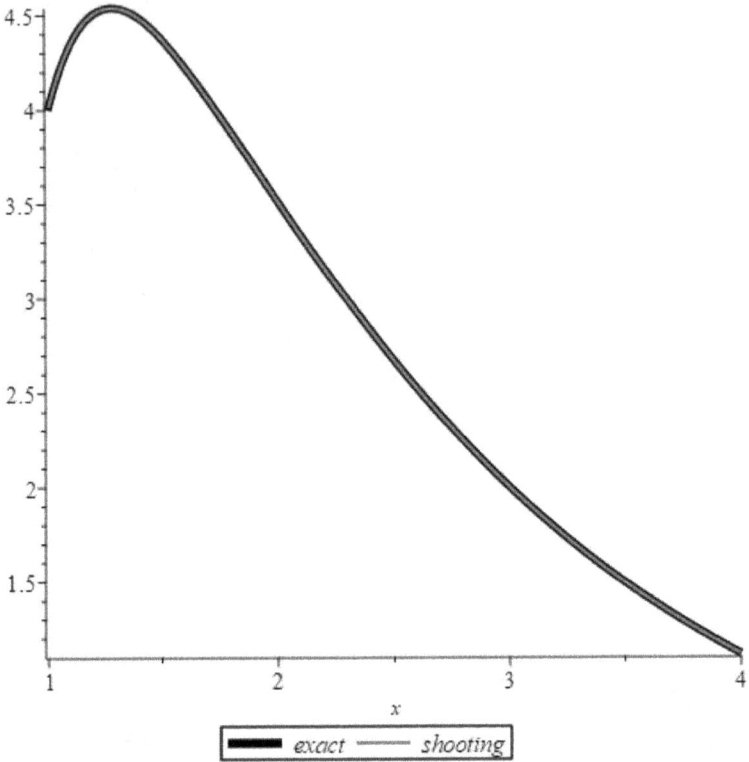

FIGURE 11.2
From Doug Meade's Maple application.

TABLE 11.1

Results

x	y
1	4
1.2	4.50146477230727
1.4	4.48158263597944
1.6	4.22801514152530
1.8	3.88305577041118
2	3.51576983759025
2.2	3.15847715233933
2.4	2.82518637233933
2.6	2.52089573490286
2.8	2.24627313799069
3.0	2

```
Meade
                    4.000000002
                    4.504126359
                    4.484247558
                    4.230005528
                    3.884365574
                    3.516566051
                    3.158929994
                    2.825425093
                    2.521007339
                    2.246312398
                    1.999999913
LINEAR SHOOTING METHOD B7 F
   I   X(I)          W(1,I)         W(2,I)
   0   1.00000000    4.00000000    -1.79770448
   1   1.20000000    3.65617675    -1.64279567
   2   1.40000000    3.34202640    -1.50070452
   3   1.60000000    3.05515867    -1.36961330
   4   1.80000000    2.79361871    -1.24683747
   5   2.00000000    2.55617835    -1.12743780
   6   2.20000000    2.34307214    -1.00086100
   7   2.40000000    2.15781150    -0.84279160
   8   2.60000000    2.01161600    -0.59536400
   9   2.80000000    1.93420000    -0.11910100
  10   3.00000000    1.99999900     0.92384000
```

In Python, using our code but changing the differential equation and conditions, we get

```
c is 1.9332593383583054
estimates for y are [1.  1.2 1.4 1.6 1.8 2.  2.2 2.4 2.6 2.8 3.
] [4.  4.1370922 3.97301569 3.6822739 3.35036208 3.01802721
 2.70364138 2.41449811 2.15250402 1.91703581 1.70636496]
```
> **for** *i* **from** 0 **to** *N* **do**

$$x := a + i \cdot h : w[i] := evalf\left(rhs\big(g1(x)[2]\big) + c \cdot rhs\big(g2(x)[2]\big)\right); \text{ } \textbf{end do}$$

$$x := 1$$

$$w_0 := 4.$$

$$x := \frac{6}{5}$$

$$w_1 := 4.50146477230727$$

$$x := \frac{7}{5}$$

$$w_2 := 4.48158263597944$$

$$x := \frac{8}{5}$$

$$w_3 := 4.22801514152530$$

$$x := \frac{9}{5}$$

$$w_4 := 3.88305577041118$$

$$x := 2$$

$$w_5 := 3.51576983759025$$

$$x := \frac{11}{5}$$

$$w_6 := 3.15847715233933$$

$$x := \frac{12}{5}$$

$$w_7 := 2.82518637148728$$

$$x := \frac{13}{5}$$

$$w_8 := 2.52089573490286$$

$$x := \frac{14}{5}$$

$$w_9 := 2.24627313799069$$

$$x := 3$$

$$w_{10} := 2.$$

>
Our approximation to $y(3)=2$.

Using the Meade programmed method, the shooting star approximation is 1.999999913.
Meade's program allows for extrapolation, to estimate at $x = 4$, we get

```
y(4) = 1.114732149.
Python's Shooting Method
# import libraries for numerical functions and plotting
import numpy as np
import matplotlib.pyplot as plt
from scipy.integrate import solve_ivp
def shooting_rhs(x, y):
    '''Function for dy/dx derivatives'''
    return [y[1], -4*y[1]-2*y[0] ]
# target boundary condition
target = 3
# pick a guess for y'(0)
guess1 = 1
sol1 = solve_ivp(shooting_rhs, [0, 2], [1, guess1])
```

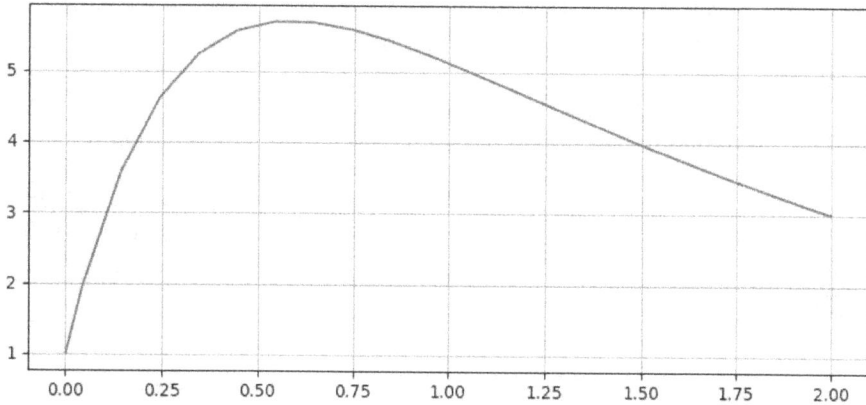

FIGURE 11.3
Example 2's plot using Python.

```
print(f'Solution 1: {sol1.y[0,-1]: .2f}')
# pick a second guess for y'(0)
guess2 = 2
sol2 = solve_ivp(shooting_rhs, [0, 2], [1, guess2])
print(f'Solution 2: {sol2.y[0,-1]: .2f}')
# now use linear interpolation to find a new guess
m = (guess1 - guess2) / (sol1.y[0,-1] - sol2.y[0,-1])
guess3 = guess2 + m * (target - sol2.y[0,-1])
print(f'Guess 3: {guess3: .2f}')
sol3 = solve_ivp(shooting_rhs, [0, 2], [1, guess3],
max_step=0.1)
print(f'Solution 3: {sol3.y[0,-1]: .2f}')
print(f'Target: {target: .2f}')
plt.plot(sol3.t, sol3.y[0,:])
plt.grid()
plt.show()
OUTPUT
Solution 1: 0.48
Solution 2: 0.59
Guess 3: 24.05
Solution 3: 3.00
Target: 3.00
```

We obtained the graphical output shown in Figure 11.3.

11.3 Linear Finite Differences Method

Each finite difference operator can be derived from Taylor expansion. Once again looking at a linear second-order differential equation $y'' = p(x)y' + q(x)y + r(x)$

on $[a, b]$ subject to boundary conditions $y(a) = \alpha$ and $y(b) = \beta$. We divide the area into evenly spaced mesh points $x_0 = a$, $x_N = b$, $x_i = x_0 + i * h$, where $h = (b - a)/N$.

We will now replace the derivatives $y'(x)$ and $y''(x)$ with the centered difference approximations $y'(x) = 1/2h(y(x_{i+1}) - y(x_{i-1})) - (h^2/12)y^3$ (ξ) and $y''(x) = 1/h^2(y(x_{i+1}) - 2y(x_i) + y(x_{i-1})) - h^2/ 6y^4$ (μi) for some $x_{i-1} \le \xi_i \mu_i \le x_{i+1}$ for $i = 1, \ldots,$ $N - 1$. We now have the equation $1/h^2 (y(x_{i+1}) - 2y(x_i) + y(x_{i-1})) = p(x_i)1/2h(y(x_{i+1}) - y(x_{i-1})) + q(x_i)y(x_i) + r(x_i)$. This is rearranged such that we have all the unknown together on the right-hand side as a system of equations so we might solve $\mathbf{Aw} = \mathbf{b}$, where

$$
\mathbf{A} = \begin{bmatrix}
2+h^2 q(x_1) & -1+\dfrac{h}{2}p(x_1) & 0 & \cdots & & 0 \\
-1+\dfrac{h}{2}p(x_2) & 2+\dfrac{h}{2}q(x_2) & -1+\dfrac{h}{2}p(x_2) & \cdots & & \cdots \\
0 & -1+\dfrac{h}{2}p(x_3) & 2+h^2 q(x_3) & \cdots & & 0 \\
\cdots & \cdots & \cdots & \cdots & -1+\dfrac{h}{2}p(x_{N-1}) \\
0 & \cdots & & \cdots & -1+\dfrac{h}{2}p(x_n) & 2+h^2 q(x_N)
\end{bmatrix}
$$

$$
\mathbf{w} = \begin{bmatrix} w_1 \\ w_2 \\ \cdots \\ w_{N-1} \\ w_N \end{bmatrix}, \text{ and}
$$

$$
\mathbf{b} = \begin{bmatrix}
-h^2 r(x_1)+(1+\dfrac{h}{2}p(x_1))w_0 \\
-h^2 r(x_2) \\
\cdots \\
-h^2 r(x_{N-1}) \\
-h^2 r(x_N)+(1+\dfrac{h}{2}p(x_N))w_{N+1}
\end{bmatrix}
$$

This method has a unique solution when functions p, q, and r are continuous, our interval $[a, b]$ and the function, $q(x) \ge 0$ on $[a, b]$, and $h < 2/L$ where $L = \max\{a \le x \le b|p(x)|\}$.

This method uses the intermediate value theorem in calculus to simplify our algorithm.

$$y''(x_i) = 1/h^2[y((x_i+1) - 2y(x_i) + y(x_i-1) - (h^2/12)y^{(4)}(e) \text{ for some } e$$
between in the interval (x_i, x_{i+1}).

$$y'(x_i) = 1/2h[y(xi + 1) - y(x_{i-1}) - h^2/6 \, y'''(n_i) \text{ for some } n_i$$
between in the interval (x_{i-1}, x_{i+1}).

In this method and the boundary conditions, $y(a) = \alpha$ and $y(b) = \beta$, we will define $w_0 = \alpha$ and $w_{N+1} = \beta$, as well as

$$\frac{2w_i - w_{i+1} - w_{i-1})}{h^2} + p(x_i)(\frac{w_{i+1} - w_{i-1}}{2h}) + q(x_i)w_i = -r(x_i) \text{ for each } I = 1, 2, \ldots, N.$$

Given the constant-coefficient BVP,

$$c_2 \, y^{(2)}(t) + c_1 \, y^{(1)}(t) + c_0 \, y(2)(t) = f$$
$$y(a) = y_a$$
$$y(b) = y_b.$$

Example 3. Simple Case

$y'' = 4y$, with boundary conditions $y(0) = 1.1752$, $y(1) = 10.0179$.
   We will let $N = 4$ for this example that we will set up and solve by hand.
   We have $p(x) = r(x) = 0$ and $q(x) = 4$.
   We have $h = 1/N = 1.4 = 0.25$.
   We substitute everything into **A**, **w**, and **b** to get the following:

$$A = \begin{bmatrix} 2.25 & -1 & 0 \\ -1 & 2.25 & -1 \\ 0 & -1 & 2.25 \end{bmatrix}, \mathbf{w} = \begin{bmatrix} w_1 \\ w_2 \\ w_3 \end{bmatrix}, \text{ and } \mathbf{b} = \begin{bmatrix} 1.1752 \\ 0 \\ 10.0179 \end{bmatrix}.$$

Previously, we have presented methods to solve these types of equations.
   Our solution to these equations is

$$w_1 = 2.14670$$
$$w_2 = 3.65488$$
$$w_3 = 6.97679.$$

Our estimates are shown in Table 11.2.
   We can solve such a system in Maple as follows see Harder (2005):

```
c[2] := 1.0;
c[1] := 4.0;
c[0] := 2.0;
f := 0.0;
n := 10;
```

TABLE 11.2

Results for Example 3 Compared to the Exact Answer

x	y	Exact	$|\text{Error}|$
0	1.1752	1.1752	0
0.25	2.1467	2.1293	0.0174
0.50	3.6549	3.6269	0.028
0.75	6.07679	6.0502	0.202659
1.0	10.0179	10.0179	0

```
a := 1.0;
b := 3.0;
ya := 4.0;
yb := 2.0;
h := (b - a) / n;
low := 2*c[2] - h*c[1];
diag := 2*h^2*c[0] - 4*c[2];
up := 2*c[2] + h*c[1];
vec := 2*h^2*f;
M := Matrix( n - 1, n - 1 );
v := Vector( n - 1 );
M[1, 1] := diag;
M[1, 2] := up;
v[1] := vec - ya*low;
for i from 2 to n - 2 do
    M[i, i - 1] := low;
    M[i, i] := diag;
    M[i, i + 1] := up;
    v[i] := vec;
end do:
M[n - 1, n - 2] := low;
M[n - 1, n - 1] := diag;
v[n - 1] := vec - yb*up;
y := LinearAlgebra[LinearSolve]( M, v );
```
Copyright ©2005 by Douglas Wilhelm Harder. All rights reserved.

We also mention that Professor Doug Meade has a Maple application for the linear finite difference method that is available from the Maple Applications website that we will also illustrate.

Example 4. $y^{(2)}(t) + 4\,y^{(1)}(t) + 2\,y(t) = 0$, $y(1) = 4$, $y(3) = 2$ **by Linear Finite Differences**

Using the algorithm from Maple from Harder (2005), we obtained the following estimates. The plot is shown in Figure 11.4.

$$y := \begin{bmatrix} 4.55349854623914 \\ 4.53051229198511 \\ 4.26177465347709 \\ 3.90307135848925 \\ 3.52630873872364 \\ 3.16333568803989 \\ 2.82701376985886 \\ 2.52133216093220 \\ 2.24624963362464 \end{bmatrix}$$

Using Doug Meade's application, we obtained

```
fd_sol1 := fsolve( {seq( eq[k], k=0..N )}, {seq( y[k], k=0..N
)} );
fd_sol1 := {y[0] = 4., y[1] = 4.553498546, y[2] = 4.530512292,
 y[3] = 4.261774623, y[4] = 3.903071358, y[5] = 3.526308739,
 y[6] = 3.163335688, y[7] = 2.827013770, y[8] = 2.521332161,
 y[9] = 2.246249634, y[10] = 2.}
```

$$\begin{bmatrix} 1 & 4. \\ \dfrac{6}{5} & 4.553498546 \\ \dfrac{7}{5} & 4.530512292 \\ \dfrac{8}{5} & 4.261774623 \\ \dfrac{9}{5} & 3.903071358 \\ 2 & 3.526308739 \\ \dfrac{11}{5} & 3.163335688 \\ \dfrac{12}{5} & 2.827013770 \\ \dfrac{13}{5} & 2.521332161 \\ \dfrac{14}{5} & 2.246249634 \\ 3 & 2. \end{bmatrix}$$

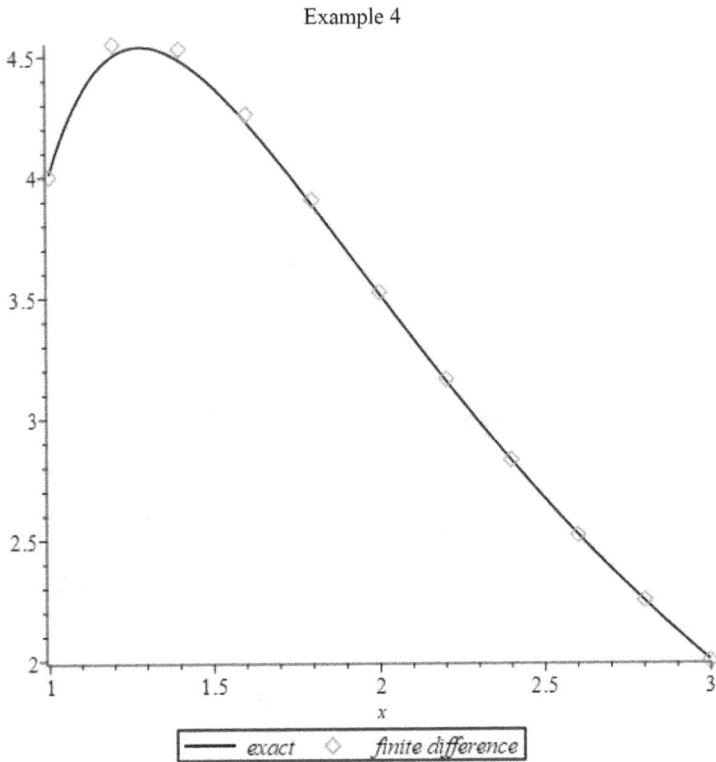

FIGURE 11.4
Plot of approximate and exact results for Example 4.

LINEAR FINITE DIFFERENCE METHOD

```
 I    X(I)         W(I)
 0 1.00000000 4.00000000
 1 1.18181818 3.68643679
 2 1.36363636 3.39752205
 3 1.54545454 3.13140834
 4 1.72727273 2.88650579
 5 1.90909091 2.66161130
 6 2.09090909 2.45622560
 7 2.27272727 2.27130563
 8 2.45454545 2.11102837
 9 2.63636364 1.98690472
10 2.81818182 1.92735711
11 3.00000000 2.00000000
```

LINEAR FINITE DIFFERENCE METHOD

```
I      X(I)        W(I)
0  1.00000000 1.00000000
1  1.10000000 1.09260052
2  1.20000000 1.18704312
3  1.30000000 1.28333687
4  1.40000000 1.38140204
5  1.50000000 1.48112026
6  1.60000000 1.58235989
7  1.70000000 1.68498902
8  1.80000000 1.78888174
9  1.90000000 1.89392110
10 2.00000000 2.00000000
```

11.4 Applications

Motorcycle Suspension

Consider a motorcycle suspension system, similar to the one shown in Figure 11.2. The suspension system can be modeled as a damped spring-mass system. We start with Newton's law: $\Sigma F = MA$, with a motorcycle with rider of total mass, M, and an acceleration, A. We have the model $mx'' = -kx - bx'$, where b is the damping constant times the instantaneous velocity (x'). Perhaps we have the choice of two suspensions. One gives a compression of a 180-pound rider of 4", while the other gives a compression of the same 180-pound rider of 8". We will solve the first compression and leave the second compression as an exercise. Let's assume the motorcycle weighs 204 pounds.

By substitution, we have

$11.8692x'' + 240x' + 1152x = 0$ that simplifies to

$x'' + 20.22x' + 97/058x = 0$,

with boundary conditions $x(0) = 4" = 1/3'$ and $x(0.2) = 0.5$.

We are looking for a smooth ride with a maximum if $f(x)$ as small as possible between [0, 0.1] time units.

First, we use the shooting point method.

$$y := \begin{bmatrix} 0.353935426034301 \\ 0.373586196761759 \\ 0.392315876661200 \\ 0.410153849053867 \\ 0.427128671878122 \\ 0.443268099539026 \\ 0.458599104197611 \\ 0.473147896513941 \\ 0.486939945867745 \end{bmatrix}$$

Output from Maple.

Our plot of the motorcycle suspension for the jump is shown in Figure 11.3.

Note also that the value of the shooting method solution at the right-hand boundary point is

evalf(eval(rhs(shoot_sol2), x=0));
   0.333333334

evalf(eval(rhs(shoot_sol2), x=.2))
   0.4999999686

This also shows good agreement between the shooting method and exact solutions.

evalf(eval(rhs(shoot_sol2), x=.1))
   0.728865405

The graphical output is shown in Figure 11.5.

In the exercise, we test the other suspension system and compare its results to these results.

Parachuting by Skydiving Free Fall

Let's consider a parachute problem where a man or woman of mass 80 kilograms jumps out of a helicopter that is 1200 meters above the earth. We will use Newton's second law: $SF = MA$, where M is mass and A is acceleration. There actually two parts to this model, the free fall and the parachute. We model and solve these separately.

Free-Fall Phase

The force acting on the jumper during the free-fall phase is the air.

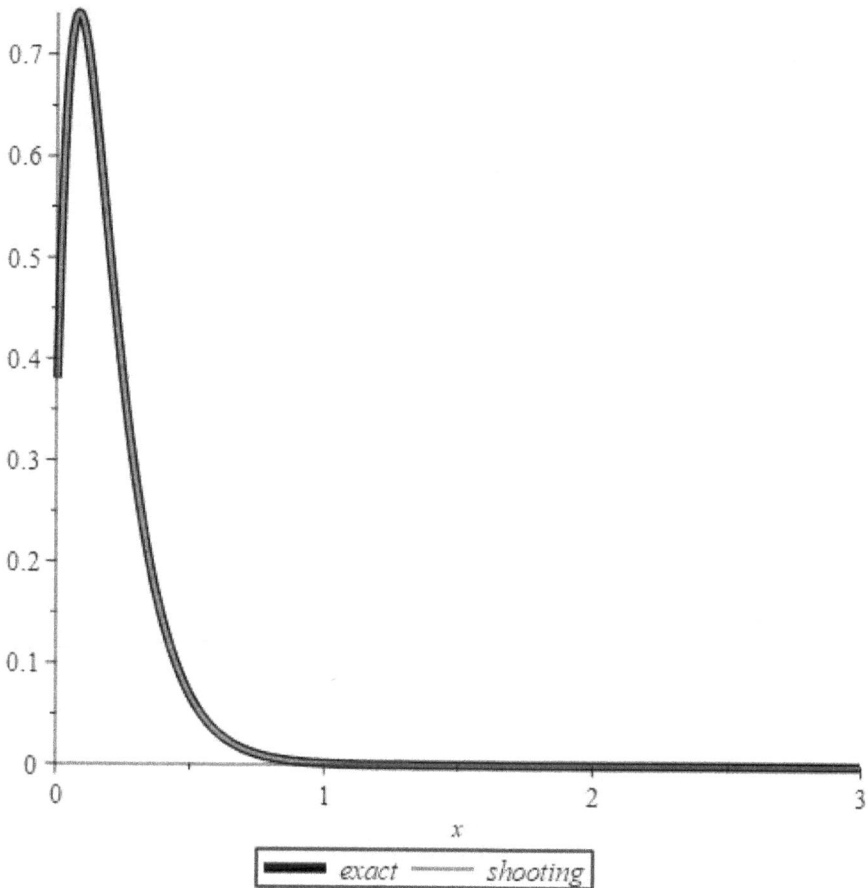

FIGURE 11.5
Motorcycle suspension for jump.

W.will use $m\,x'' = -mg - k_1x'$.

W.will use 15.79 as k_1 and 9.81 as the acceleration due to gravity.

Since down is positive, our model becomes

$80\,x'' = -80 * 9.81 - 15.79x'$.

Dividing by $m = 80$, we get

$x'' = -9.81 - 0.197375x'$.

We will use boundary condition here. We let $x(0) = 0$ and $x(10) = 929$ meters. Our output is displayed in Table 11.3, and a plot is displayed in Figure 11.6.

TABLE 11.3

Free-Fall Phase

$$
\begin{bmatrix}
0 & -0. \\
1 & 2.595342030 \\
2 & 13.65327635 \\
3 & 31.65353066 \\
4 & 55.34894378 \\
5 & 83.71640225 \\
6 & 115.9165910 \\
7 & 151.2609748 \\
8 & 189.1847109 \\
9 & 229.2244286 \\
10 & 271.
\end{bmatrix}
$$

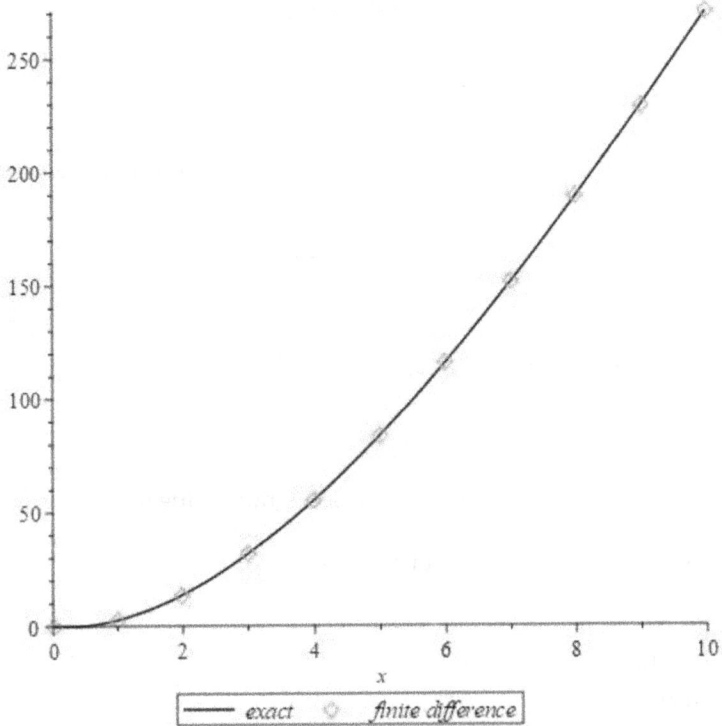

FIGURE 11.6
Plot of the free fall.

What we found from these results is that we fell 271 feet and then our parachute opened. Next, we move to the parachute-opening phase.

Parachute Phase

The forces now acting are air resistance and drag due to the open chute (we assume the chute opens instantly).

$mx'' = -mg - k_2x'$.

W.will use $k_2 = 160$.

$80\,x'' = -80 * 9.81 - 160x'$.

Dividing by $m = 80$, we get

$x'' = -9.81 - 2x'$.

Our boundary conditions here might be a little tricky. We will use $x(0) = 929$ and $x(t) = 1200$ meters for some t in the future. From historical records, we might use 180 seconds for t.

We obtain a graph in Figure 11.7.

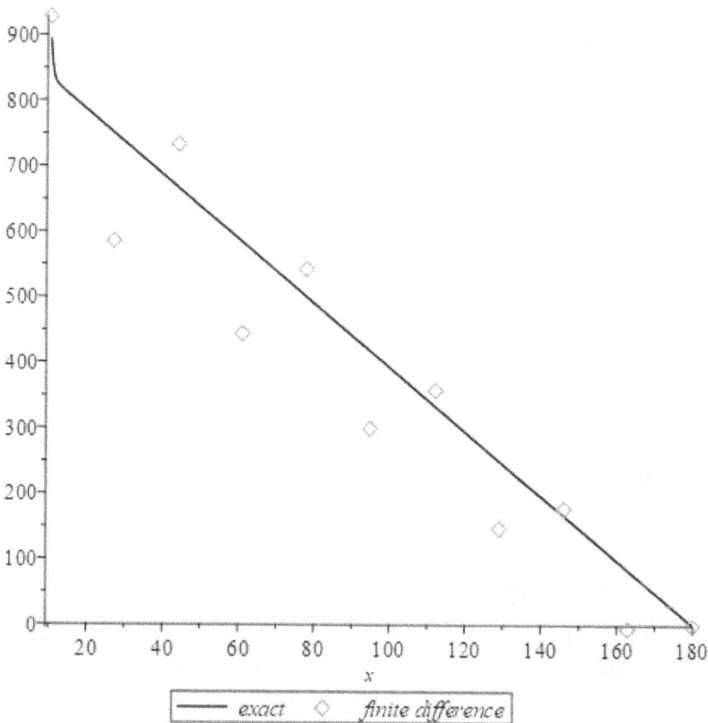

FIGURE 11.7
Chute-open phase.

We note our boundary agreement is only fair. We see that we do land in approximately 180 seconds. We hit the ground with a constant velocity of approximately 5.92 m/s.

Free Fall

A sky diver of mass, m, falls fare enough before their parachute opens with drag force equal to $k_1 v^2$ to obtain their terminal velocity in distance per seconds. When the parachute opens the drag force becomes $k_2\ v$. We can solve this boundary problem.

Newton's second law is sum $\Sigma F = MA$. Let $y(t)$ be the vertical distance downward until the parachute opens. We will call that $t = 0.\Sigma$

Our second-order ODE is $Ky'' - my' = mg$. We will use as our boundary conditions $y(0) = 0$ and $y(4.2) = 55$ meters. In this problem, we use

Mass of skydiver, $m = 70$ kilograms

Acceleration due to gravity, $g = 9.81$ m/s^2

Air-resistance constant $K = 110$ kg/s

B. substitution, $110\ y'' - 70\ y' = 70 * 9.81$ or

$y'' - 0.636636363\ y' = 6.242727.$

ysol:=(70*9.81)/110 * (t + (52/9.81-(70/110)*(1-exp((−110/70)*t))
Sol1:=(70*9.81/110)*((70/110)*ln(((52*110)/(9.81*70)−1)/.01+ (52/9.81-70/110)*1-90.01/(52*110)/(9.81*70)−1)

Bungee Two

In Chapter 7, we solved the bungee jumping problem as a first order differential equation with initial conditions in terms of velocity. As a review, in bridge jumping, a participant attaches one end of a bungee cord to themselves, attaches the other end to a bridge railing, and then drops off the bridge. In this project, the jumper will be dropping off the Royal Gorge Bridge, a suspension bridge that is 1053 feet above the floor of the Royal Gorge in Colorado. The jumper will use a 200-foot-long bungee cord. It would be nice if the jumper has a safe jump, meaning that the jumper does not crash into the floor of the gorge or run into the bridge on the rebound. In this project, you will do some analysis of the fall.

Assume the jumper weighs 160 pounds. The jumper will free-fall until the bungee cord begins to exert a force that acts to restore the cord to its natural (equilibrium) position. In order to determine the spring constant of the bungee cord, you found that that a mass weighing 4 pounds stretches the cord 8 feet. Hopefully, this spring force will help slow the descent sufficiently so that the jumper does not hit the bottom of the gorge.

Throughout this project, we will assume that DOWN is the POSITIVE direction.

Before the bungee cord begins to retard the fall the jumper, the only forces that act on the jumper are their weight and the force due to wind (air) resistance is $0.9v + 0.0009v^2$. Now, we will use a second-order problem of position with boundary conditions.

Assume that our two boundary condition are $x(0) = 0$ and $x(1.654) = 200$.

First, using Newton's law, the $\Sigma F = MA$, we get $W = MG$, so $M = W/G = 160\text{lb}/32.2 \text{ ft/s}^2 = 4.96$. We may substitute to get

$x'' = (-0.5x - 0.9x' - 32.2)/4.96$

$x'' = -0.1000806 - 0.18145x' - 6.4919,$

with boundary conditions $x(0) = 0$ and $x(1.654) = 200$.

Heat Transfer

Fins are used in many applications to increase heat transfer from surfaces. Usually the design of cooling pin fins is encountered in many applications such as the one shown in Figure 11.8

We can model the temperature distribution in a pin where the length of the fin is L, the start and start and end of the fin is $x = 0$ and $x = L$. Let the

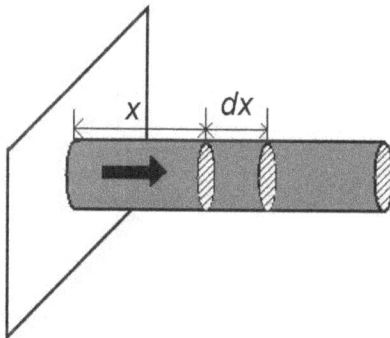

Heat Transfer through Fin with Uniform Cross-Sectional Area

FIGURE 11.8
Heat transfer.
(www.bing.com/images/search?view=detailV2&ccid=x28RXp3x&id=FD3FF080F06B8E8C6122B7
D774DACDEC4D2FB7ED&thid=OIP.x28RXp3x9fyZAEvG0NhFtgHaEB&mediaurl=https%3a%2f
%2fwww.careerride.com%2fimages%2fMechanical%2fHeat-Transfer-through-Fin-with-Uniform-
Cross-sectional-Area.png&cdnurl=https%3a%2f%2fth.bing.com%2fth%2fid%2fR.c76f115e9df1f5fc
99004bc6d0d845b6%3frik%3d7bcvTezN2nTXtw%26pid%3dImgRaw%26r%3d0&exph=283&expw
=520&q=heat+transer++from+surafces+as+an+ode&simid=608011165998719417&FORM=IRPRST
&ck=F86E9D0506323504312163B774B6F444&selectedIndex=45&ajaxhist=0&ajaxserp=0)

temperatures at the two ends be T_0 and T_L. T_s is the temperature of the surrounding environment. If we consider both convection and radiation, the steady-state temperature of the pin fin $T(x)$ can be modelled by the following modelled by equation 11.5:

Heat Transfer

Fins are used in many applications to increase heat transfer from surfaces. Usually the design of cooling pin fins is encountered in many applications such as the one shown in Figure 11.9 and Figure 11.10 (with fins).

We can model the temperature distribution in a pin where the length of the fin is L, the start and start and end of the fin is $x = 0$ and $x = L$. Let the temperatures at the two ends be T_0 and T_L. Ts is the temperature of the surrounding environment. If we consider both convection and radiation, the steady-state temperature of the pin fin $T(x)$ can be modeled by the following equation:

$$d^2Tdx^2 - \alpha_1(T - T_s) - \alpha_2(T^4 - T_s^4) = 0 \quad (11.5)$$ with the boundary conditions: $T(0) = T_0$ and $T(L) = T_L$, and α_1 and α_2 are coefficients.

Our fin has a 5-millimeter diameter and a 100-millimeter length. The thermal conductivity of our fin material is 400 $Wm^{-1}K^{-1}$. One end of the fin is

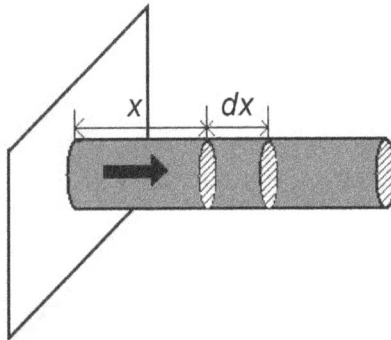

Heat Transfer through Fin with Uniform Cross-Sectional Area

FIGURE 11.9
Heat transfer.
(www.bing.com/images/search?view=detailV2&ccid=x28RXp3x&id=FD3FF080F06B8E8C6122B7
D774DACDEC4D2FB7ED&thid=OIP.x28RXp3x9fyZAEvG0NhFtgHaEB&mediaurl=https%3a%2f
%2fwww.careerride.com%2fimages%2fMechanical%2fHeat-Transfer-through-Fin-with-Uniform-
Cross-sectional-Area.png&cdnurl=https%3a%2f%2fth.bing.com%2fth%2fR.c76f115e9df1f5fc
99004bc6d0d845b6%3frik%3d7bcvTezN2nTXtw%26pid%3dImgRaw%26r%3d0&exph=283&expw
=520&q=heat+transer++from+surafces+as+an+ode&simid=608011165998719417&FORM=IRPRST
&ck=F86E9D0506323504312163B774B6F444&selectedIndex=45&ajaxhist=0&ajaxserp=0)

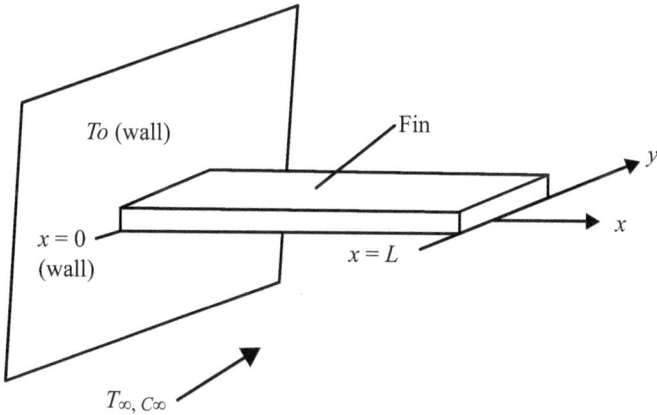

FIGURE 11.10
Heat transfer from a fin (web.mit.edu/16.unified/www/FALL/thermodynamics/notes/node128.html).

maintained at 130 °C, and the remaining surface is exposed to the ambient air at 30 °C. The convective heat transfer coefficient is 40 Wm^{-2}k^{-1}.

We consider only the initial case without radiation effects, where the second-order boundary ODE is

d^2T/dx^2-(2*H/ K R)(T-T$_o$) with boundary conditions
T(0)=130, T(100)=30.
0 ≤ x≤ 1 cm
H=heat transfer
R=radius
K=thermal conductivity
By simple substitution we have
d^2T/dx^2-((2*40)/ (400* 2.5)(T-T$_0$) =
d^2T/dx^2-0.08(T-T$_o$) with T(0)=130, T(100)=30

LINEAR SHOOTING METHOD

```
 I     X(I)           W(1,I)              W(2,I)
 0  0.00000000  130.00000000   -99.86611080
 1  0.10000000  120.00305560  -100.05677600
 2  0.20000000  109.99157810  -100.15944560
 3  0.30000000   99.97356551  -100.19014780
 4  0.40000000   89.95541298  -100.16491570
 5  0.50000000   79.94191281  -100.09977000
 6  0.60000000   69.93625610  -100.01070830
 7  0.70000000   59.94003595   -99.91370053
 8  0.80000000   49.95324941   -99.82469084
 9  0.90000000   39.97430020   -99.75960617
10  1.00000000   30.00000000   -99.73437043
```

At $x = 0.5$ cm, the temperature is 79.94191281 with the shooting point method. The exact solution at $x = 0.5$ is 80.50420126. The absolute error is 0.56218845.

$$T'' - \alpha_1 (T-Ts) - \alpha_2 (T^4 - T_s^4)$$

11.5 Beam Deflection

A common problem in civil engineering concerns the deflection of a beam subject to uniform loading, while the ends of the beam are supported so that they undergo no deflection. The differential equation approximating the physical situation is of the form

$$\frac{d^2w}{dx^2} = \frac{S}{El}w + \frac{qx}{2El}(x-1), 0 \le x \le 1,$$

where $w = w(x)$ is the deflection of a distance x from the left end of the beam and $q, E, s,$ and l represent the length of the beam, the intensity of the uniform load, the modulus of elasticity, the stress on the endpoint, and the central moment of inertia, respectively. The moment of inertia that we are concerned with is about the neutral axis perpendicular to the web at the center. Associated with the differential equation are two boundary conditions given by the assumption that no deflection occurs at the end of the beam, $w(0) = 0, w(1) = 0$.

When the beam is of uniform thickness, the product $E * I$ is constant, and the exact solution is easily obtained. In many applications, however, the thickness is not uniform, so the moment of inertia I is a function of x, and approximation techniques are required. You have been designing a catwalk for a new hotel in Atlantic City, and you are concerned with the deflection of the beam. The beam that you will be using will be a W10-type steel I-beam with the following characteristics: length $l = 120$ inches, weight is 87 lb/ft, intensity of uniform load $q = 1000$ lb/ft, modulus of elasticity $E = 3.0 \times 10^7$ lb/in.^{2}, and stress at ends $S = 1000$ pounds. The central moment of inertia, I, is given as a constant, $I = 625$ in.^{4}. We need to approximate the deflection $w(x)$ of the beam every 12 inches. If the deflection is $>1/300$ (0.0033), then it is too great. Where is the greatest deflection, and what is its value?

Here are our shooting point method results, followed by the exact solution results.

```
> for i from 0 to 120 by 10 do evalf( eval( rhs(shoot_sol2),
x=i ) ); end do;
   0.
  -0.0003
  -0.0007
  -0.0009
  -0.0012
  -0.0016
  -0.0017
```

```
-0.0018
-0.0019
-0.0016
-0.0013
-0.0009
 0.
```

$$\textbf{for } i \textbf{ from } 0 \textbf{ to } 120 \textbf{ by } 10 \textbf{ do } evalf\left(subs\left(x = i, \frac{e^{\frac{\sqrt{3}x}{7500}}\left(-312500 + 312619 e^{\frac{2\sqrt{3}}{125}}\right)}{2e^{\frac{4\sqrt{3}}{125}} - 2}\right.\right.$$

>

$$\left.\left.+ \frac{\left(312500 e^{\frac{2\sqrt{3}}{125}} - 313619\right)e^{\frac{2\sqrt{3}}{125} - \frac{\sqrt{3}x}{7500}}}{2e^{\frac{4\sqrt{3}}{125}} - 2} - \frac{x^2}{240} + \frac{x}{240} - 156250\right)\right) ; \textbf{end do};$$

```
0.0020
0.0017
0.0014
0.0011
0.0008
0.0005
0.0003
0.0002
0.0003
0.0004
0.0006
0.0012
0.0020
```

We see the minimum value graphically in Figure 11.11. We solve the optimization next to find the value.

> *with* (*Optimization*) :

> $NLPSolve\left(\frac{e^{\frac{\sqrt{3}x}{7500}}\left(-312500 + 312619 e^{\frac{2\sqrt{3}}{125}}\right)}{2e^{\frac{4\sqrt{3}}{125}} - 2}\right.$

$$\left.+ \frac{\left(312500 e^{\frac{2\sqrt{3}}{125}} - 312619\right)e^{\frac{2\sqrt{3}}{125} - \frac{\sqrt{3}x}{7500}}}{2e^{\frac{4\sqrt{3}}{125}} - 2} - \frac{x^2}{240} + \frac{x}{240} - 156250, x = 0..120\right) ;$$

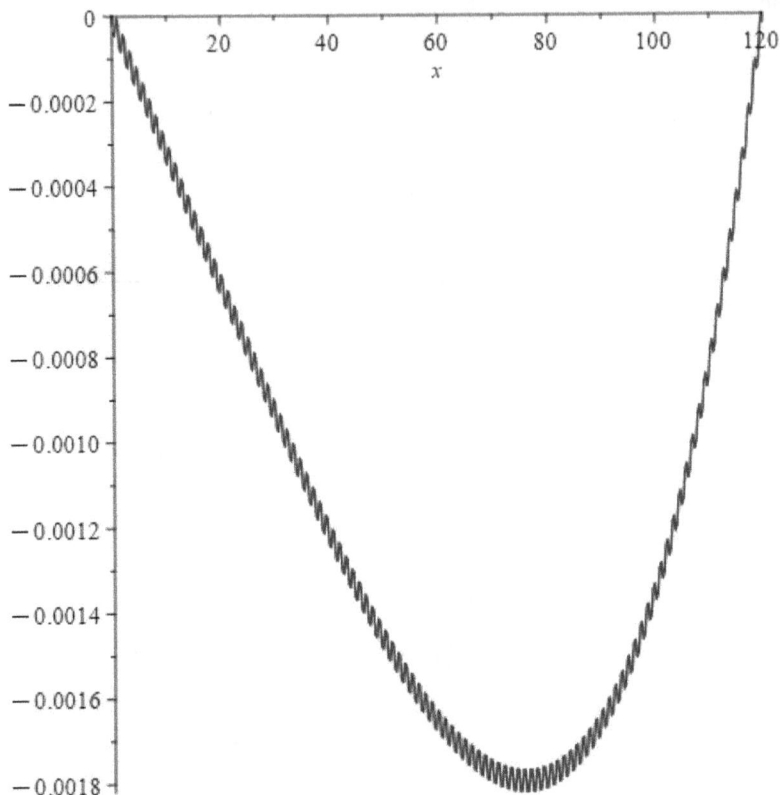

FIGURE 11.11
Graphical output from Maple for beam deflection.

```
[ - 0.00178984407102689, [x = 75.6796602742220]
>
```

The greatest deflection is 0.0017, which is <1/300. We should be all right.

LINEAR SHOOTING METHOD

```
I      X(I)         W(1,I)       W(2,I)
0   0.00000000 0.00000000 0.00003146
1 10.00000000 0.00031449 0.00003140
2 20.00000000 0.00062660 0.00003092
3 30.00000000 0.00092991 0.00002956
4 40.00000000 0.00121352 0.00002690
5 50.00000000 0.00146210 0.00002248
6 60.00000000 0.00165587 0.00001587
7 70.00000000 0.00177061 0.00000660
8 80.00000000 0.00177765 -0.00000575
```

```
 9   90.00000000 0.00164390 -0.00002163
10  100.00000000 0.00133178 -0.00004149
11  110.00000000 0.00079929 -0.00006578
12  120.00000000 0.00000000 -0.00009493
```

Exercises

1. The BVP

   $y'' = 4(y - x)$, $0 \leq x \leq 1$ with $y(0) = 0$ and $y(1) = 2$. Does this problem has an exact solution of $y(x) = e^2(e^4 - 1)^{-1}(e^{2x} - e^{-2x})+x$? Use the linear shooting method to approximate the solution and compare the results when $h = 1/2$ and $h = 1/4$.

2. The BVP

   $y'' = y' + 2y + \cos(x)$, $0 \leq x < \pi/2$ with $y(0) = -0.03$ and $y(\pi/2) = -0.01$. Does this problem has an exact solution of $y(x) = -0.1(\sin(x) + 3\cos(x)$? Use the linear shooting method to approximate the solution and compare the results when $h = \pi/2$ and $h = \pi/4$.

3. Use the linear shooting method to approximate the solution to the following BVPs:

   a. The exact solution is $-y(x) = -11.60967305 * \exp(-0.585786438 * x) + 74.84991806 * \exp(-3.414213562 * x)$ for $y' + 0.197375 * y' + 9.81 * y(x) = 0$ for $0 \leq x \leq 3$, with $y(1) = 4$, $y(3) = 2$.

   b. $y'' + 4 * y' + 13 * y = 0 = 0$ for $0 \leq x \leq 1$, with $y(0) = 1$, $y(1) = 0$, with an exact solution of $y(x) = -\exp(-2 * x) * (\cot(3) * \sin(3 * x) - \cos(3 * x))$.

   c. $y'' + 2 * y' - 10 * y(x) = 7 * \exp(-x) + 4$ for $0 \leq x \leq 1$, with $y(0) = 2$, $y(1) = -5$ and an exact solution of $y(x) = -0.435070729 * \exp(2.316624790 * x) + 3.471434365 * \exp(-4.316624790 * x) - 0.4000000000 - 0.6363636364 * \exp(-1. * x)$

1. Artillery Firing Situation

You are an ROTC cadet and you have just been assigned to an artillery unit, and you are in charge of firing a M109A3 155mm Howitzer (or for those in Navy ROTC, you have just been assigned to a battleship in charge of a 16-inch gun, or for those in Air Force ROTC, you have just been assigned to an F-16. . . .). You just have a few rounds left, and you really want to hit a stationary target roughly 5.1 kilometers away. Wind conditions are heavy but steady, and you are using a brand-new type of ammunition, so the targeting tables you have been issued do not apply since the drag characteristics are not well known. However, there is a highly accurate radar attached to your unit, and you have a laptop computer that can receive data directly from the radar, and the laptop has Maple loaded and ready for action. Having studied two-point BVPs in your numerical analysis class you realize that you can set-up a two-point boundary value problem that you can solve, for the proper angle of elevation, using the Newton secant method.

The Model: The first step is developing equations of motion that will model the flight of the projectile. The problem can be placed in a two-dimensional coordinate frame in which you are at $x = 0$, on the slope of the hill. We will assume that the projectile obeys Newton's laws of motion, $\Sigma F = MA$. The projectile experiences acceleration in the vertical direction due to gravity, $g = 9.81$ m/s^2. Drag due to air friction is in a direction opposite the direction of motion, and has a magnitude $\rho(y)|v(t)|^2$, where $v \rightarrow (t)$ is the velocity of the projectile at time t. The air friction constant $\rho(y)$ depends on altitude because air pressure decreases with increasing altitude.

With these two assumptions, the second-order equations are

$$x'' = -\rho(y) * (x')^2 \text{ and}$$
$$y'' = -\rho(y) * (y')^2 - g,$$

where the meteorologists report that the air pressure decreases linearly with altitude according to

$$\rho(y) = -4.75 * 10^{-8} y + 2.0 * 10^{-4}.$$

You are on a small mountain called "Gaussian Hill," whose terrain is described by the function

$$hill(x) = 1000e^{10^\wedge -6(x-4800)^\wedge 2}$$

You are at the fixed location $x_0 = 0$ meter on the hill, and the enemy is at the fixed location, on the far side of Gaussian Hill, $x_{target} = 5100$ meters. The mussel velocity of your gun is $|v \rightarrow (0)| = 350$ m/sec.

Mission: Hit within 1 meter of the enemy target. Given the preceding model, the only parameter the gunner can vary is α, the angle of gun elevation. Find the α to achieve your objective. Hint: $v_x = v \cos (\alpha)$ and $v = v \sin (\alpha)$.

Requirements: The following are meant to lead you through the steps of solving this two-point boundary value problem using the Newton-secant method. (1) Defining the variables $x' = vx$, and $y' = vy$, change the form of the equations of motions from two second-order differential equations to four first-order differential equations. This will allow you to directly apply Euler's solver to integrate the flight of the projectile. (2) Define a coordinate frame. (3) State the differential equations with the initial conditions, in which the initial velocities depend on α. (4) State the conditions in which the target has been hit. (5) Adapt the Euler solver to integrate our equations of motion, and place stopping condition on the algorithm so that the solutions are calculated until the projectile hits the hill. Plot a sample flight for α = your favorite angle. Discuss variations in step size h, and then choose an appropriate h that is a balance between accuracy, the speed of your machine, the accuracy of the integrator you are using, and your patience at looking at the light-bulb blink. Discuss the benefits of a higher order integration scheme such as the Runge–Kutta 4. Example: Here is my favorite angle: 45 degrees.

(6) Develop a function shoot(α), which returns the x-coordinate of the projectile when it hits dirt. (7) Develop a "range" function: range(α) = |shoot(α) – x target)| that automatically returns how close to the target a shot is given an input angle α. Plot range(α) for $0 \leq \alpha \leq 90°$. (8) Choose two initial angles, and then find the roots of range(α) by developing a Newton-secant algorithm. What should the stopping criterion be on the root finder to achieve the mission? (9) Discuss the benefits of a higher order root finder. (10) What modifications would be required of your program if the equations of motion are changed? (11) Discuss the dependence of the angle-finding part of this project, that is, the secant method, for of the model-based range(α) function.

References and Further Readings

Burden, R. and D. Faires (1997). *Numerical Analysis*. Brooks-Cole, Pacific Grove, CA.

Fox, W. P. (2014). *Mathematical Modeling with Maple*. Cengage Publishers, Boston, MA.

Fox, W. P. (2018). *Mathematical Modeling for Business Analytics*. Taylor and Francis Publishers, Boca Raton, FL.

Giordano, F., W. Fox and S. Horton (2013). *A First Course in Mathematical Modeling*, 5th ed. Cengage Publishers, Boston, MA.

Harder, D. W. (2005). https://ece.uwaterloo.ca/~dwharder/nm/ Meade, Doug, Maple Application Center. www.maplesoft.com (Doug Meade)

12

Approximation Theory and Curve Fitting

12.1 Introduction

Consider the proportionality model where we have collected data that will enable us to test our proportionality hypothesis. Proportionality is defined as two items x and y are proportional, if and only if $y = kx$, $k > 0$. This suggests that we can fit a linear relationship and find the value of the slope, k.

Example 1. Kepler's Law

To assist in further understanding the idea of proportionality, let's examine one of the famous proportionalities from Table 12.1.

Kepler's Third Law. In 1601 the German astronomer Johannes Kepler became director of the Prague Observatory. Kepler had been helping Tycho Brahe in collecting 13 years of observations on the relative motion of the planet Mars. By 1609, Kepler had formulated his first two laws: (1) Each planet moves along an ellipse with the sun at one focus. (2) For each planet, the line from the sun to the planet sweeps out equal areas in equal times. Kepler spent many years verifying these laws and formulating the third law given in Table 12.1, which relates the orbital periods and mean distances of the planets from the sun. The following data were collected from the 1993 World Almanac.

TABLE 12.1

Planets and Distances

Planet	Period (days)	Mean Distance (millions of miles)
Mercury	88.0	36
Venus	224.7	67.25
Earth	365.3	93
Mars	687.0	141.75
Jupiter	4331.8	484.8
Saturn	10,760.0	887.97
Uranus	30,684.0	1764.50
Neptune	60,188.3	2791.05

DOI: 10.1201/9781032703671-12

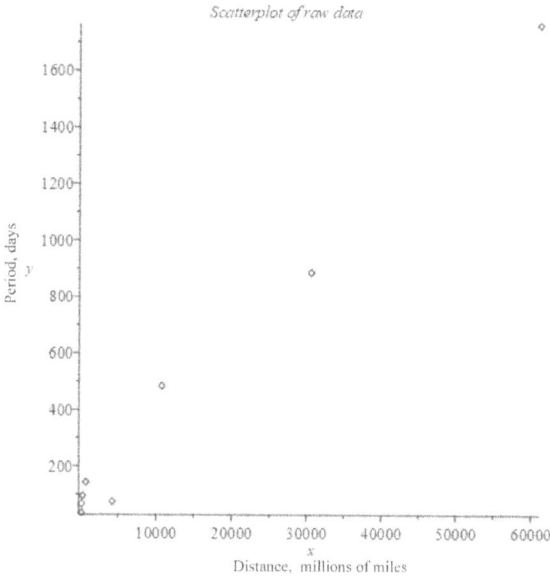

FIGURE 12.1
Scatterplot of data.

T proportional to D $^{3/2}$.

In Figure 12.2, we plot the period versus the mean distance to the 3/2 power. The plot approximates a line that projects through the origin. We can easily estimate the slope (constant of proportionality) by picking any two points that lie on the line passing through the origin.

D$^{3/2}$	Period
216	88
551.4910046	224.7
896.8595208	365.3
1687.658618	687
10,674.41091	4331.8
26,460.46306	10,760
74,119.50223	30,684
147,452.2572	60188.3
0.4072611	

The slope is $\Delta y/\Delta x = (60188.3 - 88)/(147452.2572 - 216) = 0.4072611$
The model is

$T = 0.4072611 D^{3/2}$.

But we can do better to estimate the slope, k.

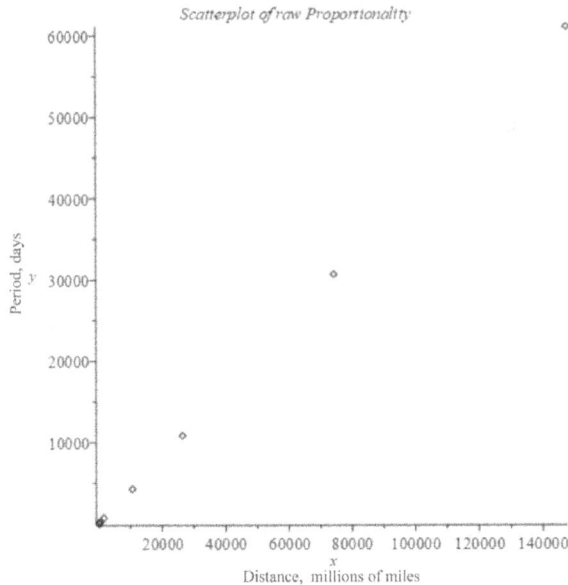

FIGURE 12.2
Kepler's third law as a proportionality plot.

Example 2. Spring-Mass System

Consider a spring-mass system, such as the one shown in Figure 12.3. We conducted an experiment to measure the stretch of the spring as a function of the mass (measured as weight) placed upon the spring. Consider the data collected for this experiment, displayed in Table 12.2. The graph showing an approximate straight line (passing through the origin) is presented in Figure 12.4

The data appear to follow the proportionality rule that elongation (e) is proportional to the mass (m), e $\propto$ m. The straight line appears to pass through the origin. This geometric understanding allows us to look at the data to determine if proportionality is a reasonable simplifying assumption from which we may estimate the slope, k. In this case, the assumption appears valid, so we estimate the constant of proportionality by picking any two points that lie on our straight line as shown in Table 12.2.

The data plot, seen in Figure 12.4, looks reasonably like a straight line through the origin. Our next step was to calculate the slope. We found the model as $F = 0.00155S$. We now want a more exact fit of our line to the data. Model fitting, especially with least squares, will be how we obtain a better fit.

This chapter describes how to determine the parameters of a model analytically, according to some criterion of "best fit," and test the adequacy of the model.

Suppose it is proposed that a parabolic model might best explain a behavior being studied, and you are interested in selecting that member of the parabolic family, say, $y = kx^2$. Recall that if y is proportional to x^2,

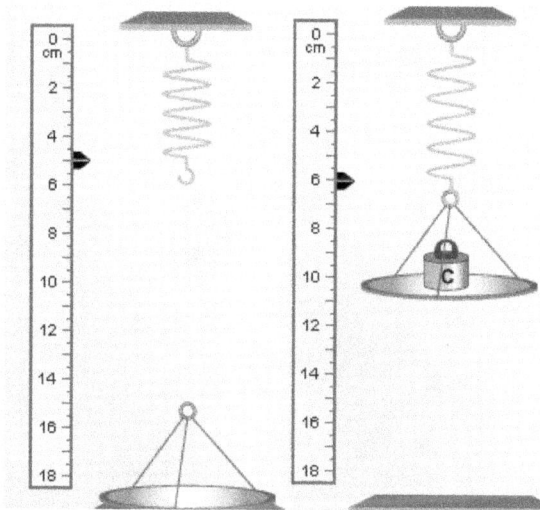

FIGURE 12.3
Spring-mass system.

TABLE 12.2

Spring-Mass System

Mass (g)	Stretch (m)
50	0.1
100	0.1875
150	0.275
200	0.325
250	0.4375
300	0.4875
350	0.5675
400	0.65
450	0.725
500	0.80
550	0.875

then we may only want to fit the model, $y = kx^2$, instead of $y = ax^2 + bx + c$. Using Maple, y versus x^2 can be plotted, and we could use the graph to assist in selecting two points to obtain an estimate of the slope of the line, as demonstrated in Chapter 2 (or Chapter 3) with proportionality as a simplifying assumption. In this chapter, we want to find the "best-fit" line using a specific criterion.

This chapter focuses on the analytical methods to arrive at a model for a given data set using a prescribed criterion. Again, from the family $y = kx^2$, the parameter k can be determined analytically by using a

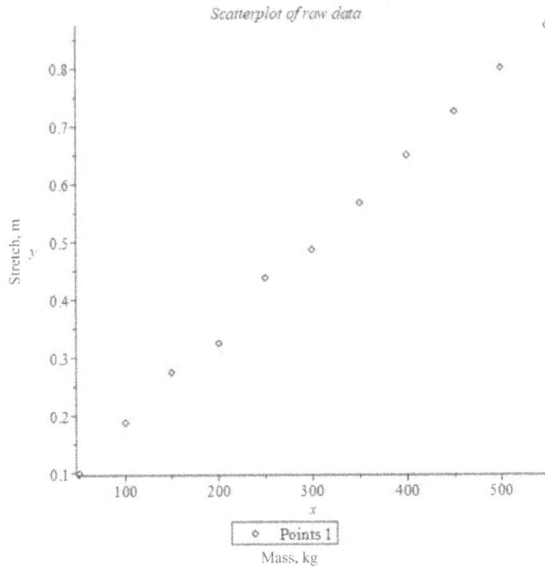

FIGURE 12.4
Plot of spring-mass data.

curve-fitting criterion, such as discrete least squares, Chebyshev's, or minimizing the sum of the absolute error, and then solving the resulting optimization problem. Although we briefly present these other criteria, we concentrate on the presentation of least squares in this chapter. We present the Maple commands that solve the least-squares optimization problem with analysis of the "goodness of the fit" of the resulting model.

12.2 Model Fitting

In this section, we will introduce several criteria for fitting data with models: least squares, Chebyshev's approximation, and minimize the sum of the absolute deviations. Each of these criteria are concerned with the error between the raw data and the model. We can define the error, e_i, as the difference between each real data value and the value the model yields for its approximation, $e_i = y_i - \hat{y}_i$. In discrete least squares we want to minimize the sum of the error, e_i^2.

Criterion 1: Discrete Least Squares

The method of least-squares curve fitting, also known as **ordinary least squares** and **linear regression**, is simply the solution to a model that minimizes the sum of the squares of the deviations between the observations and predictions. In Maple, the fit command, within the statistical package, fits a model curve to a set of data points using the least-squares methods. Least squares will find the parameters of the function, $f(x)$ that will minimize the sum of squared differences between the real data and the proposed model, shown in Equation 12.1.

$$\text{Minimize } S = \sum_{j=1}^{m} \left[y_1 - f(x_j) \right]^2 \tag{12.1}$$

For example, to fit a proposed proportionality model $y = kx^2$ to a set of data, the least-squares criterion requires the minimization of Equation 12.2. Note in Equation 12.2, k is estimated by a.

$$\text{Minimize } S = \sum_{j=1}^{5} \left[y_i - kx_j^2 \right]^2 \tag{12.2}$$

Minimizing Equation 12.2 is achieved using the first derivative, setting it equal to zero, and solving for the unknown parameter, k.

$$\frac{ds}{dk} = -2\sum x_j^2 (y_j - kx_j^2) = 0. \text{ Solving for } k: \ k = \left(\sum x_j^2 y_j \right) / \left(\sum x_j^4 \right). \tag{12.3}$$

Given the data set in Table 12.3, we will find the least squares fit to the model, $y = kx^2$.

Solving for k: $k = \left(\sum x_j^2 y_j \right) / \left(\sum x_j^4 \right) = (195.0) / (61.1875) = 3.1869$ and the model $y = kx^2$ becomes $y = 3.1869x^2$.

If our function was $f(x) = ax + b$, add normal equations.

$$n * b_0 + b_1 \sum x_i = \sum y_i$$

$$b_0 \sum x_i + b_1 \sum x_i^2 = \sum x_i y_i$$

In Maple, the fit command fits the model type $Y = b + b_1 X_1 + b_2 X_2 + \ldots + b_k X_k$ to the data set given in the k-specified columns. To fit the quadratic model $y = A_0 + A_1 x + A_2 x^2$ to a data set of x values, xv, and of y values, yv, the Maple fit command required to solve a full quadratic equation is *quadraticfit:=[leastsquare[x,y],yy=a*x^2 + b*x + c,{a,b,c}]] ([xv,yv])*, where x and y are command calls that will use the two data sets named in the command (in the case xv and yv), while a, b, and c are the coefficient variables for which a least-squares solution will be fit. To fit our example, using only, $y = Ax^2$, we use the following command:

quadraticfit:=[leastsquare[x,y],y=A*x^2 ,{A}]] ([xv,va]).

TABLE 12.3

Least-Squares Data

X	0.5	1.0	1.5	2.0	2.5
Y	0.7	3.4	7.2	12.4	20.1

Since fit is part of the statistics package, the with(stats): command must be entered once prior to using the fit command. First, we illustrate the procedure with the model, $y = ax^2 + bx + c$:

```
> with(stats) :
> with(plots) : Digits = 5 :
>
> xv := [.5, 1, 1.5, 2, 2.5];
        xv := [0.5, 1, 1.5, 2, 2.5];
> yv := [.7, 3.4, 7.2, 12.4, 20.1];
        yv := [0.7, 3.4, 7.2, 12.4, 20.1];
>
> eqfit1:= fit[leastsquare[[x,y], y=a*x^2+b*x+c, {a,b,c}]]
  ([xv,yv]);
        eqfit1 := y = 3.2607 x² - 0.22223 x + 0.12630
>
>
> f := unapply(rhs(eqfit1), x);
        f := x→ 3.2607 x² - 0.22223 x + 0.12630
> xy := {seq([xy[i], yv[i]], i = 1 ..5)} :
> plot1 := plot(xy, style = pont, symbol = diamond) :
> plot2 := plot(f(x), x = 0 ..3) :
> display({plot1, plot2}, title = 'Least Squares Fit');
```

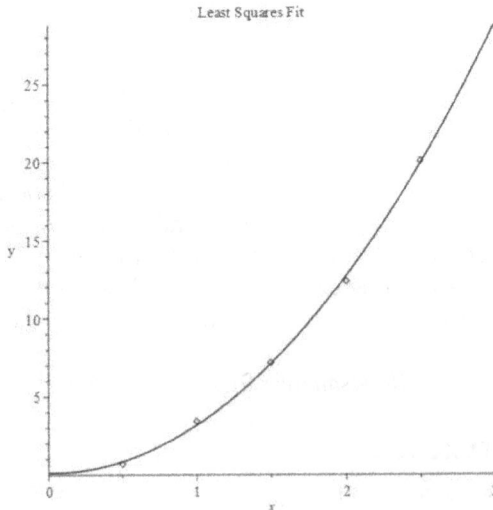

FIGURE 12.5
Least-squares fit plotted with the original data.

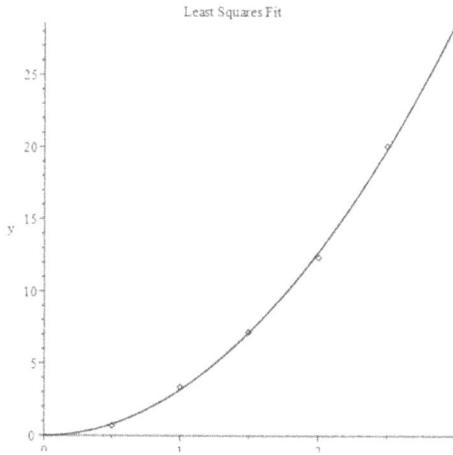

FIGURE 12.6
Least-squares kx^2 fit plotted with the data.

Next, we illustrate the fit command applied to the proportionality model $y = kx^2$ for the data set for our example. As obtained previously, the least-squares model is $y = 3.1870x^2$ (rounded to three decimal places). Since the two data sets used in this example, xv and yv, have been entered previously, they can be called without reentering the data. Having previously invoked both with(plots): and with(stats): commands, they need not be repeated in this continuation example either.

```
> eqfit2:= fit[leastsquare[[x,y], y=k*x^2, {k}]]([xv,yv]);
           eqfit2 := y = 3.1870. x²
>
>
> f := unapply(rhs(eqfit2), x);
           f := x → 3.1870 x²
> xy := {seq([xv[i], yv[i]], i = 1 ..5)} :
> plot1 := plot(xy, style = point, symbol = diamond) :
> plot2 := plot(f(x), x = 0 ..3) :
> display({plot1, plot2}, title = 'Least Squares Fit');
```

Using normal equations,

```
>solve({p1 = 0, p2 = 0}, {a, b});
```

$$
\left\{ a = \frac{n\left(\displaystyle\sum_{k=0}^{n} y(k)x(k)\right) - \left(\displaystyle\sum_{k=0}^{n} x(k)\right)\left(\displaystyle\sum_{k=0}^{n} y(k)\right) + \left(\displaystyle\sum_{k=0}^{n} y(k)x(k)\right)}{n\left(\displaystyle\sum_{k=0}^{n} x(k)^2\right) - \left(\displaystyle\sum_{k=0}^{n} x(k)\right)^2 + \left(\displaystyle\sum_{k=0}^{n} x(k)^2\right)}, b = \right.
$$

$$-\frac{\left(\displaystyle\sum_{k=0}^{n}y(k)x(k)\right)\left(\displaystyle\sum_{k=0}^{n}x(k)\right)-\left(\displaystyle\sum_{k=0}^{n}x(k)^2\right)\left(\displaystyle\sum_{k=0}^{n}y(k)\right)}{n\left(\displaystyle\sum_{k=0}^{n}x(k)^2\right)-\left(\displaystyle\sum_{k=0}^{n}x(k)\right)^2+\left(\displaystyle\sum_{k=0}^{n}x(k)^2\right)}\Bigg\}$$

Least Squares in Python

```
CODE
import pandas as pd
#create dataset
df = pd.DataFrame({'hours': [1, 2, 4, 5, 5, 6, 6, 7, 8, 10,
11, 11, 12, 12, 14],
     'score': [64, 66, 76, 73, 74, 81, 83, 82, 80, 88, 84, 82,
     91, 93, 89]})
#view first six rows of dataset
df[0:6]
import matplotlib.pyplot as plt
plt.scatter(df.hours, df.score)
plt.title('Hours studied vs. Exam Score')
plt.xlabel('Hours')
plt.ylabel('Score')
plt.show()
import statsmodels.api as sm
#define response variable
y = df['score']
#define explanatory variable
x = df[['hours']]
#add constant to predictor variables
x = sm.add_constant(x)
#fit linear regression model
model = sm.OLS(y, x).fit()
#view model summary
print(model.summary())
OUTPUT
```

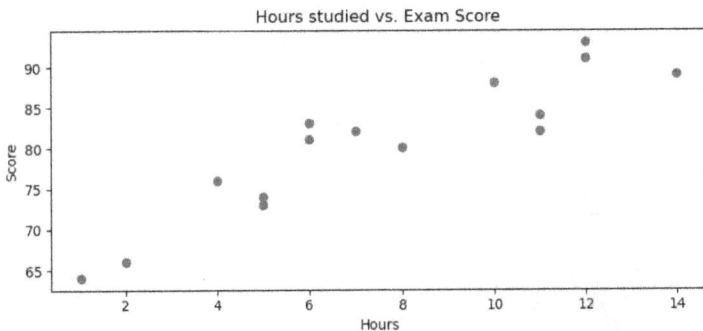

Hours studied vs. Exam Score

```
                          OLS Regression Results
==============================================================================
Dep. Variable:                  score   R-squared:                       0.831
Model:                            OLS   Adj. R-squared:                  0.818
Method:                 Least Squares   F-statistic:                     63.91
Date:                Tue, 25 Jul 2023   Prob (F-statistic):           2.25e-06
Time:                        08:07:36   Log-Likelihood:                -39.594
No. Observations:                  15   AIC:                             83.19
Df Residuals:                      13   BIC:                             84.60
Df Model:                           1
Covariance Type:            nonrobust
==============================================================================
                 coef    std err          t      P>|t|      [0.025      0.975]
------------------------------------------------------------------------------
const          65.3340      2.106     31.023      0.000      60.784      69.884
hours           1.9824      0.248      7.995      0.000       1.447       2.518
==============================================================================
Omnibus:                        4.351   Durbin-Watson:                   1.677
Prob(Omnibus):                  0.114   Jarque-Bera (JB):               .1.329
Skew:                           0.092   Prob(JB):                        0.515
Kurtosis:                       1.554   Cond. No.                         19.2
==============================================================================

Notes:
[1] Standard Errors assume that the covariance matrix of the errors is correctly specified.
```

The model is Exam score = 65.3340 + 1.9824 * hours.

Problems with Higher Order Polynomials

There are problems that might exist in higher order polynomial fitting. As we said earlier, sometimes in fitting the data the curve has oscillations and snaking occur near the endpoints. This oscillating and snaking behavior makes the polynomial fit ill fated for interpolation near the endpoints and further makes predictions outside the endpoints pointless. We provide an illustrative example in which we fit a complete polynomial. Note that we have lost the trend of the data (although we fit each datum exactly) and that at the endpoints, we have oscillations, Therefore, we suggest a close examination of all higher order polynomials before accepting any.

Here are the data and the scatterplot (Figure 12.7) suggesting a smooth curve with concave-up trends.

```
> Xvalues:=[.55,1.2,2,4,6.5,12,16];Yvalues:=[.13,.64,5.8,102,
  210,2030,3900];
  Xvalues := [0.55, 1.2, 2, 4, 6.5, 12, 16]
  Yvalues := [0.13, 0.64, 5.8, 102, 210, 2030, 3900]
> pointplot(zip((x,y)->[x,y],Xvalues,Yvalues));
```

Next, we fit a sixth-order polynomial ($n = 7$) and then plot the polynomial. We will use the *interp* command to obtain a $(n - 1)$–order polynomial.

```
> FI:=interp([.55,1.2,2,4,6.5,12,16],[.13,.64,5.8,102,
  210,2030,3900], z);
```

$$FI := -0.01383726235\, z^6 + 0.5084246673\, z^5 - 18.09506969 - 6.437923862\, z^4$$
$$+ 64.31279044\, z + 34.85731000\, z^3 - 73.99155349\, z^2$$

```
>plot(-.1383726235e-1*z^6+.5084246673*z^5-18.09506969-6.437923862*z^4
+64.31279044*z+34.85731000*z^3-73.99155349*z^2,z=0..16);
```

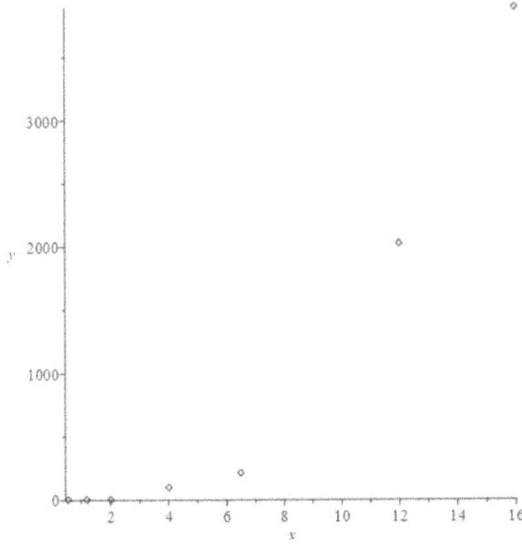

FIGURE 12.7
Scatterplot of data.

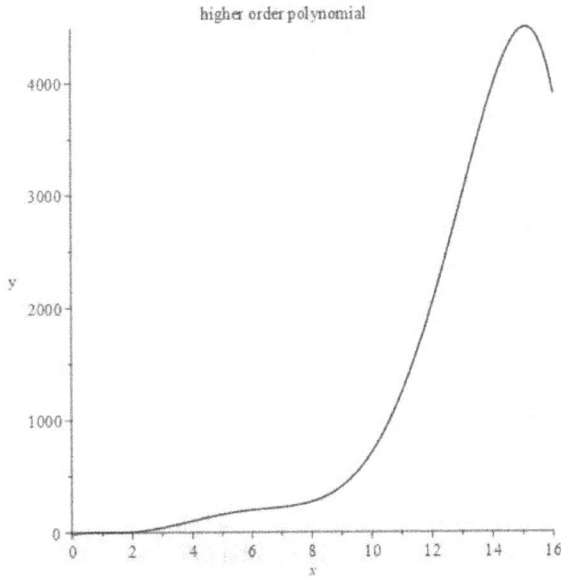

FIGURE 12.8
Higher order polynomial showing oscillations near endpoints.

Again note the oscillation at the endpoints in Figure 12.8. Thus, although the higher order $(n - 1)$–order polynomial gives a perfect fit through the data points, its oscillations make interpolation and prediction less accurate.

Example 3. Fitting a Fifth-Order Polynomial Using Least-Squares

Given a set of data points, see Table 6.12, an analyst decides to attempt to fit a curve to the data using a high-order polynomial. Table 12.4 displays the commands required, and Figure 12.9 displays the polynomial curve superimposed on the data.

```
> with (stats) with (plots) :
> xdata := [1, 2, 3, 4, 5, 6];
xdata := [1, 2, 3, 4, 5, 6]
> ydata := [305, 266, 135, -16, 125, 1230];
ydata := [305, 266, 135, -16, 125, 1230]
> xyfit := fit[leastsquare[[x, y], y = a*x^5 + b*x^4 +
    c x^3 + d*x^2 + e*x + g, {a, b, c, d, e, g}]]([xdata, ydata]);
xyfit := y = x^5 - 5x^4 - 3x^3 + 7x^2 + 5x +300
> f := unapply(rhs(xyfit), x);
f := x → x^5 - 5x^4 - 3x^3 + 7x^2 + 5x +300
> xy := {seq([xdata[i], ydata[i]], i = 1 ..6)};
xy := {[1, 305], [2, 266], [3, 135], [4, -16], [5, 125],
    [6, 1230}
> c1 := pointplot({seq([xdata[i], ydata[i]], i = 1 ..6)},
    thickness = 3) :
c2 := plot(f(x), x = 0 ..6) :
display({c1, c2});
```

TABLE 12.4

Data for Polynomial

x	1	2	3	4	5	6
y	305	266	135	-16	125	1230

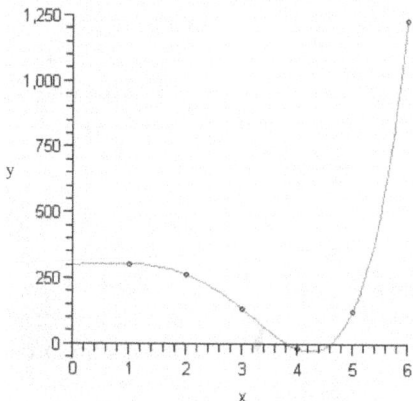

FIGURE 12.9
The plot for Example 3.

Do we have to use a fifth-order polynomial? The answer lies in the need and use of the model as well as how well the trend is captured regardless of the perfect fit.

The lesson here is perhaps a lower order polynomial that captures the trend in the data is what we might want to use.

12.3 Application of Planning and Production Control

Many optimization problems require the simultaneous consideration of a number of independent variables. In planning and producing items, one must consider many factors that impact in the process. The company might desire to maximize profit, minimize cost, maximize production levels, improve efficiency, minimize shipping time, and a host of other options. Many of these can be solved by the following techniques: differential calculus, Lagrange multipliers, linear programming, and dynamic programming. In this scenario, we will illustrate several model-fitting techniques for multivariable functions.

Least Squares Chebyshev's Criterion Minimize the sum of the absolute errors

- **Key Definitions and Variables:** z is the measure of performance. x_1, $x_2, \ldots, x_n$ are the inputs that affect z. Optimum point: The values of $x_1, x_2, \ldots, x_n$ which maximize or minimize z.
- **Optimal Value:** The value of z for the optimum point.
- **Unimodal:** Most search strategies rely on the assumption that the surface is unimodal; that is, it only has one peak over the region of concern.
- You are hired as a consultant and ask to optimize all facets of a company's planning and production. RABA manufactures 15-inch color TV sets. The company plans to improve their building of color TVs.
- It appears as though a new chip added to the circuitry will improve reception and survivability of the TV. The new chip is extremely sensitive and must be continuously monitored. The monitoring process is assumed to be modeled by the following expression:

$$f(x) = Ax + B/x,$$

where A is initially assumed to remain constant throughout the process at a value of 68. The value of B fluctuates slowly because of gradual environmental changes. The company has recently measured the process at $(x, y) = (0.5, 79)$ and estimates B at that instance to be 22.5. This seems unrealistic to you, so you collect data over a 12-week period and decide to use the least squares to estimate both A and B.

TABLE 12.5

TV Data

Week	1	2	3	4	5	6	7	8	9	10	11	12
x	1.1	2.1	3	4.5	5	6	7.1	8	9	10	11.1	12
y	76.6	78	97.5	120.5	145	170.5	196	222	248	274	301	328

Minimize $S = \Sigma(y_i - (Ax_i + (B/x_i)))^2$. The data were previously collected and are provided in Table 12.5.

The following are the normal equations we obtain

$$2\sum\left[y_i - \left(Ax_i + \left(\frac{B}{x_i}\right)\right)\right](-x_i) = 0$$

$$2\sum\left[y_i - \left(Ax_i + \left(\frac{B}{x_i}\right)\right)\right](-1/x_i) = 0.$$

And these may be simplified knowing M is the number of data pairs ($M = 12$) as

$$Ax_i^2 + BM = \Sigma x_i y_i \text{ and}$$
$$AM + \Sigma(1/x_i^2) = \Sigma(y_i/x_i).$$

We obtain the following equations after substitution:

$$\begin{bmatrix} 658.49 & 12 \\ 12 & 1.354344 \end{bmatrix}\begin{bmatrix} A \\ B \end{bmatrix} = \begin{bmatrix} 18247 \\ 388.3262 \end{bmatrix}.$$

Solving these yields $A = 26.8149$ and $B = 49.1358$.

Week	x	y	Model
1	1.1	76.6	74.1653
2	2.1	78	79.70929
3	3	97.5	96.8233
4	4.5	120.5	131.5861
5	5	145	143.9017
6	6	170.5	169.0787
7	7.1	196	197.3063
8	8	222	220.6612
9	9	248	246.7936
10	10	274	273.0626
11	11.1	301	302.072
12	12	328	325.8735

Assuming that *A* and *B* are represent settings on a machine, to minimize the sum of squared error, we would set *A* to 26.8149 and *B* to 49.1359. Maple has some nice closed routines for performing least squares.

```
Xvals:=[1.1,2.1,3,4.5,5,6,7.1,8,9,10,11.1,12];
Xvals := [1.1, 2.1, 3, 4.5, 5, 6, 7.1, 8, 9, 10, 11.1, 12] >
Yvals:=[76.6,79,97.5,120.5,145,170.5,196,222,248,274,301,328];
Yvals := [76.6, 79, 97.5, 120.5, 145, 170.5, 196, 222, 248,
   274, 301, 328] > with(plots):with(stats):with(fit):
> points:=pointplot(zip((x,y)- >[x,y],Xvals,Yvals)):
> eq:=leastsquare[[x,y],y=a*x+b/x,{a,b}]([Xvals,Yvals]);
eq := y = 26.81348281 x + 49.43330857 1/x
```

This is our least-squares model. Now, let obtain a plot of the data and the model for a visual fit shown in Figure 12.10.

The fit appears to be very good within the domain [1, 12] of the independent variable.

We take the derivative of our function and set the derivative equal to 0 and solve. We find $x = 1.496776046$ minimizes our functions.

$$df/dx = 0 = 26.8149 - 40.1359/x^2$$

There are two roots, and we select the positive root of the derivative and see the function plot in Figure 12.11.

```
NULL;
```

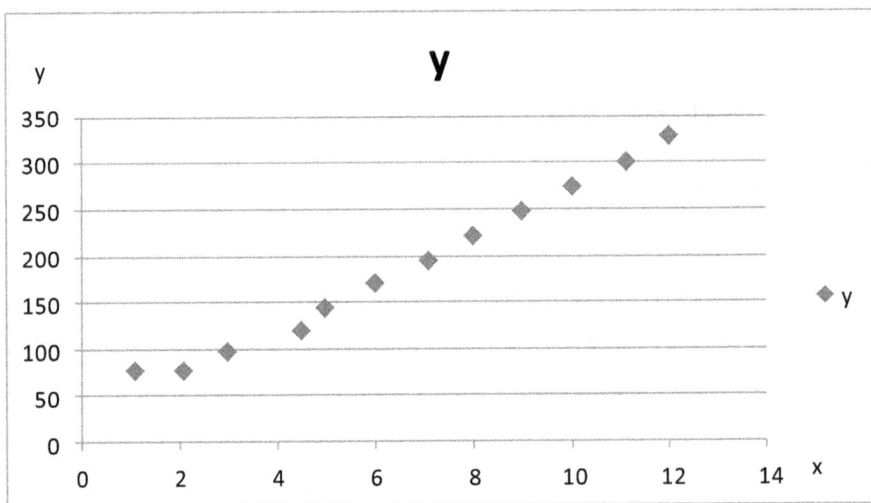

FIGURE 12.10
Plot of TV model's data.

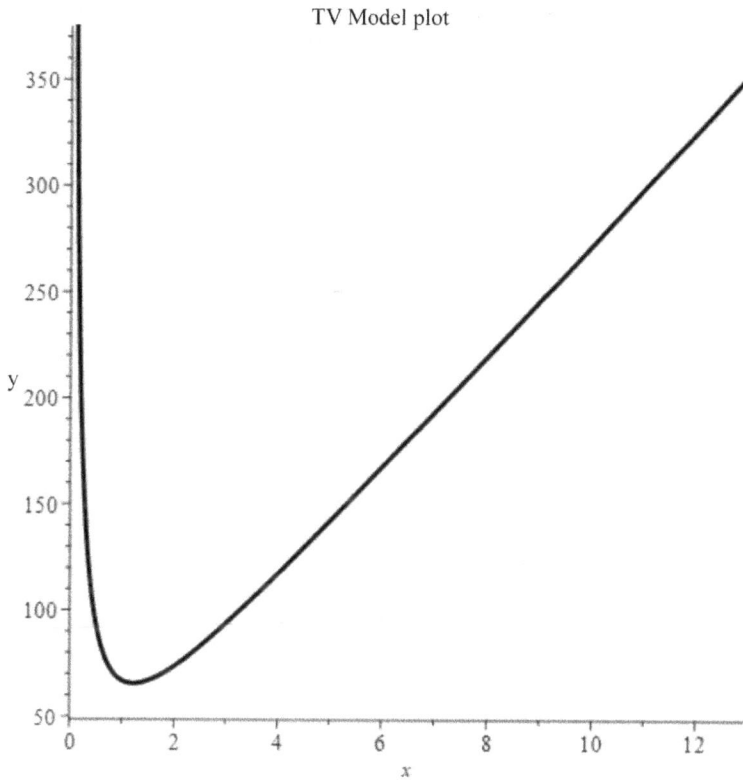

FIGURE 12.11
Model plot.

12.4 Continuous Least Squares

Suppose we have a function, f, defined on $[a, b]$, and we desire a polynomial of degree n to minimize the error.

Let $P_n(x) = a_0 + a_1 x + a_2 x^2 + \ldots + a_n x^n$, where $P_n(x)$ is an approximating polynomial.

Setting up the necessary and sufficient conditions leads to the following normal equations:

$$\sum_{k=0}^{n} a_k \int_a^b x^{j+k} \, dx = \int_a^b x^j f(x) \, dx \text{ for each } j = 0, 1, 2, \ldots, n.$$

This is then solved for the $n + 1$ unknowns, $a_0, a_1, \ldots, a_{n+1}$ to produce the coefficients for $P_n(x)$.

Example 4. Find the Continuous Least-Squares Approximating Polynomial of Degree 2 for the Function, $f(x) = \sin(\pi x)$ on the Interval [0, 1]

Our approximating polynomial is $P_2(x) = a_0 + a_1 x + a_2 x^2$.
The normal equations are

$$a_0 \sum_0^1 1\,dx + a_1 \int_0^1 x\,dx + a_2 \int_0^1 x^2\,dx = \int_0^1 \sin(\dot{A}x)dx$$

$$a_0 \sum_0^1 x\,dx + a_1 \int_0^1 x^2\,dx + a_2 \int_0^1 x^3\,dx = \int_0^1 x\sin(\dot{A}x)dx$$

$$a_0 \sum_0^1 x^2\,dx + a_1 \int_0^1 x^3\,dx + a_2 \int_0^1 x^4\,dx = \int_0^1 x^2 \sin(\dot{A}x)dx \,.$$

Using a Computer Algebra System (CAS) technology to assist, we find

```
>eq := b0 int(1, x = 0..1)+b1 int(x, x = 0..1)+b2
int(x², x = 0..1) = int(sin(pi x),x = 0..1);
```

$$eq1 := b0 + \frac{b1}{2} + \frac{b2}{3} = \frac{2}{\pi}$$

```
>eq2 := b0 int(x, x = 0..1)+b1 int(x², x = 0..1)+b2
int(x³, x = 0..1) = int(sin(pi x),x = 0..1);
```

$$eq2 := \frac{b0}{2} + \frac{b1}{3} + \frac{b2}{4} = \frac{1}{\pi}$$

```
>eq3 := b0 int(x², x = 0..1)+b1 int(x³, x = 0..1)+b2
int(x⁴, x = 0..1) = int(x² sin(pi x),x = 0..1);
```

$$eq3 := \frac{b0}{3} + \frac{b1}{4} + \frac{b2}{5} = \frac{\pi^2 - 4}{\pi^3}$$

```
>fsolve( { eq1 = 0, eq2 = 0, eq3 = 0}, {bo, b1, b2} );
   {bo = - 0.050465516, b1 = 4.122511720, b2 = - 4.122511715}
```

$$>plot\left(\left\{\sin(\text{Pi} \cdot x), -.050465516 + 4.12251172 \cdot x - 4.122511715 \cdot x^2\right\}, x = 0..1,\right.$$
$$\left. thickness = 3, color = black\right);$$

$$>plot\left(\left\{\sin(\text{Pi} \cdot x), -.050465516 + 4.12251172 \cdot x - 4.122511715 \cdot x^2\right\}, x = 0..1,\right.$$
$$\left. thickness = 3, color = black\right);$$

Our approximating polynomial is $f(x) = -0.05045516 + 4.12251172x - 4.122511715x^2$. Figure 12.12 show the function and the approximating polynomial overlaid, a "good" fit.

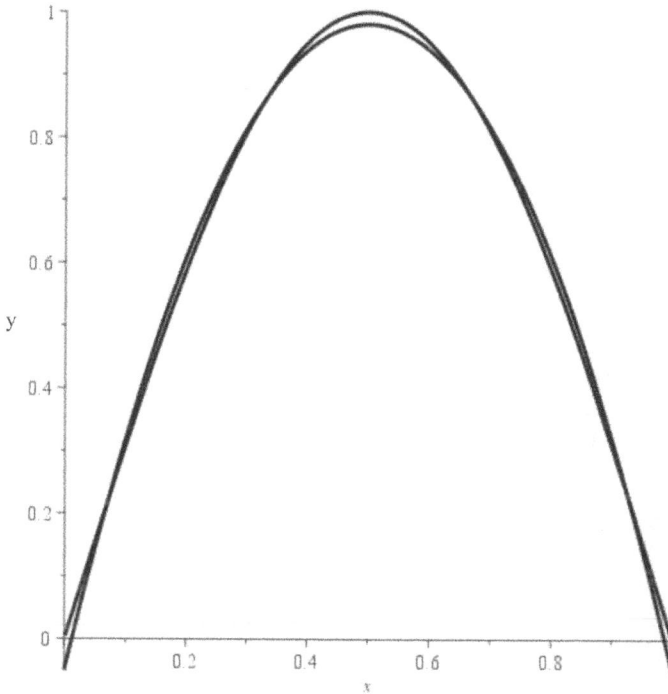

FIGURE 12.12
sin(πx) and $f(x)$ overlaid, showing a good fit.

Example 5. Find the Continuous Least-Squares Approximating Polynomial of Degree 2 for the Function, $f(x) = \cos(\pi x)$ on the Interval [−0.5, 0.5]

$$a_0 \sum_{-5}^{5} 1 dx + a_1 \int_{-5}^{5} x dx + a_2 \int_{-5}^{5} x^2 dx = \int_{-5}^{5} \cos(\dot{A}x) dx$$

$$a_0 \sum_{-5}^{5} x dx + a_1 \int_{-5}^{5} x^2 dx + a_2 \int_{-5}^{5} x^3 dx = \int_{-5}^{5} x \cos(\dot{A}x) dx$$

$$a_0 \sum_{-5}^{5} x^2 dx + a_1 \int_{-5}^{5} x^3 dx + a_2 \int_{-5}^{5} x^4 dx = \int_{-5}^{5} x^2 \cos(\dot{A}x) dx$$

> $eq11 := b0 \cdot int(1, x = -.5...5) + b1 \cdot int(x, x = -.5...5) + b2 \cdot int(x^2, x = -.5...5)$
> $= int(\cos(Pi \cdot x), x = -.5...5);$

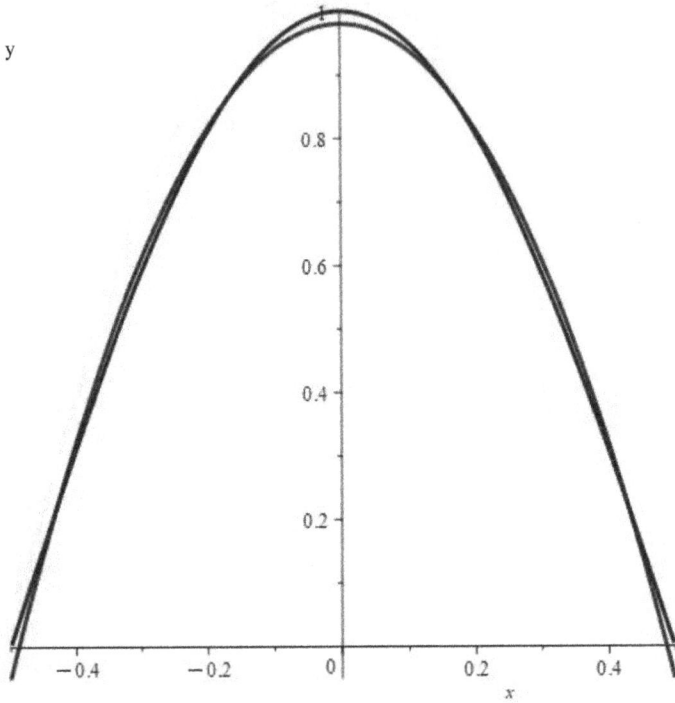

FIGURE 12.13
Approximating polynomial and cos(πx) over [−0.5, 0.5].

```
eq11 := 1.0 b0 + 12.37500000 b1 + 0.08333333333 b2 =
0.6366197724
```

$$> eq12 := b0 \cdot int(x, x = -.5...5) + b1 \cdot int(x^2, x = -.5..5) + b2 \cdot int(x^3, x = -.5...5)$$
$$= int(x \cdot \cos(\text{Pi} \cdot x), x = -.5...5);$$

```
eq12 := 41.70833333 b1 = 0.
```

$$> eq13 := b0 \cdot int(x^2, x = -.5...5) + b1 \cdot int(x^3, x = -.5..5) + b2 \cdot int(x^4, x = -.5...5)$$
$$= int(x^2 \cdot \cos(\text{Pi} \cdot x), x = -.5...5);$$

```
eq13 := 0.08333333333b0 + 156.2343750b1 +
0.01250000000b2 = 0.03014880536
```
```
>fsolve( { eq11 = 0, eq12 = 0, eq13 = 0}, {b0, b1, b2} );
{ bo = 0.9801624075, b1 = -4.122511621 }
```

Example 6. Find the Continuous Least-Squares Approximating Polynomial of Degree 3 for the Function, $f(x) = \ln(x + 2)/x$ **on the Interval [1, 3]**

```
> nf := 1/x ln(x+2);
```

$$nf := \frac{\ln(x+2)}{x}$$

```
>evalf( int( nf, x = 1 ..3 ) );

1.460466464
>
```

```
> eq11 := b0 · int(1, x = 1..3) + b1 · int(x, x = 1..3) + b2 · int(x², x = 1..3)
      + b3 · int(x³, x = 1..3) = evalf(int(nf, x = 1..3));
```

$$eq11 := 2\,b0 + 4\,b1 + \frac{26\,b2}{3} + 20\,b3 = 1.460466464$$

```
> eq12 := b0 · int(x², x = 1..3) + b1 · int(x², x = 1..3) + b2 · int(x⁴, x = 1..3)
      + b3 · int(x⁴, x = 1..3) = evalf(int(x · nf, x = 1..3));
```

$$eq12 := \frac{26\,b0}{3} + \frac{26\,b1}{3} + \frac{242\,b2}{5} + \frac{242\,b3}{5} = 2.751352693$$

```
> eq13 := b0 · int(x², x = 1..3) + b1 · int(x³, x = 1..3) + b2 · int(x⁴, x = 1..3)
      + b3 · int(x⁵, x = 1..3) = int(x² · nf, x = 1..3);
```

$$eq13 := \frac{26\,b0}{3} + 20\,b1 + \frac{242\,b2}{5} + \frac{364\,b3}{3} = \frac{3\ln(3)}{2} + \frac{5\ln(5)}{2}$$

```
> eq14 := b0 · int(x³, x = 1..3) + b1 · int(x⁴, x = 1..3) + b2 · int(x⁵, x = 1..3)
      + b3 · int(x⁶, x = 1..3) = int(x³ · nf, x = 1..3);
```

$$eq14 := 20\,b0 + \frac{242\,b1}{5} + \frac{364\,b2}{3} + \frac{2186\,b3}{7} = -\frac{26}{9} - 3\ln(3) + \frac{35\ln(5)}{3}$$

```
>fsolve( { eq11 = 0, eq12 = 0, eq13 = 0, eq14 = 0 },
 { b0, b1, b2, b3} );

 { b0 = 1.275379036, b1 = - 0.0836865830, b2 = -
0.2095854034, b3 = 0.05304307765 }
```

```
>
  plot( { 1.275379036 - 0.0836865830 · x - .2098554034 · x²
          + .0534307765 · x³, nf }, x = 1..3 );
```

The approximating polynomial is

```
f(x) =1.275379036-0.0836865830 x -0.2098554034 x² +
0.534307765 x³.
```

The plot is shown in Figure 12.14. Clearly visible, there are some errors in our approximating polynomial.

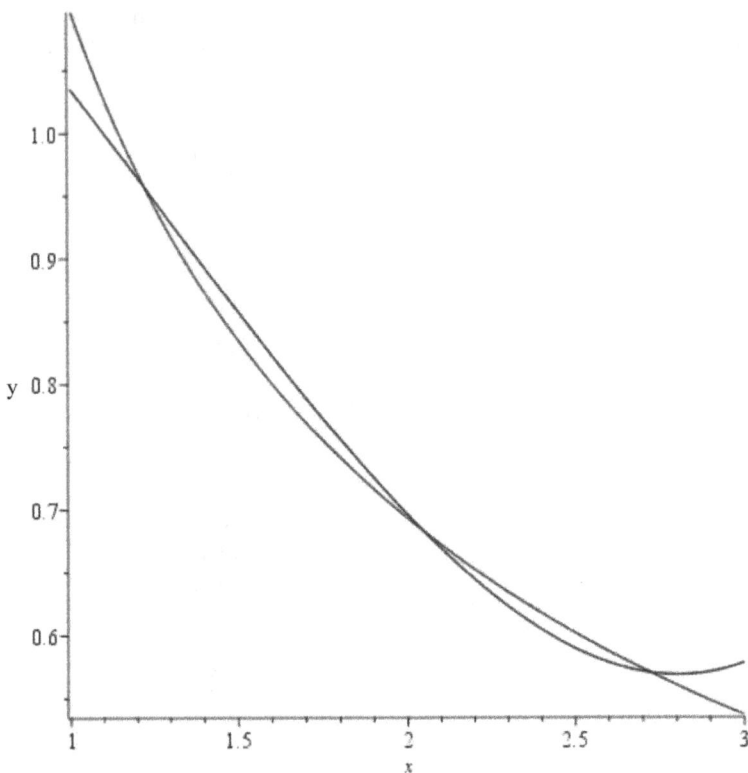

FIGURE 12.14
Plot of $\ln(x + 2)/x$ and its approximating polynomial from $[1, 3]$.

12.5 Co-Sign Out a Cosine

You are working in a research lab and you are working on pattern recognition and image processing. Your research leader believes there may be some use for the cosine generators in the old lab storage room. Your research leader would like you to use cosine functions to construct signals and break down image signals for your communications group. The storage room has a number of old "cosine wave generators" around and now let us see if we can add up combinations of cosine curves to get the signals that we need to send out. The cosine generators can generate functions of the form:

$$\cos(i\pi x) \text{ for } i = 0, 1, 2, 3, \ldots$$

We need the coefficients, b_i for $i = 0, 1, 2, 3. \ldots$ such that for $y = f(x)$ defined over a closed interval $[-1, 1]$.

We need to obtain approximations close enough to $f(x)$. It appears possible that the more cosine generators that we add up, the better our approximation of $y = f(x)$ will be on our interval $[-1, 1]$. Our first function is $f(x) = 1 - x^2$. Let's find the "best" coefficients b_i to produce the best approximations to our function when $n = 0$, 1, and 2. Then let's generalize the result.

Our model is

$$\int_{-1}^{1} b_i \left(\cos(n\dot{A}i) - (1-x^2) \right)^2 dx = 0.$$

Let $i = 0$ so

$$\int_{-1}^{1} b_0 \left(1 - (1-x^2) \right)^2 dx = 0$$

$$= b_0(x - (x - x^3/3)) = b_0[(1 - 1 + 1/3) - (1 - 1 - 1/3)] = 0, \text{ so } b_0 = 2/3.$$

$i = 1$; then

$$\int_{-1}^{1} b_0 + b1 * \cos(\pi x) - \left(1 - (1-x^2) \right)^2 dx = 0.$$

Solving yields $b_1 = 4/\pi^2$ and $b_0 = 2/3$.

For $i = 2$, we have

$$\int_{-1}^{1} b_0 + b1 * \cos(\pi x) + b2 \cos(2\pi x) - \left(1 - (1-x^2) \right)^2 dx = 0.$$

Integrating and solving, we get

$$b_2 = 1/\pi^2, b_0 = 4/3, \text{ and } b_1 = 4/\pi^2$$

If we generalize the results, $b_k = -4/n^2 p^2$ for n being odd and $b_k = 1/(n^2/2\pi^2)$ when n is even.

Exercises

Fit the following data with the models using least squares:

1.

x	1	2	3	4	5
y	1	1	2	2	4

   a. $y = b + ax$

   b. $y = ax^2$

2. Stretch of a spring data:

$x (\times 10^{-3})$	5	10	20	30	40	50	60	70	80	90	100
$y (\times 10^5)$	0	19	57	94	134	173	216	256	297	343	390

   a. $y = ax$

   b. $y = b + ax$

   c. $y = ax^2$

3. Data for the ponderosa pine:

x	17	19	20	22	23	25	28	31	32	33	36	37	39	42
y	19	25	32	51	57	71	113	140	153	187	192	205	250	260

a. $y = ax + b$

b. $y = ax^2$

c. $y = ax^3 + bx^2 + c$

4. Given Kepler's data:

Body	Period (s)	Distance from the Sun (m)
Mercury	7.60×10^6	5.79×10^{10}
Venus	1.94×10^7	1.08×10^{11}
Earth	3.16×10^7	1.5×10^{11}
Mars	5.94×10^7	2.28×10^{11}
Jupiter	3.74×10^8	7.79×10^{11}
Saturn	9.35×10^8	1.43×10^{12}
Uranus	2.64×10^9	2.87×10^{12}
Neptune	5.22×10^9	4.5×10^{12}

Fit the model $y = ax^{3/2}$.

5. Find the linear least-squares polynomial to approximate $f(x)$ on the indicated interval.

a. $f(x) = \cos(2\pi/3)$, $[-1, 1]$

b. $f(x) = x \ln(x)$, $[2, 4]$

c. $f(x) = e^x$, $[0, 2]$

d. $f(x) = e^{-2x}$, $[0, 2]$

Projects

1. Consider the function, $f(x) = 0.0625 - x^4$; find the "best" coefficients b_i to produce the best approximations with $n = 5, 10, 20$ and in general for k and an even $k = \infty$ cosine wave generators. Provide a general form for the coefficients b_i.

2. Lumber Cutters

Lumber cutters wish to use readily available measurements to estimate the number of board feet of lumber in a tree. Assume they measure the diameter of the tree in inches at waist height. Develop a model that predicts board feet as a function of diameter in inches. The following data are provided for your test:

x	17	19	20	23		28	32	38	39	40
y	19	25	32	57	71	113	123	252	259	294

The variable x is the diameter of a ponderosa pine in inches and y is the number of board feet divided by 10. Consider two separate assumptions, allowing each to lead to a model. Completely analyze each model. (1) Assume that all trees are right circular cylinders and that all trees are about the same height, so the model to fit by least squares are $y = a + bx^2$ and $y = kx^2$. (2) Assume that all trees are right circular cylinders and that the height of the tree is proportional to the diameter, leading to the model to fit by least squares are $y = a + bx^3$ and $y = kx^3$. Which of these model appears better, and why? Justify your conclusions.

4. Telemetry

Use the telemetry data from Chapter 5 with the values that were found for the missing data elements. Build a least-squares cubic model of the form $v(t) = a_0 + a_1 t + a_2 t^2 + a_3 t^3$. Compare the ability of this least-squares model to those in Chapter 5 to predict and interpolate for (a) $t = 7$ seconds, b) $v(t) = 50$ feet per second, and (c) $v(t) = 800$ feet per second.

References and Further Readings

Burden, R. and D. Faires (1997). *Numerical Analysis*. Brooks-Cole, Pacific Grove, CA.

Fox, W. P. (2014). *Mathematical Modeling with Maple*. Cengage Publishers, Boston, MA.

Fox, W. P. (2018). *Mathematical Modeling for Business Analytics*. Taylor and Francis Publishers, Boca Raton, FL.

Giordano, F., W. Fox and S. Horton (2013). *A First Course in Mathematical Modeling*, 5th ed. Cengage Publishers, Boston, MA.

13

Numerical Solutions to Partial Differential Equations

13.1 Introduction, Methods, and Applications

Problems that involve more than one variable, such as x and t are often expressed as partial differentiation equations. Here we present a brief introduction of some of the techniques used to solve the basic standard heat equation.

Methods

The three methods that we will present for parabolic problems are the forward-difference method, the backward-difference method, and the Crank–Nicolson method.

Forward-Difference Method

Given the partial differential equation

$$\frac{\partial u}{\partial t}\left(x_i, t_j\right) - \alpha^2 \frac{\partial^2 u}{\partial x^2}\left(x_i, t_j\right) = 0$$

so that the difference quotient is

$$\frac{w_{i,j+1} - w_{i,j}}{k} - \alpha^2 \frac{w_{i+1,j} - 2w_{i,j} + w_{i-1,j}}{h^2} = 0,$$

where $w_{i,j}$ approximates $u(x_i, t_j)$. The error for the difference equation is

$$\tau_{ij} = \frac{k}{2}\frac{\partial^2 u}{\partial t^2}\left(x_i, u_j\right) - \alpha^2 \frac{h^2 \partial^2 u}{12 \partial x^4}\left(\varepsilon_i, t_j\right).$$

DOI: 10.1201/9781032703671-13

We will use linear algebra to assist us:

The explicit nature of the differences gives rise to a $m-1$ by $m-1$ matrix, A.

$$A = \begin{bmatrix} (1-2\lambda) & \lambda & 0 & \cdots & & 0 \\ \lambda & (1-2\lambda) & \lambda & & & \cdots \\ 0 & \lambda & (1-2\lambda) & \lambda & & \\ \cdots & 0 & \lambda & (1-2\lambda) & \lambda & \\ 0 & 0 & 0 & \lambda & (1-2\lambda) \end{bmatrix},$$

where $\lambda = \alpha^2(k)/h^2$.

If we set $\mathbf{w}^{(0)} = (f(x_1), f(x_2), \ldots f(x_{m-1}))^t$

and let

$\mathbf{w}^{(j)} = (w_{1,j}, w_{2,j}, \ldots, w_{m-1,j})^t$ **for each j=1,2,...**

We can approximate the solution with matrix multiplication

$\mathbf{w}^{(j)} = \mathbf{A}\,\mathbf{w}^{(j-1)}$ **for each j = 1,2,...**

Example 1. Forward-Difference for Heat Equation

Given the following heat equation,

$\frac{\partial u}{\partial t}(x,t) - \frac{\partial^2 u}{\partial t^2}(x,t) = 0$, for $0 \le x \le 1$ and $t \ge 0$ with boundary conditions

$u(0, t) + 0u(1, t) = 0$ for $t > 0$ and
$u(x, 0) = \sin(\pi x)$ for $0 \le x \le 1$.

```
>restart :
```
Forward-difference method
```
>with(linalg) :
>digits := 20;
```

$$digits := 20$$

```
> l:=1:alpha:=1:m:=h := 1/m ;k:=0.0005;
```

$$h := \frac{1}{10}$$

$$k := 0.0005$$

```
> f := x → sin(Pi·x)
```

$$f := x \mapsto \sin(\pi \cdot x)$$

```
>A := matrix(m- 1,m- 1,0);
```

$$A := \begin{bmatrix} 0 & 0 & 0 & 0 & 0 & 0 & 0 & 0 & 0 \\ 0 & 0 & 0 & 0 & 0 & 0 & 0 & 0 & 0 \\ 0 & 0 & 0 & 0 & 0 & 0 & 0 & 0 & 0 \\ 0 & 0 & 0 & 0 & 0 & 0 & 0 & 0 & 0 \\ 0 & 0 & 0 & 0 & 0 & 0 & 0 & 0 & 0 \\ 0 & 0 & 0 & 0 & 0 & 0 & 0 & 0 & 0 \\ 0 & 0 & 0 & 0 & 0 & 0 & 0 & 0 & 0 \\ 0 & 0 & 0 & 0 & 0 & 0 & 0 & 0 & 0 \\ 0 & 0 & 0 & 0 & 0 & 0 & 0 & 0 & 0 \end{bmatrix}$$

>$w := vector(\text{m} - 1, 0);$

$$w := \begin{bmatrix} 0 & 0 & 0 & 0 & 0 & 0 & 0 & 0 & 0 \end{bmatrix}$$

> $u := vector(m - 1, 0);$

$$u := \begin{bmatrix} 0 & 0 & 0 & 0 & 0 & 0 & 0 & 0 & 0 \end{bmatrix}$$

> for i from 1 to $m - 1$ do $w[i] := evalf\left(f(i \cdot h)\right);$ end do;

$$w_1 := 0.3090169944$$
$$w_2 := 0.5877852524$$
$$w_3 := 0.8090169944$$
$$w_4 := 0.9510565165$$
$$w_5 := 1.$$
$$w_6 := 0.9510565165$$
$$w_7 := 0.8090169944$$
$$w_8 := 0.5877852524$$
$$w_9 := 0.3090169944$$

> lambda $:= \dfrac{\text{alpha}^2 \cdot k}{h^2};$

$$\lambda := 0.0500$$

> $A[1,1] := 1 - 2 \cdot \text{lambda};$

$$A_{1,1} := 0.9000$$

> $A[1,2] := \text{lambda};$

> for i from 2 to $m - 2$ do $A[i, i-1] := \text{lambda}; A[i, i+1] := A[i, i]$
$:= 1 - 2 \cdot \text{lambda}$: end do

$$A_{2,1} := 0.0500$$
$$A_{2,3} := 0.0500$$
$$A_{2,2} := 0.9000$$
$$A_{3,2} := 0.0500$$
$$A_{3,4} := 0.0500$$
$$A_{3,3} := 0.9000$$
$$A_{4,3} := 0.0500$$
$$A_{4,5} := 0.0500$$

$$A_{4,4} := 0.9000$$
$$A_{5,4} := 0.0500$$
$$A_{5,6} := 0.0500$$
$$A_{5,5} := 0.9000$$
$$A_{6,5} := 0.0500$$
$$A_{6,7} := 0.0500$$
$$A_{6,6} := 0.9000$$
$$A_{7,6} := 0.0500$$
$$A_{7,8} := 0.0500$$
$$A_{7,7} := 0.9000$$
$$A_{8,7} := 0.0500$$
$$A_{8,9} := 0.0500$$
$$A_{8,} := 0.9000$$

```
>A[m - 1, m -2] := lambda;
```

$$A_{9,8} := 0.0500$$

> $A[m-1, m-1] := 1 - 2 \cdot \text{lambda};$

$$A_{9,9} := 0.9000$$

```
>print(A);
```

$$\begin{bmatrix} 0.9000 & 0.0500 & 0 & 0 & 0 & 0 & 0 & 0 & 0 \\ 0.0500 & 0.9000 & 0.0500 & 0 & 0 & 0 & 0 & 0 & 0 \\ 0 & 0.0500 & 0.9000 & 0.0500 & 0 & 0 & 0 & 0 & 0 \\ 0 & 0 & 0.0500 & 0.9000 & 0.0500 & 0 & 0 & 0 & 0 \\ 0 & 0 & 0 & 0.0500 & 0.9000 & 0.0500 & 0 & 0 & 0 \\ 0 & 0 & 0 & 0 & 0.0500 & 0.9000 & 0.0500 & 0 & 0 \\ 0 & 0 & 0 & 0 & 0 & 0.0500 & 0.9000 & 0.0500 & 0 \\ 0 & 0 & 0 & 0 & 0 & 0 & 0.0500 & 0.9000 & 0.0500 \\ 0 & 0 & 0 & 0 & 0 & 0 & 0 & 0.0500 & 0.9000 \end{bmatrix}$$

> $B := Matrix(A)^{1000};$

B := [[0.000141315047920884, 0.000268797111482164. 0.000369967306866671
0.000434922390455969, 0.000457304157315686, 0.000434921807031926,
0.000369966362866739, 0.000268796167482232, 0.000141314464496840],
10.000268797111482164. 0.000511282354787554, 0.000703719501938133,
0.000827271464182357, 0.000869844197487895, 0.000827270520182425,
0.000703717974514157, 0.000511280827363578, 0.000268796167482231],
[0.000369967306866671, 0.000703719501938133, 0.000968586512103240,
0.00113864130897006, 0.00119723782704910. 0.00113864036497013,
0.000968584984679265, 0.000703717974514157, 0.000369966362866739],
[0.000434922390455969, 0.000827271464182357, 0.00113864130897006,
0.00133855287496998, 0.00140743747645229, 0.00133855229154594,
0.00113864036497013, 0.000827270520182425, 0.000434921807031926 j,
[0.000457304157315686, 0.000869844197487894, 0.00119723782704910,

0.00140743747645229, 0.00147986733946682, 0.00140743747645229,
0.00119723782704910, 0.000869844197487895,0.000457304157315686],
[0.000434921807031926. 0.000827270520182425, 0.00113864036497013,
0.00133855229154594, 0.00140743747645229,0.00133855287496998,
0.00113864130897006, 0.000827271464182358, 0.000434922390455970],
[0.000369966362866739. 0.000703717974514157, 0.000968584984679265,
0.00113864036497013, 0.00119723782704910, 0.00113864130897006,
0.000968586512103242, 0.000703719501938134, 0.000369967306866671],
[0.000268796167482232, 0.000511280827363578, 0.000703717974514157,
0.000827270520182425, 0.000869844197487895, 0.000827271464182358,
0.000703719501938134, 0.000511282354787555, 0.000268797111482164],
[0.000141314464496840, 0.000268796167482231, 0.000369966362866739,
0.000434921807031926, 0.000457304157315686, 0.000434922390455970,
0.000369967306866671, 0.000268797111482164, 0.000141315047920884]]

> $u := multiply(B, w)$;

$u := [0.00228652078684004, 0.00434922098793708, 0.00598618913593038,$
$0.00703718738306659, 0.00739933669818067, 0.00703718738306660,$
$0.00598618913593038, 0.00434922098793709, 0.00228652078684004]$

>

We note that if we use $k = 0.01$, our results are not as good. Using the same routine but with $k = 0.01$, we get the following:

u := Matrix(1, 9, [[−0.3886081114e13, 0.7391765332e13, −0.1017389253e14, 0.1196012549e14,−0.1257562912e14,0.1196013230e14,−0.1017389593e14, 0.7391763759e13, −0.3886077575e13]])

The errors at $w_{i,50}$ are so large we do not even include them in Table 13.1

TABLE 13.1
Output for Example 1

x	$u(x_i, 0.5)$	$w_{i,1000}$	|error|	$w_{i,50}$
0	0	0		0
0.1	0.00222241	0.00228652	6.41×10^{-5}	−0.388608e13
0.2	0.00422728	0.00434922	1.219×10^{-4}	0.739176e13
0.3	0.00581836	0.0598619	1.678×10^{-4}	−0.101738e14
0.4	0.00683989	.00703719	1.973×10^{-4}	0.1196013e14
0.5	0.00719188	0.00739934	2.075×10^{-4}	−0.125756e14
0.6	0.00683989	0.00703718	1.973×10^{-4}	0.1196013e14
0.7	0.00581836	0.00598619	1.678×10^{-4}	−0.101738e14
0.8	0.00422728	0.00434922	1.219×10^{-4}	0.739176e13
0.9	0.00222241	0.00228652	6.41×10^{-5}	−0.388608e14
1.0	0	0		0

Backward-Difference Method

This method involves using mesh points and involves approximating points marked with an X in Figure 13.1.

As before, we let $\lambda = \alpha^2(k)/h^2$, then the backward difference becomes

$$(1 + 2\lambda)w_{i,j} = \lambda w_{i+1,j} - \lambda w_{i-1,j} = w_{i,j-1}.$$

Our A matrix in this case is

$$A = \begin{bmatrix} (1+2\lambda) & -\lambda & 0 & \cdots & & 0 \\ -\lambda & (1+2\lambda) & -\lambda & & & \cdots \\ 0 & -\lambda & (1+2\lambda) & -\lambda & & \\ \cdots & 0 & -\lambda & (1+2\lambda) & -\lambda \\ 0 & 0 & 0 & -\lambda & (1+2\lambda) \end{bmatrix}.$$

And solve $\mathbf{Aw}^{(j)} = \mathbf{w}^{(j-1)}$.

The $j - 1$ term is equal to t/k.

Example 2. Backward Difference

```
>restart:
Backward-difference Method
>with(linalg) :
> l:=1:alpha:=1:m:=10:h:=1/m;k:=0.01;
```

$$h := \frac{1}{10}$$
$$k := 0.01$$

```
> f:=x→sin(Pi·x)
```

$$f : x \mapsto \sin(\pi \cdot x)$$

```
>A := matrix(m - 1, m - 1, 0);
```

$$A := \begin{bmatrix} 0 & 0 & 0 & 0 & 0 & 0 & 0 & 0 & 0 \\ 0 & 0 & 0 & 0 & 0 & 0 & 0 & 0 & 0 \\ 0 & 0 & 0 & 0 & 0 & 0 & 0 & 0 & 0 \\ 0 & 0 & 0 & 0 & 0 & 0 & 0 & 0 & 0 \\ 0 & 0 & 0 & 0 & 0 & 0 & 0 & 0 & 0 \\ 0 & 0 & 0 & 0 & 0 & 0 & 0 & 0 & 0 \\ 0 & 0 & 0 & 0 & 0 & 0 & 0 & 0 & 0 \\ 0 & 0 & 0 & 0 & 0 & 0 & 0 & 0 & 0 \\ 0 & 0 & 0 & 0 & 0 & 0 & 0 & 0 & 0 \end{bmatrix}$$

```
>w := vector(m -1, 0);
```

$$w := \begin{bmatrix} 0 & 0 & 0 & 0 & 0 & 0 & 0 & 0 & 0 \end{bmatrix}$$

```
>u := vector(m -1, 0);
```

$$u := \begin{bmatrix} 0 & 0 & 0 & 0 & 0 & 0 & 0 & 0 & 0 \end{bmatrix}$$

```
> for i from 1 to m − 1 do w[i]:= evalf(f(i·h));end do;
```

$$w_1 := 0.3090169944$$
$$w_2 := 0.5877852524$$
$$w_3 := 0.8090169944$$
$$w_4 := 0.9510565165$$
$$w_5 := 1.$$
$$w_6 := 0.9510565165$$
$$w_7 := 0.8090169944$$
$$w_8 := 0.5877852524$$
$$w_9 := 0.3090169944$$

```
> lambda := \frac{alpha^2 · k}{h^2};
```

$$\lambda := 1.00$$

```
>A[1, 1] := 1 + 2
```

$$A_{1,1} := 3.00$$

```
>A[1, 2] := -lambda
> for i from 2 to m-2 do A[i, i-1] := -lambda; A[i, i+1] :=
  -lambda; A[i,i] := 1+2 lambda : end do
```

$$A_{2,1} := -1.00$$
$$A_{2,3} := -1.00$$
$$A_{2,2} := 3.00$$
$$A_{3,2} := -1.00$$
$$A_{3,4} := -1.00$$
$$A_{3,3} := 3.00$$
$$A_{4,3} := -1.00$$
$$A_{4,5} := -1.00$$
$$A_{4,4} := 3.00$$
$$A_{5,4} := -1.00$$
$$A_{5,6} := -1.00$$
$$A_{5,5} := 3.00$$
$$A_{6,5} := -1.00$$
$$A_{6,7} := -1.00$$
$$A_{6,6} := 3.00$$
$$A_{7,6} := -1.00$$
$$A_{7,8} := -1.00$$
$$A_{7,7} := 3.00$$
$$A_{8,7} := -1.00$$
$$A_{8,9} := -1.00$$
$$A_{8,8} := 3.00$$

```
>A[m-1, m-2] :=lambda;
```

$$A_{9,8} := -1.00$$

```
>A[m-1, m-1] :=1+2 lambda;
```
$$A_{9,9} := 3.00$$
```
>B :=augment(A, w);
```
   $B := [[3.00, -1.00, 0, 0, 0, 0, 0, 0, 0, 0.3090169944],$
     $[-1.00, 3.00, -1.00, 0, 0, 0, 0, 0, 0, 0.5877852524],$
     $[0, -1.00, 3.00, -1.00, 0, 0, 0, 0, 0, 0.8090169944],$
     $[0, 0, -1.00, 3.00, -1.00, 0, 0, 0, 0, 0.9510565165],$
     $[0, 0, 0, -1.00, 3.00, -1.00, 0, 0, 0, 1.],$
     $[0, 0, 0, 0, -1.00, 3.00, -1.00, 0, 0, 0.9510565165],$
     $[0, 0, 0, 0, 0, -1.00, 3.00, -1.00, 0, 0.8090169944],$
     $[0, 0, 0, 0, 0, 0, -1.00, 3.00, -1.00, 0.5877852524],$
     $[0, 0, 0, 0, 0, 0, 0, -1.00, 3.00, 0.3090169944]]$
```
>M := gausselim(B);
```
   $M := [[3.00, -1.00, 0., 0., 0., 0., 0., 0., 0., 0.3090169944],$
     $[0, 2.666666667, -1.00, 0., 0., 0., 0., 0., 0., 0.6907909172],$
     $[0., 0, 2.625000000, -1.00, 0., 0., 0., 0., 0., 1.068063588],$
     $[0., 0., 0, 2.619047619, -1.00, 0., 0., 0., 0, 1.357937883],$
     $[0., 0., 0., 0, 2.618181818, -1.00, 0., 0., 0., 1.518485374],$
     $[0., 0., 0., 0., 0, 2.618055556, -1.00, 0., 0., 1.531033569]$
     $[0., 0., 0., 0., 0., 0, 2.618037135, -1.00, 0., 1.393814962]$
     $[0., 0., 0., 0., 0., 0., 0, 2.618034448, -1.00, 1.120174554]$
     $[0., 0., 0., 0., 0., 0., 0., 0, 2.618034056, 0.7368855256]]$
```
> v := backsub(M);
```
   $v := [0.2814652178, 0.5353786589, 0.7368855067, 0.8662608669,$
    $0.9108405782, 0.8662608671, 0.7368855068, 0.5353786590,$
    $0.2814652177]$

The result is $w^{(1)}$.

We loop 50 times; for j to 50, do
```
    B := augment(A, v);
    M := gausselim(B);
    v := backsub(M);
    j := j + 1;
end do
```

Our result after looping 50 times is

```
v := [0.02725737253, 0.05184660352, 0.07136072773,
  0.08388956673,
0.08820671044, 0.08388956670, 0.07136072774, 0.05184660351,
  0.02725737252]
```

Crank–Nicholson Method

Since both the forward and backward differences have error issues, this method employs an average of the two methods to improve the error.

 As before, we let $\lambda = \alpha^2(k)/h^2$.

We create two matrices, **A** and **B** as

$$A = \begin{bmatrix} (1+\lambda) & -\lambda/2 & 0 & \cdots & 0 \\ -\lambda/2 & (1+\lambda) & -\lambda/2 & & \cdots \\ 0 & -\lambda/2 & (1+\lambda) & -\lambda/2 & \\ \cdots & 0 & -\lambda/2 & (1+\lambda) & -\lambda/2 \\ 0 & 0 & 0 & -\lambda/2 & (1+\lambda) \end{bmatrix}$$

and

$$B = \begin{bmatrix} (1-\lambda) & \lambda/2 & 0 & \cdots & 0 \\ \lambda/2 & (1-\lambda) & \lambda/2 & & \cdots \\ 0 & \lambda/2 & (1-\lambda) & \lambda/2 & \\ \cdots & 0 & \lambda/2 & (1-\lambda) & \lambda/2 \\ 0 & 0 & 0 & \lambda/2 & (1-\lambda) \end{bmatrix}$$

We establish $\mathbf{w}^{(0)}$ first by using $f(xi * h)$.

We use matrix multiplication to obtain a vector **u**.

Augment **A** with **u**.

Use Gaussian elimination on the augments matrix, and find the vector **z** that represents the updated vector, **w**.

We repeat the process until we have found k/h^2 iteration.

Example 3. Using the Crank–Nicolson Method

CRANK-NICOLSON METHOD

```
I  X(I)           W(X(I),5.000000e-01)
1 0.10000000    0.00230512
2 0.20000000    0.00438461
3 0.30000000    0.00603489
4 0.40000000    0.00709444
5 0.50000000    0.00745954
6 0.60000000    0.00709444
7 0.70000000    0.00603489
8 0.80000000    0.00438461
9 0.90000000    0.00230512
```

Application Scenario

One of the most important partial differential equations is the heat equation, illustrated in Figure 13.1. As a consequence of the balance of energy and Fourier's law of heat conduction, the flow of heat in a thin, laterally

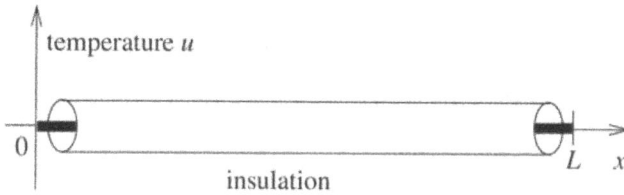

FIGURE 13.1
Generic heat problem.

insulated homogeneous rod is modeled by the one-dimensional heat equation:

$$\frac{\partial u}{\partial t} = k \frac{\partial^2 u}{\partial x^2}.$$

Together with initial conditions, $u(x, 0) = f(x)$ and homogeneous boundary condition. If the temperatures at either end are to be held at zero degrees, then

$$u(0, t) = u(L, t) = 0.$$

Since the rate at which heat flows through the rod depends upon the material that makes up the rod, the constant K is related to the thermal diffusivity of the material. Several situations may be considered: (a) The two ends of the rods are held at constant temperature, (b) the ends may be insulated, and (c) a combination of these conditions.

13.2 Solving the Heat Equation with Homogeneous Boundary Conditions

BACKWARD-DIFFERENCE METHOD

```
I X(I)          W(X(I),1.000000e+00)
1 0.25000000   -0.7081068
2 0.50000000   -1.0000000
3 0.75000000   -0.7071068
4 #1
```

CRANK–NICOLSON METHOD

```
I X(I)          W(X(I),1.000000e+00)
1 0.25000000   -0.68849397
```

```
2 0.50000000  -0.97367751
3 0.75000000  -0.68849397
```

CRANK–NICOLSON METHOD

```
I X(I)         W(X(I),1.000000e+00)
1 0.25000000  -0.68849397
2 0.50000000  -0.97367751
3 0.75000000  -0.68849397
```

Exact

$$
\begin{aligned}
s[0] &:= 0 \\
s[0.25] &:= -0.7071067813 \\
s[0.50] &:= -1. \\
s[0.75] &:= -0.7071067813 \\
s[1.00] &:= 4.102067616\ 10^{-10}
\end{aligned}
$$

The first problem that we consider is the one previously modeled where the temperature at the ends of the insulated rod is constant and held at zero. The initial temperature of the rod is represented as a function, $f(x)$. Let's assume we have a thin insulated rod whose initial temperature in the rod is modeled by $f(x) = \sin(\pi x/L)$, where L is the length of the rod and $L = 1.5$ centimeters. We want to solve this heat equation:

$$
\frac{\partial u}{\partial t} - \alpha^2 \frac{\partial^2 u}{\partial x^2} = 0, 0 < x < L, t > 0
$$

$$
u(0,t) = u(L,t) = 0,\ t > 0
$$

$$
u(x,0) = f(x),\ 0 \le x \le L
$$

We will solve using the three methods from Section 13.1 and compare the results.

Backward Differences

BACKWARD-DIFFERENCE METHOD

```
I X(I)         W(X(I),5.000000e-01)
1 0.10000000  0.00286276
2 0.20000000  0.00544530
3 0.30000000  0.00749482
4 0.40000000  0.00881070
5 0.50000000  0.00926414
6 0.60000000  0.00881074
7 0.70000000  0.00749487
8 0.80000000  0.00544535
9 0.90000000  0.00286279
```

Crank–Nicholson

This method uses a combination of the forward- and backward-difference methods with a few minor changes.

CRANK-NICOLSON METHOD

```
I X(I)          W(X(I),5.000000e-01)
1 0.10000000   0.00227709
2 0.20000000   0.00433129
3 0.30000000   0.00596151
4 0.40000000   0.00700817
5 0.50000000   0.00736883
6 0.60000000   0.00700817
7 0.70000000   0.00596151
8 0.80000000   0.00433129
9 0.90000000   0.00227709
```

The second type of problem that we consider is one in which the temperatures at the ends of the rod are at fixed (but different) temperatures. We can use the heat equation to compute the temperature $u(t, x)$ at position x and t. Let's use the following setup:

$$\partial u / \partial t - \alpha^2 \partial^2 / \partial x = 0 \; 0 < x < 2, t > 0$$
$$u(0, t) = 0, u(l,t) = 2, t > 0$$
$$u(x, 0) = 10x - 10x^4 + 2x^2, 0 \le x \le 2$$

CRANK-NICOLSON METHOD

```
I X(I)          W(X(I),2.000000e+00)
1 0.20000000   -2.07383162
2 0.40000000   -4.08531838
3 0.60000000   -5.46006486
4 0.80000000   -5.23916497
5 1.00000000   -2.07843434
6 1.20000000    5.75266492
7 1.40000000   20.37240765
8 1.60000000   44.28849774
9 1.80000000   80.40006966
```

13.3 Methods with Python

We resolve Example 1 for the forward-difference, backward-differences, and Crank–Nicolson methods using Python.

```
#Forward Finite Difference
import numpy as np
import matplotlib.pyplot as plt
from scipy.integrate import solve_ivp
plt.style.use('seaborn-poster')
import math
%matplotlib inline

L=1
alpha=1
m=10
h=1/m
k=0.0005
def f(x):
  return np.sin(np.pi*x)
A=np.empty((m-1,m-1))
print('The shape of matrix is :', A.shape)
w=np.empty(m-1)
u=np.empty(m-1)
for i in range (1,m-1):
    w[i]=f(i*h)
print(w)
l1=0.05
A[0,0]=1-2*l1
A[0,1]=l1
for j in range (1,m-2):
  A[j,j-1]=l1
  A[j,j+1]=l1
  A[j,j]=1-2*l1
A[8,7]=l1
A[8,8]=1-2*l1
print("Matrix",A)
B=np.linalg.matrix_power(A, 1000)
u=np.dot(B,w)
print("Estimates",u)

The shape of matrix is : (9, 9)
[3.00067734e-48 3.09016994e-01 5.87785252e-01 8.09016994e-01
 9.51056516e-01 1.00000000e+00 9.51056516e-01 8.09016994e-01
 5.87785252e-01]

Matrix [[0.9  0.05 0.   0.   0.   0.   0.   0.   0.  ]
        [0.05 0.9  0.05 0.   0.   0.   0.   0.   0.  ]
        [0.   0.05 0.9  0.05 0.   0.   0.   0.   0.  ]
        [0.   0.   0.05 0.9  0.05 0.   0.   0.   0.  ]
```

```
[0.    0.    0.    0.05  0.9   0.05  0.    0.    0.  ]
[0.    0.    0.    0.    0.05  0.9   0.05  0.    0.  ]
0.    0.    0.    0.    0.    0.05  0.9   0.05  0.  ]
[0.    0.    0.    0.    0.    0.    0.05  0.9   0.05]
[0.    0.    0.    0.    0.    0.    0.    0.05  0.9 ]]
```

```
Estimates [0.00217461 0.00413635 0.0056932 0.00669276
   0.00703719 0.00669276
0.00569321 0.00413636 0.00217461]
```

Backward Differences

```python
#Backward Difference
importnumpyasnp
importmatplotlib.pyplotasplt
plt.style.use('seaborn-poster')
%matplotlib inline
alpha=1
m=10
h=1/m
k=0.0005
def f(x):
  return np.sin(np.pi*x)
A=np.empty((m-1,m-1))
print('The shape of matrix is :', A.shape)
w=np.empty(m-1)
u=np.empty(m-1)
for i in range (1,m-1):
    w[i]=f(i*h)
print(w)
l1=0.05
A[0,0]=1-2*l1
A[0,1]=l1
for j in range (1,m-2):
  A[j,j-1]=l1
  A[j,j+1]=l1
  A[j,j]=1-2*l1
A[8,7]=l1
A[8,8]=1-2*l1
print("Matrix",A)
y=np.linalg.solve(A,w)
print("estimates",y)
y=np.linalg.solve(A,w)
print(y)
for k in range(1,50):
 y=np.linalg.solve(A,y)
```

```
#print("estimates",y)
print(y)
[0.00275362 0.00523713 0.00720692 0.00847107 0.00890597
0.00846884
 0.00720261 0.00523187 0.00275008]
Crank-Nicolson Method
#Crank Nicolson Method
importnumpyasnp
importmatplotlib.pyplotasplt
plt.style.use('seaborn-poster')
%matplotlib inline
alpha=1
m=10
h=1/m
k=0.0005
def f(x):
  return np.sin(np.pi*x)
A=np.empty((m-1,m-1))
B=np.empty((m-1,m-1))
print('The shape of matrix is :', A.shape)
w=np.empty(m-1)
u=np.empty(m-1)
for i in range (1,m-1):
  w[i]=f(i*h)
print(w)
l1=0.05
A[0,0]=1+l1
A[0,1]=-l1/2
B[0,0]=1-l1
B[0,1]=l1/2
for j in range (1,m-2):
  A[j,j-1]=-l1/2
  A[j,j+1]=-l1/2
  A[j,j]=1+l1
B[j,j-1]=l1/2
  B[j,j+1]=l1/2
  B[j,j]=1-l1
A[8,7]=-l1/2
A[8,8]=1+l1
B[8,7]=l1/2
B[8,8]=1-l1
u=np.dot((B,w))
print("Matrix",A)
y=np.linalg.solve(A,u)
print("estimates",y)
  y=np.linalg.solve(A,w)
```

```
print(y)
for k in range(1,50):
  y=np.linalg.solve(A,y)
#print("estimates",y)
print(y)
```

Our estimates:

```
[0.02272277 0.04334514 0.05987324 0.07059584 0.07432846
0.07063913
 0.05994318 0.04341494 0.02276583]
```

Exercises

In Exercises 1 and 2, use the backward-difference method.

1. $\dfrac{\partial u}{\partial t} - \dfrac{\partial^2 u}{\partial t^2} = 0, 0 < x < 2, t > 0$

   - $u(0,t) = u(2,t) = 0 \, u > t$

   $u(x, 0) = \sin(\pi x/2) \; 0 \leq x \leq 2$

   Use $m = 4$, $T = 0.1$, and $N = 2$.

   The actual answer is $u(t, x) = e^{-\left(\frac{\pi^2}{4}\right)t} \sin\left(\dfrac{\pi x}{2}\right)$

2. $\partial u/\partial t - 1/16(\partial^2 u)/(\partial t^2) = 0, 0 < x < 1, t > 0$

   $u(0,t) = u(1,t) = 1, t > 0$

   $u(x,0) = 2\sin(2\pi x) \; 0 \leq x \leq 1$

   Use $m = 3$, $T = 0.1$, and $N = 2$.

   The actual answer is $u(t, x) = 2e^{-\left(\frac{\pi^2}{4}\right)t} \sin(2\pi x)$

In Exercises 3 and 4, use the forward-difference method.

3. $\dfrac{\partial u}{\partial t} - \dfrac{\partial^2 u}{\partial t^2} = 0, 0 < x < 2, t > 0$

   - $u(0,t) = u(2,t) = 0 \, u > t$

   $u(x, 0) = \sin(2\pi x) \; 0 \leq x \leq 2$

   a. Use $h = 0.4$ and $k = 0.1$ and compare your answer to $t = 0.5$ to the actual answer:

   $u(x, t) = e^{-4\pi^2 t} \sin(2\pi x)$.

   b. Then use $h = 0.4$ and $k = 0.05$ and compare answers again.

4. $\dfrac{\partial u}{\partial t} - \dfrac{\partial^2 u}{\partial t^2} = 0, 0 < x < \pi, t > 0$

$u(0t) = u(\pi, t) = 0u > t$

$u(x, 0) = \sin(x)\ 0 \le x \le \pi$

Use $h = 0.1$ and $k = 0.05$ and compare your answer to $t = 0.5$ to the actual answer:

$$u(x, t) = e^{-t}\sin(x)\ at\ t = 05.$$

5. Repeat Exercises 1 and 2 using the forward-differences and Crank–Nicolson methods.

6. Repeat Exercises 3 and 4 using backward-differences and Crank–Nicolson methods.

Projects

1. The temperature $u(x, t)$ of a long thin rid of constant cross section and homogeneous conducting material is governed by a one-dimensional heat equation. If heat is generated in the material, for example by resistance to current, the heat equation becomes

$$\frac{\partial^2 u}{dx^2} + \frac{Kr}{\rho C} = k\frac{du}{dt}$$

for $0 < x < 1, t > 0$. If l is the length, r is the density, C is the specific heat, and K is the thermal diffusivity of the rod. The function $r = r(x, t, u)$ represents the heat generated per unit volume. Suppose $l = 1.5$ cm, $K = 1.04$ cal/cm *(deg*s), p = 10.6 g/cm³, and $C = 0.056$ cal/g *deg.

   a. Approximate the temperature in the long thin rid with $h = 0.15$ and $k = 0.0225$ for 0.225 seconds using all three algorithms.

   b. Complete an error analysis using your approximates to compare to each other.

   c. Using the Crank–Nicholson method, use $h = 0.1$, $T = 1.0$, and $k = 0.01$ to generate approximations for $u(x, t)$ over the domain. $0 < x < 1.5, 0 < t < 1$.

   d. Obtain a plot.

   e. Approximate $u(x, 0.5)$.

2. The temperature $u(x, t)$ of a long thin rid of constant cross section and homogeneous conducting material is governed by a one-dimensional

heat equation. If heat is generated in the material, for example, by resistance to current, the heat equation becomes

$$\frac{\partial^2 u}{dx^2} + \frac{Kr}{\rho C} = k\frac{du}{dt}$$

for $0 < x < 1$, $t > 0$. If l is the length, r is the density, C is the specific heat, and K is the thermal diffusivity of the rod. The function $r = r(x, t, u)$ represents the heat generated per unit volume. Suppose $l = 1.5$ cm, $K = 1.04$ cal/cm *(deg*s), $\rho = 10.6$ g/cm³, and $C = 0.056$ cal/g *deg, and suppose additionally $u(x,0) = 4x^3(1 - x)^2 -$, $0 < x < l$.

a. Solve the parabolic equation using the Crank–Nicholson method.

b. Approximate the temperature in the long thin rid with $h = 0.15$ and $k = 0.0225$ for 0.225 seconds using all three algorithns.

c. Complete error analysis using your approximates to compare to each other.

d. Using the Crank–Nicholson method, use $h = 0.1$, $T = 1.0$, and $k = 0.01$ to generate approximations for $u(x,t)$ over the domain. $0 < x < 1.5, 0 < t < 1$.

e. Obtain a plot.

f. Approximate $u(x, 0.5)$.

```
% Solution of the Heat Equation Using a Forward Difference
Scheme

% Initialize Data
%  Length of Rod, Time Interval
%  Number of Points in Space, Number of Time Steps

L=1;
T=0.1;
k=1;
N=10;
M=50;
dx=L/N;
dt=T/M;
alpha=k*dt/dx^2;
% Position

for i=1:N+1
  x(i)=(i-1)*dx;
end

% Initial Condition

for i=1:N+1
  u0(i)=sin(pi*x(i));
end
```

```
% Partial Difference Equation (Numerical Scheme)

for j=1:M
  for i=2:N
    u1(i)=u0(i)+alpha*(u0(i+1)-2*u0(i)+u0(i-1));
  end
  u1(1)=0;
  u1(N+1)=0;
  u0=u1;
end

% Plot solution
plot(x, u1);
function U=forward(f,c0,cl,a,b,c,n,m)
% Input  -- f=u(x,0) as a string 'f'
%        -- c0=u(0,t) and cl=u(a,t)
%        -- a and b right end points of [0,a] and [0,b]
%        -- c=alpha^2 the thermal diffusivity in heat equation
%        -- n and m number of grid points over [0,a] and [0,b]
% Output -- U solution matrix;
% Initialize parameters and U
h=a/(n-1);
k=b/(m-1);
r=c*k/h^2;
s=1-2*r;
U=zeros(n,m);
% Boundary conditions
U(1,1:m)=c0;
U(n,1:m)=cl;
% Generate the first row
U(2:n-1,1)=feval(f,h:h:(n-2)*h)';
% Generate remaining rows of U
for j=2:m
    for j=2:n-1
        U(i,j)=s*U(i,j-1)+r*(U(i-1,j-1)+U(i+1,j-1));
    end
end
U=U';

function U=crnich(f,c0,cl,a,b,c,n,m)
% Input  -- f=u(x,0) as a string 'f'
%        -- c0=u(0,t) and cl=u(a,t)
%        -- a and b right end points of [0,a] and [0,b]
%        -- c=alpha^2 the thermal diffusivity in heat equation
%        -- n and m number of grid points over [0,a] and [0,b]
% Output -- U solution matrix;
% Initialize parameters and U
h=a/(n-1);
k=b/(m-1);
r=c*k/h^2;
```

```
s1=2+2/r;
s2=2/r-2;
U=zeros(n,m);
% Boundary conditions
U(1,1:m)=c0;
U(n,1:m)=c1;
% Generate the first row
U(2:n-1,1)=feval(f,h:h:(n-2)*h)';
% Form the diagonal and off-diagonal elements of A and
% the constant vector B and solve triagonal system AX=B
Vd(1,1:n)=s1*ones(1,n);
Vd(n)=1;
Vd(1)=1;
Va=-ones(1,n-1);
Va(n-1)=0;
Vc=-ones(1,n-1);
Vc(1)=0;
Vb(1)=c1;
Vb(n)=c2;
for j=2:m
    for i=2:n-1
        Vb(i)=U(i-1,j-1)+U(i+1,j-1)+s2*U(i,j-1);
    end
    X=trisys(Va,Vd,Vc,Vb);
    U(1:n,j)=X';
end
U=U'
```

References and Further Readings

Burden, R. and D. Faires (1997). *Numerical Analysis*. Brooks-Cole, Pacific Grove, CA.

Fox, W. P. (2014). *Mathematical Modeling with Maple*, Cengage Publishers, Boston, MA.

Fox, W. P. (2018). *Mathematical Modeling for Business Analytics*. Taylor and Francis Publishers, Boca Raton, FL.

Giordano, F., W. Fox and S. Horton (2013). *A First Course in Mathematical Modeling*, 5th ed. Cengage Publishers, Boston, MA.

Answers to Selected Exercises

Chapter 1

Section 1.5

1. $\lim\limits_{x \to 4}\left(x^3 + 2x - 21\right) = 51$

2. $\lim\limits_{x \to \infty} \dfrac{x^2 + x}{x} = \infty$

3. $\lim\limits_{t \to 1} \dfrac{t^2 - 1}{t - 1} = 2$

4. $\lim\limits_{y \to 0} \dfrac{\tan(y)}{y} = 1$

5. undefined

7. $2 \times \cos(x^2)$

8. Differentiate $y = \left(\sin^2 x\right) = 2\sin(x)\cos(x)$

9. Find all first- and second-order partial derivatives for $f(x,y) = \exp\left(xy^2\right)$ so $f_x = y^2 e^{xy\wedge 2}$,

   $f_y = 2\,y\ xe^{xy\wedge 2}$, $f_{xx} = y^2 e^{xy\wedge 2}$, $f_{yy} = 2x\ e^{xy\wedge 2} + 2y\ e^{xy\wedge 2}$, $f_{xy} = f_{yx} = 2{}^*y{}^*\exp(x{}^*y\wedge 2)$ $+ 2{}^*y\wedge 3{}^*x{}^*\exp(x{}^*y\wedge 2)$

11. $f''(x) = 6x - 18$, so when $x > 3$ and $f''(x) > 0$, so f is concave-up, and when $x < 3$, $f''(x) < 0$, so f is concave-down.

12. $f''(x) = e^x$, which is always > 0.

Section 1.6

Using the Reimann sums, trapezoidal, and Simpson's methods on each of the following to approximate the area under the curve.

1. Simpson's method 0.8414710140, trapezoidal method 0.8407696420, Reimann sum (midpoint) 0.841821700 for $f(x) = \cos(x)$ $0 \le x \le 1$.

Section 1.7

1. Expand $f(x) = e^x$ around $x = 0$. $1 + x + 1/2x^2 + 1/6x^3$.

2. Expand $f(x) = e^{2x}$ around $x = 1$. $-7.389056 + 14.7781122x + 14.781122(x - 1)^2 + 9.852074796(x - 1)^3$

Chapter 3

Consider the model $a(n + 1) = ra(n)(1 - a(n))$. Let $a(0) = 0.2$. Determine the numerical and graphical solution for the following values of r. Find the pattern in the solution.

1. From the graph the DDS has an equilibrium at 0.5.
3. Chaos

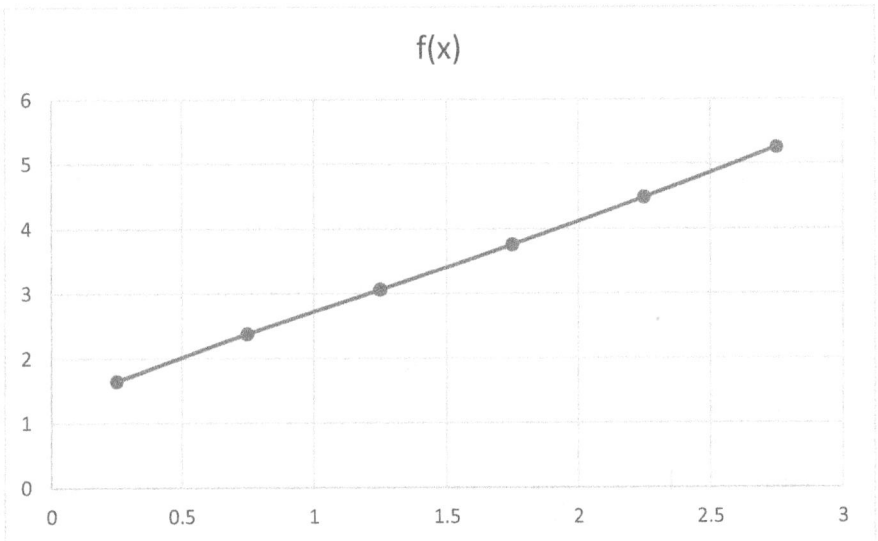

f(x)

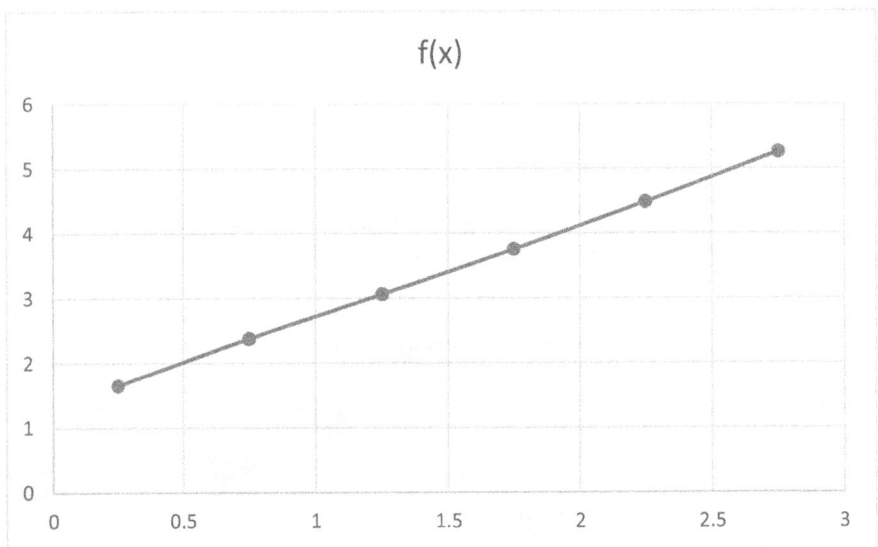

f(x)

For problems 5–8, find the equilibrium value by iteration and determine if it is stable or unstable.

5. $a(n + 1) = 1.7a(n) - 0.14a(n)^2$. EV are 0 (unstable), 5 (stable).

7. $a(n + 1) = 0.2a(n) - 0.2a(n)^3$. EV are 0 (stable), –2 (unstable).

10. Equilibrium values of the bass and trout model presented in Section 3.2 are (0, 0), (150, 200). They cannot be maintained naturally.

Chapter 4

1. The root(s) for $f(x) = x^3 - 2$ using

   a. the bisection method.

   Bisection(f, x = [1, 2], tolerance = 10^(-2));
                1.257812500

   ```
   Bisection(f, x = [1, 2], tolerance = 10^(-2), output =
   sequence);
      [1., 2.], [1., 1.500000000], [1.250000000, 1.500000000],
      [1.250000000, 1.375000000], [1.250000000, 1.312500000],
      [1.250000000, 1.281250000], [1.250000000, 1.265625000],
      1.257812500
   ```

   b. Newton's method.
   ```
   Newton(f, x = 1, output = sequence);
      1., 1.333333333, 1.263888889, 1.259933493, 1.259921050
   ```

   c. the secant method.
   ```
   Secant(f, x = [0, 2], output = sequence);
   0., 2., 0.500000000, 0.8571428571, 1.826714802, 1.100211976,
   1.202120997, 1.268187169, 1.259531737, 1.259918506, 1.259921051
   ```

2. Find the root(s) accurate to a tolerance of 10^{-4} using technology for each method, bisection, Newton's, and secant, over the specified interval.

   a. $f(x) = x+1- e^{-x^2}, 0.5 \le x \le 1$
   ```
   Bisection(f1, x = [0.5, 1], output = sequence);
      [0.5, 1.], [0.5, 0.7500000000], [0.6250000000, 0.7500000000],
      [0.6875000000, 0.7500000000], [0.7187500000, 0.7500000000],
      [0.7343750000, 0.7500000000], [0.7421875000, 0.7500000000],
      [0.7460937500, 0.7500000000], [0.7460937500, 0.7480468750],
      [0.7460937500, 0.7470703125], [0.7465820312, 0.7470703125]
   ```

```
Bisection(f1, x = [0.5, 1], output = sequence);
  [0.5, 1.], [0.5, 0.7500000000], [0.6250000000, 0.7500000000],
  [0.6875000000, 0.7500000000], [0.7187500000, 0.7500000000],
  [0.7343750000, 0.7500000000], [0.7421875000, 0.7500000000],
  [0.7460937500, 0.7500000000], [0.7460937500, 0.7480468750],
  [0.7460937500, 0.7470703125], [0.7465820312, 0.7470703125]

Secant(f1, x = [0.5, 1], output = sequence);
0.5, 1., 0.6155863532, 0.6837590238, 0.7728475023,
  0.7429322690,
  0.7466529769, 0.7468838286, 0.7468817417
```

Chapter 5

1. Interpolating polynomials:

   with(Student[NumericalAnalysis]);

```
xy := [[0, 26], [1, 7], [4, 25]];
   xy := [[0, 26], [1, 7], [4, 25]]
L := PolynomialInterpolation(xy, independentvar = x, method
   = lagrange);
Nev := PolynomialInterpolation(xy, independentvar = x, method
   = neville);
New := PolynomialInterpolation(xy, independentvar = x, method
   = newton);

expand(Interpolant(L));
25/4 x² +101/4 x + 26

expand(Interpolant(Nev));
25/4 x² +101/4 x + 26

expand(Interpolant(New));
25/4 x² +101/4 x + 26
```

3. Find the two natural cubic splines for

x	−1	0	1
y	13	7	9

```
Splines
2x³+6x²-2x+7  x <0
-2x³+6x²-2x+7 otherwise
```

4. Find a natural cubic spline between $f(8.3) = 17.5649$, $f(8.4) = 18.1$, and $f(8.6) = 18.50515$. Use your cubic spline to estimate $f(8.5)$. $f(8.5)=$ 18.38572 using the function

   $27.70986615 * x^3 - 714.9145468 * x^2 + 6149.182510 * x - 17614.41116$.

Chapter 6

1. Given the function, $f(t) = -4 * t * \exp(-t) * \cos(2 * t)$,
    a. plot the function and decide on the positive and negative regions.
    b. use a root finding technique and find the roots.
    c. use the Riemann sum, trapezoidal, and Simpson's methods to approximate the area in the nonnegative region from $t = 1$ to $t = 2$.
    d. Find the error bounds for each method in part c.

a.

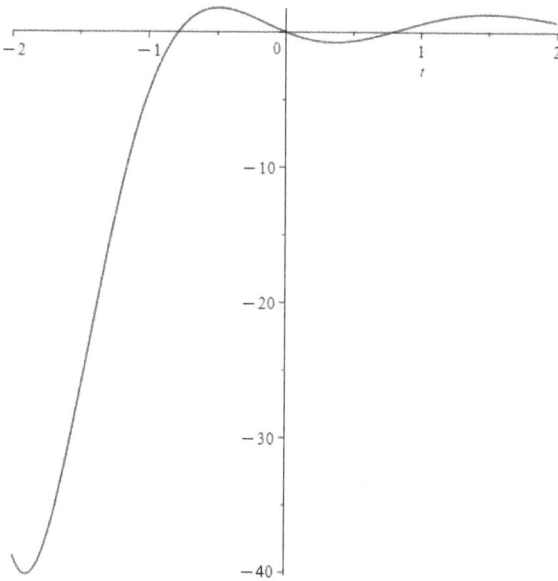

```
Roots
fsolve(f16 = 0, t);
            0.
Bisection(f16, t = [-2, -0.5], output = sequence);
[-2., -0.5], [-1.250000000, -0.5], [-0.8750000000, -0.5],

[-0.8750000000, -0.6875000000], [-0.8750000000,
-0.7812500000],

[-0.8281250000, -0.7812500000], [-0.8046875000,
-0.7812500000],

[-0.7929687500, -0.7812500000], [-0.7871093750,
-0.7812500000],
```

```
    [-0.7871093750,      -0.7841796875],      [-0.7856445313,
    -0.7841796875]
Bisection(f16, t = [0.5, 1]);
                      0.7853393555

with(Student[Calculus1]);
evalf(ApproximateInt(f16, t = 1 .. 2, method = simpson));
               1.092521309

evalf(ApproximateInt(f16, t = 1 .. 2, method =
    trapezoid));
               1.088625750

evalf(ApproximateInt(f16, t = 1 .. 2, method = midpoint));
               1.094469088
```

2. 2.6 ft/s

Chapter 6

Section 6.6

7. Use cubic spline interpolation to obtain the cubic equations for the following data:

 a. $-9 + 39/2x - 9/4x^2 + 3/4x^3 x < 2$

   $3 + 3/2x + 27/4x^2 - 3/4x^3$ otherwise.

Chapter 9

Section 9.3

1. 2000 by road
2. $f' = 0$ at $x = 0$; max is at $x = 1, f(1) = 1$.
3a. The extreme points are at 0 and 10.6667. The point $x = 0$ yields at relative maximum, and $x = 10.667$ yields a relative minimum.
4. Maximum at $x = 1$.
5. The function is maximized at 0 and $2/3r_0$. This confirms the conjecture.

Section 9.6

1. f:= x->-x^2-2*x;
   x -x²⁻²x

```
DICHOTOMOUS(f,-2,1,.2,.01);
```
The interval [a,b] is [-2.00, 1.00]and user-specified
tolerance level is 0.20000.
The first 2 experimental endpoints are x1= -0.510 and
x2 = -0.490.

Iteration	x(1)	x(2)	f(x1)	f(x2)	Interval
1	-0.5100	-0.4900	0.7599	0.7399	[-2.0000, 1.0000]
2	-1.2550	-1.2350	0.9350	0.9448	[-2.0000, -0.4900]
3	-0.8825	-0.8625	0.9862	0.9811	[-1.2550, -0.4900]
4	-1.0688	-1.0488	0.9953	0.9976	[-1.2550, -0.8625]
5	-0.9756	-0.9556	0.9994	0.9980	[-1.0688, -0.8625]

The midpoint of the final interval is -0.965625 and
f(midpoint) = 0.999.

The maximum of the function is 0.995 and the x value =
-1.068750.

```
FIBSearch(f,-2,1,.6);
```
The interval [a,b] is [-2.00, 1.00]and user-specified
tolerance level is 0.60000.
The first 2 experimental endpoints are x1= -0.800 and x2 =
-0.200.

Iteration	x(1)	x(2)	f(x1)	f(x2)	Interval
2	-1.4000	-0.8000	0.9600	0.3600	[-2.0000, -0.2000]
3	-0.8000	-0.8000	0.8400	0.9600	[-1.4000, -0.2000]

The midpoint of the final interval is-0.800000 and f(midpoint)
= 0.960.

The maximum of the function is 0.990 and the x value =
-1.100000

GOLD(f,-2,1,.6);

The interval [a,b] is [-2.00, 1.00] and user-specified
tolerance level is 0.60000.
The first 2 experimental endpoints are x1= -0.854 and x2 =
-0.146.

Iteration	x(1)	x(2)	f(x1)	f(x2)	Interval
2	-1.2918	-0.8540	0.9787	0.2707	[-2.0000, -0.146]
3	-0.8540	-0.5837	0.9149	0.9787	[-1.2918, -0.146]
4	-1.0213	-0.8540	0.9787	0.8267	[-1.2918, -0.5837]
5	-1.1245	-1.0213	0.9995	0.9787	[-1.2918, -0.8540]

The midpoint of the final interval is -1.072886 and
f(midpoint) = 0.995.

The maximum of the function is 0.9995 and the x value =
-1.1245

NEWTON's
Newton(f,-5, 10);

$$-0.5$$
$$-1.000000000$$
$$-1.000000000$$

2. f:=x->-x^2-3*x;

$$x - x^2 - 3x$$

DICHOTOMOUS(f,-3,1,.2,.01);
The interval [a,b] is [-3.00, 1.00] and user-specified
tolerance level is 0.20000.
The first 2 experimental endpoints are x1= -1.010 and x2 =
-0.990.

Iteration	x(1)	x(2)	f(x1)	f(x2)	Interval
1	-1.0100	-0.9900	2.0099	1.9899	[-3.0000, 1.0000]
2	-2.0050	-1.9850	1.9950	2.0148	[-3.0000, -0.9900]
3	1.5075	-1.4875	2.2499	2.2498	[-2.0050, -0.9900]
4	-1.7562	-1.7362	2.1843	2.1942	[-2.0050, -1.4875]
5	-1.6319	-1.6119	2.2326	2.2375	[-1.7562, -1.4875]

```
6              -1.5697  -1.5497  2.2451  2.2475  [-1.6319,  -1.4875]
```

The midpoint of the final interval is -1.559688 and
f(midpoint) = 2.246.
The maximum of the function is 2.250 and the x value =
-1.487500
FIBSearch(f, -3, 1, 6);
The interval [a,b] is [-3.00, 1.00] and user-specified
tolerance level is 0.60000.
The first 2 experimental endpoints are x1= -1.500 and x2 =
-0.500.

Iteration	x(1)	x(2)	f(x1)	f(x2)	Interval
2	-2.0000	-1.5000	2.2500	1.2500	[-3.0000, -0.5000]
3	-1.5000	-1.0000	2.0000	2.2500	[-2.0000, -0.5000]
4	-1.5000	-1.5000	2.2500	2.0000	[-2.0000, -1.0000]

The midpoint of the final interval is -1.500000 and
f(midpoint) = 2.250.

The maximum of the function is 2.250 and the x value =
-1.500000
GOLD(f,-3,1,.6);
The interval [a,b] is [-3.00, 1.00] and user-specified
tolerance level is 0.60000.
The first 2 experimental endpoints are x1= -1.472 and x2 =
-0.528.

Iteration	x(1)	x(2)	f(x1)	f(x2)	Interval
2	-2.0557	-1.4720	2.2492	1.3052	[-3.0000, -0.5280]
3	-1.4720	-1.1116	1.9412	2.2492	[-2.0557, -0.5280]
4	-1.6950	-1.4720	2.2492	2.0991	[-2.0557, -1.1116]
5	-1.4720	-1.3345	2.2120	2.2492	[-1.6950, -1.1116]

The midpoint of the final interval is-1.403312 and f(midpoint)
= 2.241.

The maximum of the function is 2.243 and the x value =
-1.583638

Newton(f,1,10);

$$1$$
$$-1.500000000$$
$$-1.500000000$$

3. f:= x->-x^2-2*x;

$$f := x \rightarrow -x^2 - 2x$$

DICHOTOMOUS(f,-2,1,.2,.01);
The interval [a,b] is [-2.00, 1.00] and user-specified
tolerance level is 0.20000.
The first 2 experimental endpoints are x1= -0.510 and x2 =
-0.490.

Iteration	x(1)	x(2)	f(x1)	f(x2)	Interval	
1	-0.5100	-0.4900	0.7599	0.7399	[-2.0000,	1.0000]
2	-1.2550	-1.2350	0.9350	0.9448	[-2.0000,	-0.4900]
3	-0.8825	-0.8625	0.9862	0.9811	[-1.2550,	-0.4900]
4	-1.0688	-1.0488	0.9953	0.9976	[-1.2550,	-0.8625]
5	-0.9756	-0.9556	0.9994	0.9980	[-1.0688,	-0.8625]

The midpoint of the final interval is-0.965625 and f(midpoint)
= 0.999.

The maximum of the function is 0.995 and the x value
= -1.068750.

FIBSearch(f,-2,1,.6);
The interval [a,b] is [-2.00, 1.00] and user-specified
tolerance level is 0.60000.
The first 2 experimental endpoints are x1= -0.800 and x2 =
-0.200.

Iteration	x(1)	x(2)	f(x1)	f(x2)	Interval	
2	-1.4000	-0.8000	0.9600	0.3600	[-2.0000,	-0.2000]
3	-0.8000	-0.8000	0.8400	0.9600	[-1.4000,	-0.2000]

The midpoint of the final interval is-0.800000 and f(midpoint)
= 0.960.

The maximum of the function is 0.990 and the x value =
-1.100000.

GOLD(f,-2,1,.6);
The interval [a,b] is [-2.00, 1.00] and user-specified
tolerance level is 0.60000.
The first 2 experimental endpoints are x1= -0.854 and x2 =
-0.146.

Iteration	x(1)	x(2)	f(x1)	f(x2)	Interval	
2	-1.2918	-0.8540	0.9787	0.2707	[-2.0000,	-0.1460]
3	-0.8540	-0.5837	0.9149	0.9787	[-1.2918,	-0.1460]
4	-1.0213	-0.8540	0.9787	0.8267	[-1.2918,	-0.5837]
5	-1.1245	-1.0213	0.9995	0.9787	[-1.2918,	-0.8540]

The midpoint of the final interval is-1.072886 and f(midpoint)
= 0.995.

The maximum of the function is 0.915 and the x value =
-1.291772.

Newton(f, -3, 20)

$$-3$$
$$-1.$$
$$-1.$$

4. f:= x->x-exp(x);

$$x \rightarrow x - e^x$$

DICHOTOMOUS(f,-1,3,.2,.01);
The interval [a,b] is [-1.00, 3.00] and user-specified
tolerance level is 0.20000.
The first 2 experimental endpoints are x1= 0.990 and x2 = 1.010.

Iteration	x(1)	x(2)	f(x1)	f(x2)	Interval	
1	0.9900	1.0100	-1.7012	-1.7356	[-1.0000,	3.0000]

```
2              -0.0050   0.0150 -1.0000 -1.0001 [-1.0000,  1.0100]
3              -0.5025  -0.4825 -1.1075 -1.0997 [-1.0000,  0.0150]
4              -0.2538  -0.2338 -1.0296 -1.0253 [-0.5025,  0.0150]
5              -0.1294  -0.1094 -1.0080 -1.0058 [-0.2538,  0.0150]
6              -0.0672  -0.0472 -1.0022 -1.0011 [-0.1294,  0.0150]
```
The midpoint of the final interval is -0.057188 and
f(midpoint) = -1.002.

The maximum of the function is -1.000 and the x value =
0.015000.
FIBSearch(f,-1,3,.1);
The interval [a,b] is [-1.00, 3.00] and user-specified
tolerance level is 0.10000.
The first 2 experimental endpoints are x1= 0.527 and x2 =
1.473.


```
Iteration   x(1)     x(2)     f(x1)     f(x2)        Interval
2          -0.0545   0.5273 -1.1670 -2.8884 [-1.0000,  1.4727]
3          -0.4182  -0.0545 -1.0015 -1.1670 [-1.0000,  0.5273]
4          -0.0545   0.1636 -1.0764 -1.0015 [-0.4182,  0.5273]
5          -0.2000  -0.0545 -1.0015 -1.0141 [-0.4182,  0.1636]
6          -0.0545   0.0182 -1.0187 -1.0015 [-0.2000,  0.1636]
7           0.0182   0.0909 -1.0015 -1.0002 [-0.0545,  0.1636]
8           0.0182   0.0182 -1.0002 -1.0043 [-0.0545,  0.0909]
```


The midpoint of the final interval is 0.018182 and f(midpoint)
= -1.000.

The maximum of the function is -1.001 and the x value =
-0.054545.

GOLD(f,-1,3,.1);
The interval [a,b] is [-1.00, 3.00] and user-specified
tolerance level is 0.10000.
The first 2 experimental endpoints are x1= 0.528 and x2 = 1.472.


```
Iteration   x(1)     x(2)     f(x1)     f(x2)        Interval
2          -0.0557   0.5280 -1.1675 -2.8859 [-1.0000,  1.4720]
3          -0.4163  -0.0557 -1.0015 -1.1675 [-1.0000,  0.5280]
```

```
4        -0.0557   0.1673  -1.0758  -1.0015  [-0.4163,  0.5280]
5        -0.1934  -0.0557  -1.0015  -1.0148  [-0.4163,  0.1673]
6        -0.0557   0.0295  -1.0175  -1.0015  [-0.1934,  0.1673]
7         0.0295   0.0821  -1.0015  -1.0004  [-0.0557,  0.1673]
8        -0.0031   0.0295  -1.0004  -1.0035  [-0.0557,  0.0821]
9        -0.0231  -0.0031  -1.0000  -1.0004  [-0.0557,  0.0295]
```
The midpoint of the final interval is -0.013095 and
f(midpoint) = -1.000.

The maximum of the function is -1.002 and the x value =
-0.055696.
Newton (f, -1, 20);

$$-1$$
$$0.718281828$$
$$0.2058711269$$
$$0.0198090911$$
$$0.00019491102$$
$$1.90103 \ 10^{-8}$$
$$1.030036 \ 10^{-11}$$
$$1.030036 \ 10^{-11}$$

The value of *x* is essentially 0.

Chapter 10

1. a. f:=2*x1*x2-2*x1^2-x2^2;

$$f := 2\,x1\,x2 - 2\,x1^2 - x2^2$$

```
>
> (kt,MP,z1,z2,z3):=STEEPEST(50,.05,1.0,1.0,f):
```

```
              Initial Condition: ( 1.0000, 1.0000)
Iter  Gradient                 Vector G    G x[k]              Step
                               magnitude                       Length
 1    (-2.0000, 0.0000)        2.0000      (1.0000,1.0000)     0.2500
 2    (0.0000, -1.0000)        1.0000      (0.5000,1.0000)     0.5000
```

3	(-1.0000, 0.0000)	1.0000	(0.5000,0.5000)	0.2500
4	(0.0000, -0.5000)	0.5000	(0.2500,0.5000)	0.5000
5	(-0.5000, 0.0000)	0.5000	(0.2500,0.2500)	0.2500
6	(0.0000, -0.2500)	0.2500	(0.1250,0.2500)	0.5000
7	(-0.2500, 0.0000)	0.2500	(0.1250,0.1250)	0.2500
8	(0.0000, -0.1250)	0.1250	(0.0625,0.1250)	0.5000
9	(-0.1250, 0.0000)	0.1250	(0.0625,0.0625)	0.2500
10	(0.0000, -0.0625)	0.0625	(0.0312,0.0625)	0.5000
11	(-0.0625, 0.0000)	0.0625	(0.0312,0.0312)	0.2500
12	(0.0000, -0.0312)	0.0312	(0.0156,0.0312)	

```
------------------------------------------------------------------
         Approximate Solution: ( 0.0156, 0.0312)
            Maximum Functional Value: -0.0005
            Number gradient evaluations: 13
            Number function evaluations: 12
------------------------------------------------------------------
```

1. b. Newtons(f,10,.5,55,55);

```
Hessian: [ -4.000   2.000 ]
         [  2.000  -2.000 ]
eigenvalues: -5.236 -0.764
pos def: false
new x= 0.000 new y= 0.000
Hessian: [ -4.000   2.000 ]
         [  2.000  -2.000 ]
eigenvalues: -5.236 -0.764
pos def: false
new x= 0.000 new y= 0.000
final new x= 0.000 final new y= 0.000
final fvalue is 0.000
```

2. a. $f := 3*x1*x2 - 4*x1^2 - 2*x2^2;$

$$f := 3\,x1\,x2 - 4x1^2 - 2x2^2$$

```
>.(kt, MP, z1, z2, z3):=STEEPEST(50,.05, 1.0, 1.0, f):
```

```
------------------------------------------------------------------
            Initial Condition: ( 1.0000, 1.0000)
```

Iter	Gradient	Vector G magnitude G	x[k]	Step Length
1	(-5.0000, -1.0000)	5.0990	(1.0000,1.0000)	0.1494
2	(0.5287, -2.6437)	2.6960	(0.2529,0.8506)	0.1884
3	(-1.7625, -0.3525)	1.7974	(0.3525,0.3525)	0.1494

4	(0.1864, -0.9319)	0.9503	(0.0891, 0.2998)	0.1884
5	(-0.6212, -0.1242)	0.6336	(0.1242, 0.1242)	0.1494
6	(0.0657, -0.3285)	0.3350	(0.0314, 0.1057)	0.1884
7	(-0.2190, -0.0438)	0.2233	(0.0438, 0.0438)	0.1494
8	(0.0232, -0.1158)	0.1181	(0.0111, 0.0373)	0.1884
9	(-0.0772, -0.0154)	0.0787	(0.0154, 0.0154)	0.1494
10	(0.0082, -0.0408)	0.0416	(0.0039, 0.0131)	

```
         Approximate Solution: ( 0.0039, 0.0131)
            Maximum Functional Value: -0.0003
            Number gradient evaluations: 11
            Number function evaluations: 10
```
---

2. b. $f:=3x1\ x2 - 4x1^2$

$$-2x2^2$$

```
Newtons(f,100,.5,2,2);
Hessian: [ -8.000   3.000 ]
         [  3.000   0.000 ]
eigenvalues:  -9.000  1.000
pos def: false
new x= 0.000   new y= 0.000
Hessian: [ -8.000   3.000 ]
         [  3.000   0.000 ]
eigenvalues: -9.000  1.000
pos def: false
new x= 0.000   new y= 0.000
final new x= 0.000    final new y= 0.000
final fvalue is 0.000
```

3. a. f:=-x1^3+3*x1+84*x2-6*x2^2;

$$f := -x1^3 + 3x1 + 84x2 - 6x2^2$$

```
Newtons(f,100,.01,1,1);#4: init (0.5,1)
Hessian: [ -6.000      0.000 ]
         [  0.000    -12.000 ]
eigenvalues: -12.000  -6.000
pos def: false
new x= 1.000 new y= 7.000
Hessian: [ -6.000      0.000 ]
         [  0.000    -12.000 ]
eigenvalues: -12.000  -6.000
pos def: false
new x= 1.000   new y= 7.000
final new x= 1.000    final new y= 7.000
final fvalue is 296.000
```

Chapter 11

1. The BVP

    $y'' = 4(y - x)$, $0 \leq x \leq 1$, with $y(0) = 0$ and $y(1) = 2$. If this problem has an exact solution of $y(x) = e^2(e^4 - 1)^{-1}(e^{2x} - e^{-2x}) + x$. Use the linear shooting method to approximate the solution and compare the results when $h = 1/2$ and $h = 1/4$.

    Problem 1

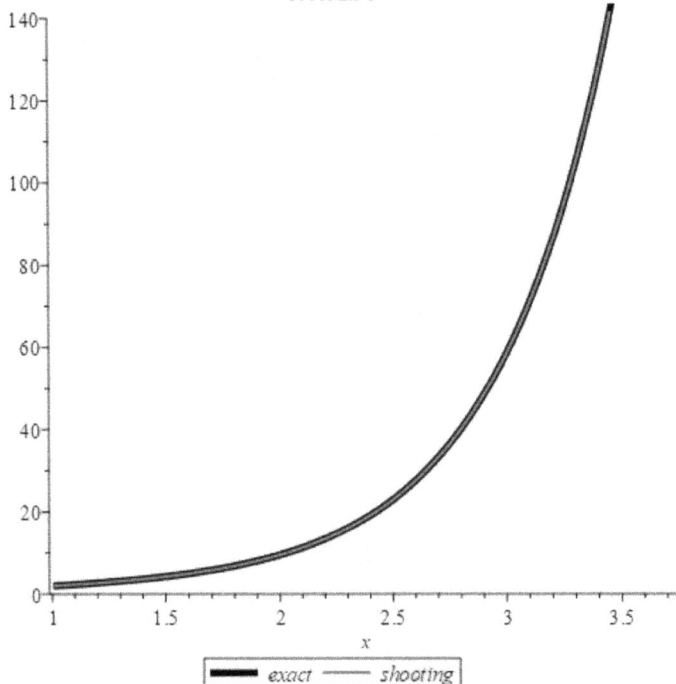

    The answer is 2.

3. Use the linear shooting method to approximate the solution to the following BVPs:

    a. The exact solution is $-y(x) = -11.60967305 * \exp(-0.585786438 * x) + 74.84991806 * \exp(-3.414213562 * x)$ for $y' + 0.197375 * y' + 9.81 * y(x) = 0$ for $0 \leq x \leq 3$, with $y(1) = 4$, $y(3) = 2$.

    b. $y'' + 4 * y' + 13 * y = 0$ for $0 \leq x \leq 1$ with $y(0) = 1$, $y(1) = 0$, with an exact solution of $y(x) = -\exp(-2 * x) * (\cot(3) * \sin(3 * x) - \cos(3 * x))$.

    c. $y'' + 2 * y' - 10 * y(x) = 7 * \exp(-x) + 4$ for $0 \leq x \leq 1$, with $y(0) = 2$, $y(1) = -5$, and an exact solution of $y(x) = -0.435070729 * \exp(2.316624790 * x) + 3.471434365 * \exp(-4.316624790 * x) - 0.4000000000 - 0.6363636364 * \exp(-1. * x)$.

Chapter 12

1. a. $y = 0.7x - 0.1$, $r^2 = 0.816$.

   b. $y = 0.158325x^2$

   5. a.

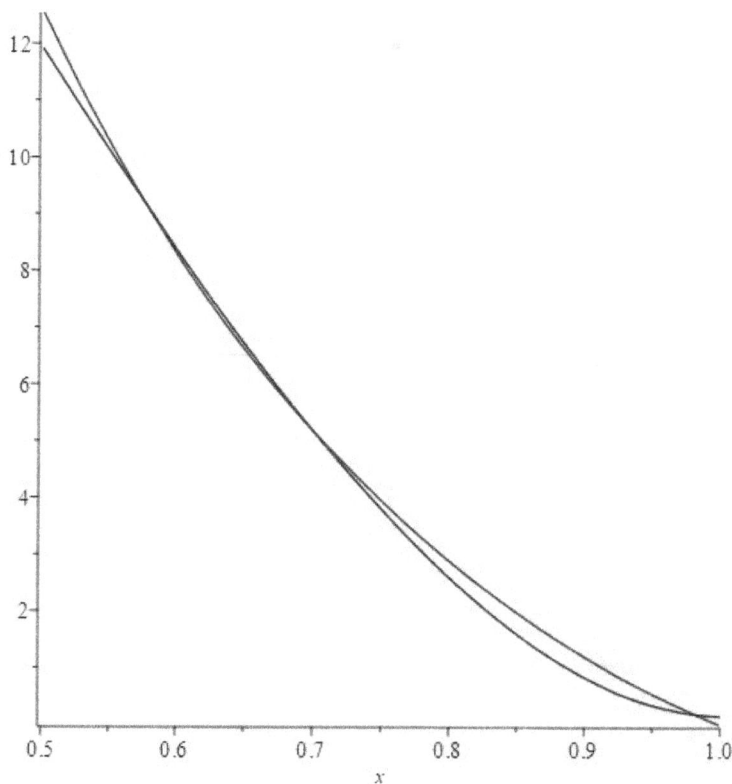

$$fsolve\left(\{eq11 = 0, eq12 = 0, eq13 = 0, eq14 = 0\}, \{b0, b1, b2, b3\}\right);$$

$$\{b0 = 26.47677854, b1 = -11.82255137, b2 = -54.30819506, b3 = 40.05016095\}$$

$$plot\left(\{26.47677854 - 11.82255137x - 54.30819506x^2 + 40.05016095x^3\right.$$

$$\left. -0.0836865830x - .2098554034x^2 + .0534307765x^3, f5\}, x = .5..1\right);$$

Index

For Product Safety Concerns and Information please contact our EU
representative GPSR@taylorandfrancis.com
Taylor & Francis Verlag GmbH, Kaufingerstraße 24, 80331 München, Germany

www.ingramcontent.com/pod-product-compliance
Lightning Source LLC
Chambersburg PA
CBHW060748220326
41598CB00022B/2369

* 9 7 8 1 0 3 2 7 0 3 6 8 8 *